Studies in Computational Intelligence

Volume 903

Series Editor

Janusz Kacprzyk, Polish Academy of Sciences, Warsaw, Poland

The series "Studies in Computational Intelligence" (SCI) publishes new developments and advances in the various areas of computational intelligence—quickly and with a high quality. The intent is to cover the theory, applications, and design methods of computational intelligence, as embedded in the fields of engineering, computer science, physics and life sciences, as well as the methodologies behind them. The series contains monographs, lecture notes and edited volumes in computational intelligence spanning the areas of neural networks, connectionist systems, genetic algorithms, evolutionary computation, artificial intelligence, cellular automata, self-organizing systems, soft computing, fuzzy systems, and hybrid intelligent systems. Of particular value to both the contributors and the readership are the short publication timeframe and the world-wide distribution, which enable both wide and rapid dissemination of research output.

The books of this series are submitted to indexing to Web of Science, EI-Compendex, DBLP, SCOPUS, Google Scholar and Springerlink.

More information about this series at http://www.springer.com/series/7092

Akash Kumar Bhoi · Pradeep Kumar Mallick ·
Chuan-Ming Liu · Valentina E. Balas
Editors

Bio-inspired Neurocomputing

Springer

Editors
Akash Kumar Bhoi
Sikkim Manipal Institute of Technology
Rangpo, Sikkim, India

Pradeep Kumar Mallick
KIIT Deemed to be University
Bhubaneswar, Odisha, India

Chuan-Ming Liu
National Taipei University of Technology
Taipei, Taiwan

Valentina E. Balas
Aurel Vlaicu University of Arad
Arad, Romania

ISSN 1860-949X ISSN 1860-9503 (electronic)
Studies in Computational Intelligence
ISBN 978-981-15-5494-0 ISBN 978-981-15-5495-7 (eBook)
https://doi.org/10.1007/978-981-15-5495-7

This Springer imprint is published by the registered company Springer Nature Singapore Pte Ltd.
The registered company address is: 152 Beach Road, #21-01/04 Gateway East, Singapore 189721, Singapore

Preface

This book volume comprises new theoretical and practical findings in solving problems related to the healthcare and allied fields using bio-inspired neurocomputing-based approaches. Moreover, this edited book (bio-inspired neurocomputing) also emphasizes the ongoing evolutionary techniques and their applications for addressing real-time problems and possible solutions in various areas. The initial chapters discuss on hybridized and deep learning-based medical image processing techniques for precise analysis. Subsequent chapters deliver about the AI and IoT for healthcare systems, an in-depth review on brain tumour detection and classification, and healthcare transformation using electronics health record by integrating deep learning techniques. The next chapter introduces background of sleep, electrophysiological parameters of sleep, individual sleep stages behaviour, and different types of sleep-related disorders and its treatment solutions. Ricky Mohanty et al. have modelled an adaptive neuro-fuzzy inference system for disease prediction which is prevalent in the health industry. Sandeep Raj has presented a time-frequency-based method for continuous analysis of heartbeats by combining several robust methods such as Pan-Tompkins (PT) technique, double density complex wavelet transform (DDCWT), and support vector machine (SVM). The next chapter discusses on how to identify the relationship between stress, depression, and self-differentiation among Korean college students and to verify the effect of self-differentiation in the relationship between stress and depression. Jalluri Gnana Siva Sai et al. have discussed an approach for segmentation of MR images of brain, where it started with pre-processing, using the adaptive bilateral filter (ABF) and followed by the binary thresholding and fuzzy recurrent neural network (FR-Net). Diseases such as cancer, diseases of lung, rheumatoid arthritis, diabetic retinopathy, diseases of heart, Alzheimer's disease, hepatitis, dengue, liver disease, and Parkinson's disease were detected and diagnosed using deep learning approaches and were discussed in the chapter presented by Sushruta Mishra et al. The remaining chapters describe the applied and allied technologies and their applications *such as* deep neural networks, hDEBSA global optimization method, ultrasonography measurement muscle, SVM, bio-inspired chicken swarm optimization, moderated mediation effect, mobile data source framework, mobile computing, machine learning algorithms to reduce power consumption in wearables, Harris hawks optimization algorithm, systematic review on computational neuroscience models, and appearance management behaviour.

We wish our readers will get informative findings and deliberation through these chapters and take further lead to explore the future research possibilities and scope on the discussed topics. We would like to express our thanks and gratitude to all the authors involved during the submission, revision process, providing us the camera-ready chapters and spared time and energy during compiling this book. Moreover, we would like to express our sincere thanks to the higher authorities of our institutes and universities who have given their all kind of support in all fronts and motive us to do research.

Rangpo, India — Dr. Akash Kumar Bhoi
Bhubaneswar, India — Dr. Pradeep Kumar Mallick
Taipei, Taiwan — Prof. Dr. Chuan-Ming Liu
Arad, Romania — Prof. Dr. Valentina E. Balas

Contents

About the Editors

Dr. Akash Kumar Bhoi [PhD (Sikkim Manipal University), M.Tech (Karunya University), B.Tech (TAT, BPUT)] working as Assistant Professor in the Department of Electrical and Electronics Engineering at Sikkim Manipal Institute of Technology (SMIT), SMU, India. He has specialized in the field of Biomedical Engineering. His areas of research are biomedical signal processing, Medical Image Processing, Computational Approaches and Pattern Recognition etc.. He is also a regular reviewer of journals of repute namely IEEE, Springer, Elsevier, Taylor and Francis, Inderscience etc. He has published several papers in national and international journals and conferences. He has also served on numerous organizing panels for the international conferences and workshops. He is currently editing several books with Springer Nature and serving as Guest editor for special issues of the journal like Springer Nature and Inderscience.

Dr. Pradeep Kumar Mallick is currently working as Associate Professor in the School of Computer Engineering, Kalinga Institute of Industrial technology (KIIT) Deemed to be University, Odisha, India. He was a postdoctoral fellow in Kongju National University South Korea. He received his PhD from Siksha O Anusandhan University, M.Tech. (CSE) from Biju Patnaik University of Technology (BPUT), and MCA from Fakir Mohan University Balasore, India. Besides academics, he is also involved in various administrative activities - he is a Member of Board of Studies, Member of Doctoral Research Evaluation Committee, and Admission Committee. His areas of research include algorithm design and analysis, data mining, image processing, soft computing, and machine learning. He has published 5 books and more than 55 research papers in national and international journals and conference proceedings.

Dr. Chuan-Ming Liu is a professor in the Department of Computer Science and Information Engineering (CSIE), National Taipei University of Technology (Taipei Tech), TAIWAN, where he was the Department Chair from 2013-2017. Currently, he is pointed to be the Head of the Extension Education Center at the same school. Dr. Liu received his Ph.D. in Computer Science from Purdue University in 2002 and joined the CSIE Department in Taipei Tech in the spring of 2003. In 2010 and 2011, he has held visiting appointments with Auburn University, Auburn, AL, USA, and the Beijing Institute of Technology, Beijing, China. He has services in many journals, conferences and societies as well as published more than 100 papers in many prestigious journals and international conferences. Dr. Liu was the co-recipients of many conference best paper awards, including ICUFN 2015 Excellent Paper Award, ICS 2016 Outstanding Paper Award, and WOCC 2018 Best Paper Award. His current research interests include big data management and processing, uncertain data management, data science, spatial data processing, data streams, ad-hoc and sensor networks, location-based services.

Dr. Valentina E. Balas is currently Full Professor in the Department of Automatics and Applied Software at the Faculty of Engineering, "Aurel Vlaicu" University of Arad, Romania. She holds a Ph.D. in Applied Electronics and Telecommunications from Polytechnic University of Timisoara. Dr. Balas is author of more than 300 research papers in refereed journals and international conferences. Her research interests include intelligent systems, fuzzy control, soft computing, smart sensors, information fusion, modeling and simulation. She is the Editor-in Chief of the International Journal of Advanced Intelligence Paradigms and the International Journal of Computational Systems Engineering. She also serves as an editorial board member of several national and international journals and is the Director of the Department of International Relations and Head of Intelligent Systems Research Centre in Aurel Vlaicu University of Arad. She is a member of EUSFLAT, SIAM and a Senior Member IEEE, member in TC–Fuzzy Systems (IEEE CIS), chair of the TF 14 in TC – Emergent Technologies (IEEE CIS), member in TC – Soft Computing (IEEE SMCS).

Performance Measurement of Various Hybridized Kernels for Noise Normalization and Enhancement in High-Resolution MR Images

P. Naga Srinivasu, Valentina E. Balas, and Norita Md. Norwawi

Abstract In this article, a focus is laid on the hybridization of various noise removal kernels that are used in the normalization of the noise in the medical MR images which is acquainted into the images during the processes of image rendering caused due to inappropriate calibration of the equipment and poor illumination of the subject. In the process of statistical study of various kernels that include Otsu-based Adaptive Weighted Bilateral Kernel (AWBK), Adaptive Contourlet Transform (ACT), Adaptive Fuzzy Hexagonal Weighted Mean (AFHWM) Kernel, and Adaptive Multiscale Data Condensation Kernel (AMDC), the experimentation is carried over images that are corrupted at distinct noise levels. During the recovery of the noisy image, the performances of the various included approaches have been evaluated and presented in this article. Upon practical implementation, it is observed that each of those hybridized kernels outperformed the type of noise on which they are experimented. The mean computational time of each kernel is also been presented in the results.

Keywords Noisy image · Hybrid kernel · Poisson noise · Gaussian noise · Speckle noise

1 Introduction

In the field of Computer-Aided Diagnosis, there are various imaging techniques that are being widely used in the identification of abnormalities in the human body, among all the approaches Magnetic Resonance Imaging (MRI) is most predominantly

P. Naga Srinivasu (✉)
Department of CSE, GIT, GITAM Deemed to be University, Visakhapatnam, India
e-mail: parvathanenins@gmail.com

V. E. Balas
Department of Automation and Applied Informatics, Aurel Vlaicu University of Arad, Arad, Romania
e-mail: balas@drbalas.ro

N. Md. Norwawi
Faculty of Science and Technology, Universiti Sains Islam Malaysia, Nilai, Malaysia
e-mail: norita@usim.edu.my

A. K. Bhoi et al. (eds.), *Bio-inspired Neurocomputing*, Studies in Computational Intelligence 903, https://doi.org/10.1007/978-981-15-5495-7_1

accepted due to its non-invasive nature and limited ionization. There are various types of MRI scanners that are being universally accepted, which are categorized based on the strength of the generated magnetic field for radio waves, and these machines are categorized as $5T$–$3T$ where T designates the Tesla that is used in measuring the strength of the magnetic field. For all these types of MR scanning machines, the common challenge is the noise that is affixed in the image at the time of rendering caused due to inappropriate calibration of the equipment or poor illumination of the region of interest or due to unbalanced and irregular emission and capturing of protons while capturing the image.

It is very crucial to address the noise issue well before the image is given as the input for statistical analysis of the abnormality in the human body. At times, when a raw image is directly fed as the input for the algorithm, the proposed architecture/approach might misinterpret the noisy data and would lead to inaccurate and poor results. Hence at most care must be taken while processing the image for noise removal and image enhancement as mentioned by Naga Srinivasu et al. [1]; Hemanth et al. [2]. In this chapter, various types of noises in MR imaging technology have been discussed.

There are various types of noises that are generally seen in the medical MR images as reported by Aja-Fernández and Vegas-Sánchez-Ferrero [3] and Isa et al. [4], they are categorized based on nature of noise, few of them includes the salt and pepper noise is also well known as Impulse noise. It is a very usually observed type of noise in the MR images caused due to instantaneous disruption in the signal that would appear on the image like dots of extreme white and dark pixels on the image. From the existing studies, it is very much clear that the salt and pepper type of noise is usually caused while performing the analog to digital conversion in digital imaging technology, and in some cases, salt and pepper type of noise can also be caused at the time of data transfer. There are various images de-noising techniques like harmonic mean and contra-harmonic mean kernels that are used in addressing the issues, as stated by Al-Khaffaf et al. [5] in their article on noise removal from document images.

Poisson noise is the other very common type of noise we come across in the medical-related images, and sometimes it is also called Shot noise, which is generally represented as a Poisson distribution function that mainly focuses on the randomness of the data over the presumed mathematical space. When the Poisson noise is presiding when there are a considerable number of elements that are capable of transferring, the energy is minimal, which would obviously lead to a Poisson distribution that will lead to randomness in the individual items that would lead to a noise in the image. Kirti et al. [6] in his article removal of Poisson noise from the X-Ray image, has been presented where the author has focused on the Harris corner point detector that is capable enough to pinpoint the erroneous pixel for processing it through a responsive median filter.

Gaussian noise is also a very usual noise in medical MR images, which is also called amplifier noise, which is usually caused due to the impact of the inadequate illumination, and in some cases, it could also be caused due to the temperature at the scanning machine. In the field of medical imaging, the Gaussian noise is being addressed through the spatial filter. However, the spatial filter is not a better solution for noise removal as it blurs the image, which could miss some crucial data from the image. Ali [7], in his work on image de-noising based on the fundamental filter, has evaluated the median filter, Wiener filter, and Adaptive median filter. However, in the majority of the cases, the pixel values are assessed from the neighboring pixels, and in the majority of the cases, the approximated value would smoothen the image. In other words, we can also identify it as blurring the image for noise removal.

The other very common type of noise is Speckle noise, that is acquainted into the MR image during the process of the image rendering due to the presence of interference that significantly affects the quality of the image, where this is a very common type of noise that is caused due to the presence of arbitrary inferences that lies in between the issues of the coherent return from objects and the identifier where plenty of the scatterers present in the surface that causes some extreme dark and bright spots on the image.

All the aforementioned noises have been successfully addressed through various hybridization filters at varying noise levels. The performances of each of these hybrid filters have been assessed through diverse performance evaluation metrics like Peak Signal to Noise Ratio (PSNR), which is one among the widely accepted type of metric that regulate between a value with respect to the pixel's intensity to the weight of the corrupted signal generally represented as a decimal value, Mean Square Error (MSE) that determines the average of the squares of mistaken intensities with respect to the actual pixel intensity, Root Mean Square Error (RMSE) is the rooted value of the approximated mean square error, Image Quality Index (IQI) is the other predominantly used image quality evaluation metric that is measured as the ratio among the misstate pixel intensity due to noise and the value of the same pixel in the reference image and Kappa Coefficient (KC) is a probabilistic measure that determines the most happening intensity of the pixel whose values are being estimated would be the exact value that is being corrupted by chance.

2 The Intent of the Chapter

In this chapter, various types of noises that corrupt the Magnetic Resonance Images have been discussed, and experimentation has been performed over the corrupted images at varying noise levels. It is observed that the hybridization of the noise removal kernels has exhibited better performance over the simple linear kernels. On the practical implementation of the hybridized kernels, the statistical analysis of various approaches has been presented in the results and discussion session of the chapter.

3 Incorporation of Hybridization Kernel

The process of the Hybridization of the kernels is one such novel approach for de-noising the MR image by using the existing linear kernels in conjunction with the non-linear kernels. The resultant output pixel approximation is made based on the input image, i.e., the value is of pixels is determined by the values of its neighboring pixels. But, the main challenging task by using the linear kernels-based image enhancement is that the image will smoothen as the pixel value is the cumulative average of its neighboring pixels due to which some crucial like edge and boundary information may be missing. Hence, the linear kernels could be implemented in coherence with non-linear kernels. The resultant value of the corresponding pixels is not dependent on the input value. Moreover, the performance of the non-linear filters significantly affects the quality of the resultant image.

The non-linear pixels are conscientiously stronger than the linear pixels, and they are completely designed in the mathematical models that would work in the spatial domain. Generally, the linear pixels usually deal with the additive noise where the unwanted signal U can be easily discarded from the actual signal A. In the sense, the value of A could be effortlessly restored by using a linear kernel like Laplacian filter utilized by Ranjbaran et al. [8] and Talebi and Milanfar [9] Wiener filter as stated by Kazubek [10], Mean filter as stated by Srinivas et al. [11]; Srinivasu et al. [12] and song et al. [13] in their paper on image de-noising, Low-pass filter, High-pass filter, Kalman filter by song Conte et al. [14], Bandpass filter and Differentiator filter as stated by Verma and Saini [15] in their article on forward and backward processing technique for image de-noising.

The non-linear kernels need not be like the filters that are seen earlier; they could be some value which is related to phase secured iterations, or the relational formulas that can glimpse the modulated radio frequency, or the mathematical models that are frequency mixers that can generate new modulated values, the mathematical models of the various median filters, and other statistical inclination that are a selective feature. Few of the researchers like Beagum et al. [10] have stated the estimations that are made through the non-linear kernels as a part of image enhancements are being used with the linear kernels for better randomness in choosing the appropriate value for replacing the noisy pixel. The outcome of the hybrid kernel would always be a better value than the linear kernels alone due to incorporating the weighted nonparametric feature for assessing the noisy pixel intensity value. Moreover, the non-linear parameters could preserve the edge information related to the pixel that is crucial in the identification of the boundaries of the objects.

4 Evaluating the Optimal Size of the Kernel

The size of the kernel would be the major constraint in the process of image enhancement and noise removal, At most care must be taken while choosing the size of the kernel, sometimes when the kernel is chosen to be too large would be cumbersome and the mean value or the median would be misleading when the pixel of the different region is covered as the part of the kernel that might mislead the assessment of the noise pixel or when the kernel is too small some of the pixels that are crucial in deciding the optimal value of the noise pixel may not be considered for the evaluation of the noise pixel that might lead to inappropriate evaluation of the pixel value. Hence at most care must be taken while choosing the size of the kernel.

The optimal size of the kernel could be decided through the equation stated below

$$\min_k = \sqrt{\frac{T_{nf} - 1}{1 - \sigma}} \tag{1}$$

In the above equation, the variable $\min_k$ designates the minimum size of the kernel, and the variable T_{nf} designates the total number of noise-free pixels in the considered region, and the standard deviation values have been evaluated from the equation $\sigma = \frac{T_n}{i \times j}$ From the above equation (i, j) is the number of rows and the columns in the MR image. T_n designates the total number of noise pixels in the image.

5 Otsu-Based Adaptive Weighted Bilateral Kernel (AWBK)

Adaptive Weighted Bilateral Filter (AWBF), as mentioned by Zhang and Allebach [16] in their article on image de-noising have stated that the non-linear kernels that preserve the boundary information(edge) of the target present in the image and can achieve noise depletion by aggregating the noise locations in the image with respect to its closely surrounded pixels that are carried out using the Otsu thresholding approach as stated by sha et al. [17]. By approximating the intensity value of the distorted pixel caused due to noise, the intensity value is assessed through Contra-Harmonic mean of the pixel intensities in the corresponding noisy region in the MR image. It is a general presupposition that the pixels that are surrounded closely would be having extremely likely pixel intensities and the weight has noticeable importance, and the significance of the weight keeps reducing as we move away from the centroid of the pixel, and it is would be closely negligible when it moves far away.

The noise that has been appended would be analogously weakly correlated where the pixel intensities mean of the neighboring pixels have been upheld where the noise has been ignored. At the stage of image enhancement, when a low-pass kernel is utilized for computing the average of the pixel intensities, it will smoothen the image, i.e., it would blur the region of the image. But to maintain the edge-related

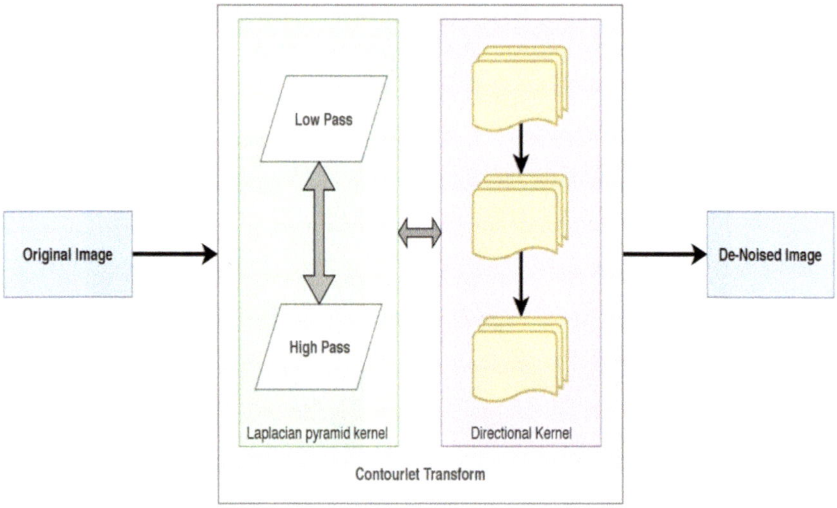

Fig. 3 Represents the working procedure and components of ACT kernel

frequency model would highlight the sensitive and pivotal details of corresponding MR image. Laplacian kernel is implemented that measures the dissimilarity between the taken MR image and the resultant image of low-pass kernel. The MR image has been blurred to discard the considerably low-frequency signal in the image, and the directional kernel is employed to reconstruct the original image by wedge-like frequency partitioning which is employed to conserve the boundary and edge associated data in the original MR image by evaluating the first-order dervion of the pixel's intensities of all the pixels that remain within the window which could be used in any of the directions (Fig. 3).

By using the suggested hybrid kernel for the noise normalization, the noise in the pixel data is being predicted by using the equation stated below

$$p(i, j) = Gp_{int}(i, j) - \left\{ 4 \sum_{m=-2}^{2} \sum_{n=-2}^{2} \omega(p, q) Gp_{int} \left(\frac{i-p}{2}, \frac{j-q}{2} \right) \right\} \quad (5)$$

In the equation mentioned above, the variable $Gp_{int}(i, j)$ designates the Gaussian kernel linked with the coordinates (i, j). $\omega(p, q)$ designates the value of standard deviation among the image dimensional coordinates m and n (Fig. 4).

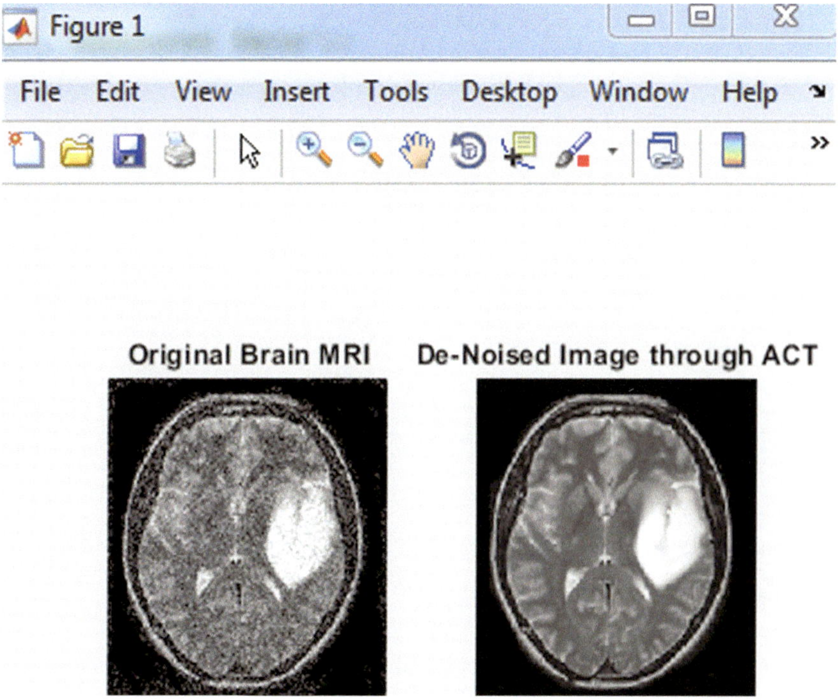

Fig. 4 Represents the resultant outcome of ACT kernel on image restoration

7 Adaptive Fuzzy Hexagonal Weighted Mean (AFHWM) Kernel

Adaptive Fuzzy Hexagonal Weighted Mean Kernel-based image de-noising and enhancement of the image that includes local order kernel, Bilateral kernel alongside with the statistical features as stated by Kala and Deepa [21] in their article on brain MRI de-noising.

7.1 *Statistical Features*

Statistical features of an image pixel that include some of the features like standard deviation, Local and Global Mean are being considered. There is a special significance for the local and Global Mean in estimating the noisy pixel and segregating the foreground and background regions in the image by computing the Local Mean (lm) of the surrounded neighborhood and the Global Mean of the noisy pixels in the image for evaluating the fuzzy membership of the pixel.

The standard deviation is being assessed through the approximated mean value of the background region's pixel in the MR image, where the background pixels are being recognized through Otsu-based thresholding, as stated by Yang et al. [22] which is evaluated as follows

$$\sigma = \sqrt{\frac{b_m}{2}} \tag{6}$$

From the above equation, b_m designates the background mean of which is evaluated for an input MR image $f(p,\ q)$ of size $i \times j$ through the formula stated below

$$b_m = \frac{1}{i \times j} \sum_{p,q=0}^{i,j} f(p,q) \tag{7}$$

7.2 *Local Kernel*

Local mean/median filter is a very commonly used kernel that is being widely used in addressing the issue of inflated corrupted noisy pixels like spikes in the image as stated by Banerjee et al. [23]. Moreover, it is statistically identical to the minimal (min) and maximal (max) filter that takes the pixel intensities of the neighboring pixels into consideration. It is used with a mask of size M which is determined by the formula stated below

$$M = 2 \times L_{\text{med}} + 1 \tag{8}$$

From the above equation, L_{med} designates the local median of the pixel that lies within the mask, and almost the value is set to 1 in the majority of the cases. And it used a search space window W size that is determined by the formula stated below

$$W = 2 \times W_{\text{med}} + 1 \tag{9}$$

The corrupted pixels in the image are being reconstructed through the local kernel using the formula stated below

$$L_k = \text{local kernel}(r,\ M) \tag{10}$$

From the above equation, the variable L_k denotes the local kernel, and r represents the radius of the locally squared neighborhood pixels.

7.3 Customized Bilateral Kernel

The bilateral kernel is also most widely used in image de-noising, which is a locally implemented non-linear kernel that takes the pixel intensities of the surrounded pixels into consideration while assessing the noisy pixel by preserving the edge related information. By using the bilateral filter alone, there might be a chance that few of the crucial information might be gone missing due to the aggregation of the sub-bands of the image decomposed that are being evaluated at each level of wavelet revamp as stated by Gavaskar et al. [24]. The weights of the image pixels that decomposes by distance from the centroid of the region, the kernel is stated as follows

$$w(i,j) = \frac{1}{2\tau\sigma^2} e^{-\frac{(i^2+j^2)}{2\sigma^2}} \tag{11}$$

The estimated weights are being used with the bilateral kernel in evaluating the appropriate value of the pixel. The bilateral filter is being practically implemented using the equation stated below

$$\text{Pix}_{\text{Intensity}}(x) = \frac{1}{p} \sum_{r \in y} w_p(i,j)(||x-y||) w_q(i,j)(|I(x)-I(y)|) I(y) \tag{12}$$

From the equation $w_p(i,j)$ designates the closeness of the pixel with respect to the particular region

$$w_p(i,j)(||x-y|| = e^{-\frac{(x-y)^2}{2\sigma_p^2}} \tag{13}$$

The variable $w_q(i,j)$ in the above Eq. (12)

$$w_q(i,j)(||x-y|| = e^{-\frac{|(\text{int}(x)-\text{int}(y)|^2}{2\sigma_q^2}} \tag{14}$$

In the above Eq. (12), the variable p designates the grey level similarity coefficient, and the values ||x–y|| is the Euclidean distance measure.

8 Adaptive Fuzzy Hexagonal Weighted Mean

Through the proposed approach, all the aforementioned kernels, i.e., Local kernel, Bilateral kernel, and the obtained statistical features, to de-noise the MR image. The objective function for evaluating the degree of membership that determines the correlation of each pixel in the original MR image with respect to the membership values that lie between 0 and 1 as stated by Sharif et al. [25] that also. The Adaptive Fuzzy Hexagonal Weighted Mean would also consider the non-local order and the distribution function p_d can be determined by the following asymptotically strong formula

$$p_d = \frac{s_p}{N} \times \frac{1}{f_{sr(z,sp)}}$$

The aforementioned equation is applied over a 2D MR image of size $i_{m \times n,}$ and $N = m \times n$, that determines the total number of pixels in the image. On performing the aforementioned task, the proposed approach can recognize the opaque seed points of corresponding segments. The difference between the actual seed point and the approximated seed point would diverge from the estimated value of the probability density function in concerning to the presumed seed point (Fig. 6).

The mechanism that evaluates the distance between each seed point through Minkowski metric is determined as follows:

$$d_m = \sqrt[q]{\left(\sum_{i=1}^{n} |x_{mi} - x_{ti}|^q\right)}$$

when $q = 1$, it resembles the city-block distance and when $q = 2$ would give the Euclidean measure, for $q = \infty$ the Chebyshev distance is evaluated as follows:

$$d_m = \max_j |x_{mi} - x_{ti}|$$

The correlation approximation among the two point i, j is determined through the equation stated below

$$C(i, j) = e^{-\alpha \times dij}$$

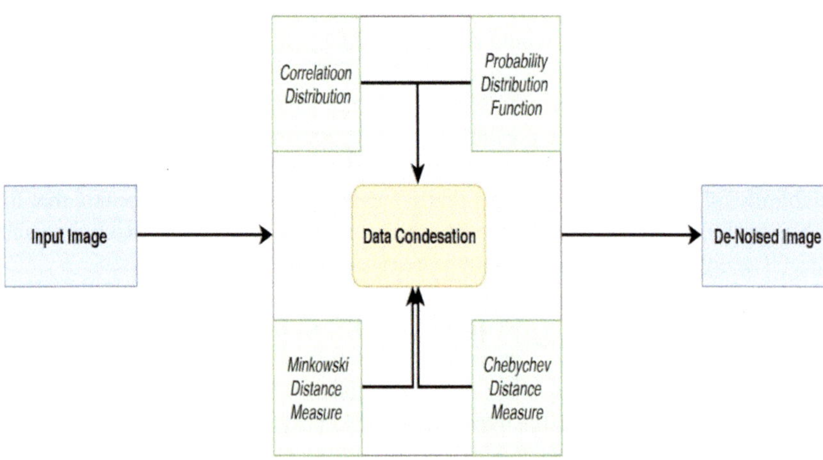

Fig. 6 Represents the working procedure and elements in the AMDC kernel

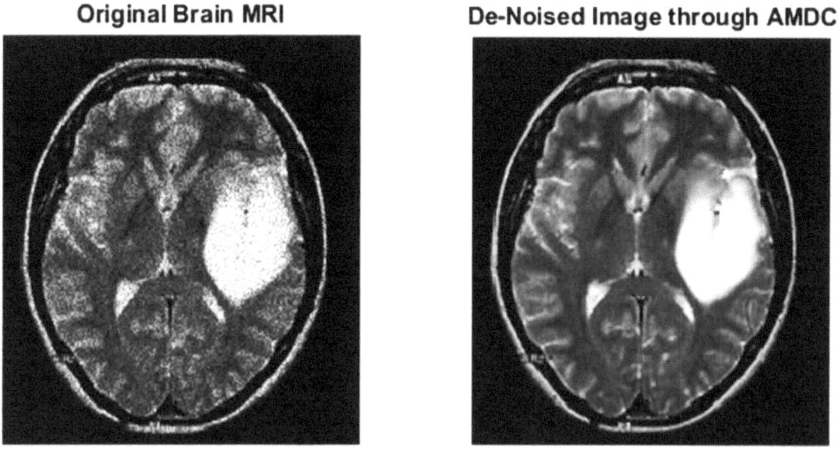

Fig. 7 Represents the resultant outcome of the AMDC kernel on image restoration

The value of the $\alpha = \frac{-(\ln 0.5)}{d} > 0$ where d designates the mean distance among the two seed points in the distance matrix (Fig. 7).

10 Results and Discussion

The experimentation setup has been carried over both clinical real-time images and synthetic images, and the performance of the aforementioned approaches have been assessed in qualitative and quantative perspective. The experimentation is performed over a core i7 working in windows environment personal computer using Matlab R2019 version software, and the execution time estimated is concerning the same experimentation setup. The experimentation has been performed over the image of size 256 × 256 and 512 × 512. The quality of the image that has been restored is assessed through the performance evaluation metric like Peak Signal to Noise Ratio (PSNR) which is expected to be high with lower noise variance, Mean Square Error (MSE) is expected to be proportional to the value of the noise variance which is a non-negative number that represents the square of the difference between the actual value and the approximated value, Root Mean Square Error (RMSE) is the rooted value of the MSE, Image Quality Index (IQI) would be high for the lower noise variance and would decrease with increase in the noise variance and Structural Similarity Index (SSIM) is a value that would be proportional to the value of the IQI, that determines how similar does the restored image in concern to the analogs reference image and the resultant evaluated metrics have been tabled in Table 1 through Table 6.

The experiment is being performed on the image that is being corrupted by the Poisson noise that can be seen from Tables 1 and 2 for variable size images, the experimentation is also performed on the image corrupted by Speckle noise that is

Table 3 The performance analysis table for varies Hybridized Kernel approaches for Speckle Noise

De-noising of Speckle noise from MR Image of size 256 × 256								
Noise variance	Otsu-based Adaptive Weighted Bilateral Kernel				Adaptive Contourlet Transform			
	PSNR	MSE	IQI	SSIM	PSNR	MSE	IQI	SSIM
1	43.24	2.45	0.892	0.801	44.32	2.34	0.881	0.821
2	42.98	3.97	0.832	0.765	43.16	3.86	0.803	0.793
4	40.62	6.54	0.748	0.7	40.09	6.2	0.714	0.722
6	38.56	11.75	0.696	0.654	38.43	10.24	0.685	0.683
8	36.41	13.01	0.534	0.594	37.64	12.75	0.591	0.624
9	35.22	15.92	0.498	0.489	35.24	15.12	0.487	0.589
10	34.54	18.01	0.401	0.452	34.32	17.23	0.403	0.502
Mean exe. time	1.8922				2.0012			
De-noising of Speckle noise from MR Image of size 256 × 256								
Noise variance	Adaptive Fuzzy Hexagonal Weighted Mean				Adaptive Multiscale Data Condensation			
	PSNR	MSE	IQI	SSIM	PSNR	MSE	IQI	SSIM
1	43	2.63	0.854	0.803	42.27	4.42	0.866	0.806
2	41.34	4.23	0.81	0.743	41.62	8.53	0.842	0.79
4	38.23	7.45	0.714	0.681	38.75	11.72	0.711	0.725
6	36.64	12.02	0.603	0.554	35.53	14.53	0.601	0.689
8	34.53	15.83	0.516	0.486	34.43	17.5	0.489	0.599
9	33.45	18.65	0.422	0.387	33.32	19.35	0.392	0.572
10	32.25	19.71	0.397	0.324	32.05	20.66	0.342	0.536
Mean exe. time	2.2089				2.1642			

Thereafter Adaptive Fuzzy Hexagonal Weighted Mean(AFHWM) has also exhibited considerable good performance but it need computationally more computational recourses for normalization of the image.

Gaussian noise effected MR images have been experimented for evaluating the performance of the hybrid mechanisms, it is observed that the Adaptive Multiscale Condensation (AMC) showcased a better functionality over the other approaches, followed by the Adaptive Fuzzy Hexagonal Weighted Mean(AFHWM) approach is proven to show superior performance in noise normalization. On the other hand, Otsu-based Adaptive Weighted kernel(AWK) and approach has presented a reasonable fair production of noise-free MR images. The performances have been experimented over a high-resolution MR images of size 256 × 256 and 512 × 512, And the performances of the afore mention approaches are much better for smaller size images when compared to that of the larger size images and moreover, the computational time is considerably high for larger size MR images.

Table 4 The performance analysis table for varies Hybridized Kernel approaches for Speckle Noise

De-noising of Speckle noise from MR Image of size 512 × 512								
Noise variance	Otsu-based Adaptive Weighted Bilateral Kernel				Adaptive Contourlet Transform			
	PSNR	MSE	IQI	SSIM	PSNR	MSE	IQI	SSIM
1	40.21	4.32	0.824	0.789	41.24	3.4	0.868	0.803
2	38.43	6.75	0.718	0.754	39.74	5.22	0.796	0.789
4	36.25	7.84	0.654	0.644	37.52	9.43	0.7	0.701
6	35.31	11.32	0.573	0.615	34.67	12.74	0.651	0.645
8	32.57	14.13	0.498	0.575	31.49	14.56	0.564	0.59
9	27.22	18.74	0.352	0.5	27.75	17.29	0.43	0.534
10	26.12	20.12	0.339	0.482	26.34	19.12	0.392	0.497
Mean exe. time	2.3223				2.1412			
De-noising of Speckle noise from MR Image of size 512 × 512								
Noise variance	Adaptive Fuzzy Hexagonal Weighted Mean				Adaptive Multiscale Data Condensation			
	PSNR	MSE	IQI	SSIM	PSNR	MSE	IQI	SSIM
1	40.23	3.98	0.822	0.795	40.78	3.96	0.802	0.753
2	38.82	6.08	0.792	0.723	38.22	6.32	0.788	0.702
4	36.47	10.24	0.704	0.654	36.34	11.43	0.693	0.632
6	34.16	14.02	0.641	0.569	34.53	14.56	0.572	0.598
8	31.86	16.72	0.552	0.464	31.44	17.25	0.5	0.503
9	27.34	19.93	0.438	0.392	28.02	21.75	0.443	0.48
10	25.46	21.05	0.388	0.366	26.83	22.63	0.321	0.465
Mean exe. time	2.4936				2.2024			

Table 5 The performance analysis table for varies Hybridized Kernel approaches for Gaussian noise

De-noising of Gaussian noise from MR Image of size 256 × 256								
Noise variance	Otsu-based Adaptive Weighted Bilateral Kernel				Adaptive Contourlet Transform			
	PSNR	MSE	IQI	SSIM	PSNR	MSE	IQI	SSIM
1	42.39	3.05	0.865	0.836	42.48	2.73	0.831	0.843
2	41.86	3.86	0.801	0.771	41.19	3.43	0.784	0.789
4	39.78	5.09	0.704	0.696	39.43	7.32	0.659	0.704
6	37.95	8.24	0.653	0.608	36.93	11.98	0.56	0.655
8	35.23	11.55	0.572	0.564	34.41	14.09	0.488	0.593
9	34.47	14.3	0.489	0.482	33.29	16.24	0.414	0.527
10	33.54	17.22	0.422	0.411	32.1	19.54	0.378	0.488
Mean exe. time	1.9824				2.0211			

(continued)

Table 5 (continued)

De-noising of Gaussian noise from MR Image of size 256 × 256								
Noise variance	Adaptive Fuzzy Hexagonal Weighted Mean				Adaptive Multiscale Data Condensation			
	PSNR	MSE	IQI	SSIM	PSNR	MSE	IQI	SSIM
1	42.96	3.65	0.852	0.842	43.11	3.12	0.851	0.871
2	40.59	4.21	0.821	0.787	42.27	6.01	0.811	0.812
4	37.65	8.2	0.79	0.699	40.47	9.19	0.78	0.775
6	35.89	13.88	0.682	0.574	37.81	12.98	0.681	0.696
8	33	16.97	0.532	0.475	35.75	15.87	0.563	0.59
9	32.97	19.71	0.485	0.396	33.87	17.05	0.421	0.525
10	32.01	20.01	0.389	0.331	33.08	18.9	0.392	0.487
Mean exe. time	2.2065				2.0921			

Table 6 The performance analysis table for varies Hybridized Kernel approaches for Gaussian noise

De-noising of Gaussian noise from MR Image of size 512 × 512								
Noise variance	Otsu-based Adaptive Weighted Bilateral Kernel				Adaptive Contourlet Transform			
	PSNR	MSE	IQI	SSIM	PSNR	MSE	IQI	SSIM
1	40.05	4.71	0.842	0.752	41.77	3.57	0.844	0.833
2	37.86	7.34	0.798	0.714	39.48	5.24	0.776	0.795
4	35.48	9.02	0.703	0.696	37.81	10.24	0.692	0.721
6	33..16	12.78	0.644	0.671	35.38	14.05	0.592	0.665
8	31.07	15.24	0.561	0.59	32.01	17.42	0.501	0.589
9	26.1	19.91	0.475	0.511	28.29	21.34	0.467	0.506
10	25.25	22.41	0.417	0.423	27.47	23.59	0.382	0.417
Mean exe. time	2.4212				2.2397			
De-noising of Gaussian noise from MR Image of size 512 × 512								
Noise variance	Adaptive Fuzzy Hexagonal Weighted Mean				Adaptive Multiscale Data Condensation			
	PSNR	MSE	IQI	SSIM	PSNR	MSE	IQI	SSIM
1	40.29	3.79	0.845	0.743	42.34	3.22	0.881	0.765
2	37.95	6.53	0.759	0.701	39.2	6.43	0.767	0.702
4	35.38	11.21	0.683	0.639	37.53	11.59	0.694	0.645
6	33.2	14.49	0.601	0.411	35.26	14.25	0.589	0.577
8	30.98	17.84	0.514	0.394	32.04	17.53	0.552	0.512
9	26.23	19.4	0.473	0.332	29.47	21.42	0.443	0.481
10	25.03	22.97	0.392	0.302	27.83	22.59	0.342	0.428
Mean exe. time	2.4814				2.1864			

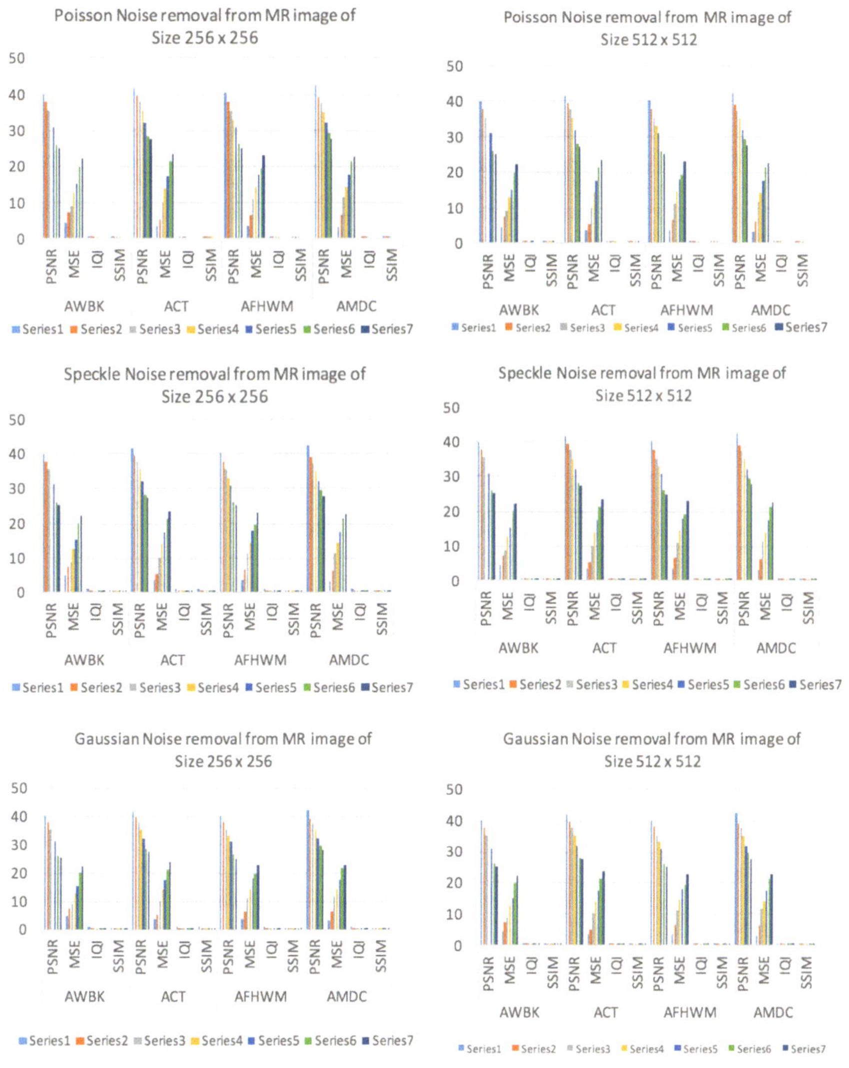

Fig. 8 Representing the performance evaluation charts of various approaches

11 Conclusion

The process of image restoration is a significant process that must be performed prior to the image analysis in medical image processing as every minute region in the medical MR images are very crucial in the process of abnormality diagnosis. It is observed that on experimentation of various hybrid kernels, the resultant outcomes

26. Ray, D., Dutta Majumder, D., Das, A.: Noise reduction and image enhancement of MRI using adaptive multiscale data condensation. In: 1st International Conference on Recent Advances in Information Technology (RAIT), Dhanbad, pp. 107–113 (2012)
27. Jin, F., Fieguth, P., Winger, L., Jernigan, E.: Adaptive Wiener filtering of noisy images and image sequences. In: Proceedings 2003 International Conference on Image Processing (Cat. No.03CH37429), pp. III–349, Barcelona, Spain (2003)
28. Kazubek, M.: Wavelet domain image denoising by thresholding and Wiener filtering. IEEE Signal Process. Lett. **10**(11), 324–326 (2003)
29. Sánchez, M.G., Vidal, V., Verdú, G., Mayo, P., Rodenas, F.: Medical image restoration with different types of noise. Conf. Proc. IEEE Eng. Med. Biol. Soc. **2012**, 4382–4385 (2012). PubMed PMID: 23366898

A Precise Analysis of Deep Learning for Medical Image Processing

Sushruta Mishra, Hrudaya Kumar Tripathy, and Biswa Acharya

Abstract Recently, usage of image processing in machine learning (ML) is growing fast. Medical image processing, image segmentation, computer-aided diagnosis, image transformation, image fusion combined with AI play a crucial role in the healthcare field. Other industries are different from the healthcare sector. This is the people's highest priority sector for those whose expectation levels of the people about care and services are high at a decent cost. It consumes a huge percentage of budgets, but still, it does not affect the social expectations. Many times it is observed that explanations provided by medical experts seem to be ambiguous. Few experts are able to effectively explain the details of medical images due to its complexity and subjective nature; interpreters and fatigue exist in different extensives. Later, the achievement of deep learning concept in varieties of real-time application domains also provides thrilling solutions with a good accuracy percentage in a medical image which will help the medical society soon. Deep learning (DL) methods are a set of algorithms in machine learning (ML), which provides an effective way to analyse medical images automatically for diagnosis/assessment of a disease. DL enables a higher level of abstraction and provides better prediction from data sets. Therefore, DL has a great impact and becomes popular in recent years. In this chapter, we discuss different states of deep learning architecture, image classifications and the medical image segmentation optimizer. This chapter provides a detailed analysis of algorithms based on deep learning that is used in clinical image with regards to recent works and their future approaches. It provides some important knowledge and the way of approaching deep learning concept in the field of healthcare image analysis. Afterwards, we will discuss the challenges that are faced when it is applied to medical images and some open research issues. In the end, a successful medical image processing is presented where implementation is done by deep learning.

S. Mishra (✉) · H. K. Tripathy · B. Acharya
School of Computer Engineering, KIIT University, Odisha, India
e-mail: mishra.sushruta@gmail.com

H. K. Tripathy
e-mail: hrudayakumar@gmail.com

B. Acharya
e-mail: acharyabiswa85@gmail.com

A. K. Bhoi et al. (eds.), *Bio-inspired Neurocomputing*, Studies in Computational Intelligence 903, https://doi.org/10.1007/978-981-15-5495-7_2

Keywords Deep learning · Machine learning · Disease diagnosis · Healthcare sector · Image processing · Brain images

1 Introduction to Deep Learning in Medical Image Processing

Nowadays, medical image analysis which is done on a computer is playing an important part at different stages like workflow in clinics starting from population screening as well as performing diagnoses to do treatment more effectively and monitoring everything. For this study, the accuracy of the method increases and the price is moderate which helps to balance the whole health society. Some years back, a vital method is a development which is deep learning and convolutional neural networks adopted by various analyses of medical image and intervention of computer-assisting jobs.

Deep learning shows a nested configuration of numerous basic tasks (generally extensive integrations like, scalar nonlinear approaches, convolutions and moment normalizations) characterized by data samples. Architecture is the function of a specific configuration, which is defined by a parametric function (customarily consisting of a bulk number of parameters) so that 'loss' function can be minimized or optimized, especially by the use of some gradient descent.

Initially, use of medical image analysis in the neural networks was done twenty years ago; in the last five years, their demands were expanded by orders of magnitude. Recently, many reviewers have highlighted the importance of deep learning in analysis of medical images and several ongoing research are being carried out in this field which include image segmentation, image classification, image detection, image registration, image reconstructing and image enhancement in critical anatomical regions such as brain, heart, lung, abdomen, breast, prostate and musculature. Each application has its specific feature; sometimes, it overlaps in software pipelines which are been implemented by many researchers.

Medical image analysis for deep learning pipelines has several associated components which are typical in deep learning methodologies:

- segregating records into training, testing and validation sets;
- during training random sample;
- data loading and a sample of an image;
- data augmentation;
- defining the architecture of a network as the arrangement of several simple functions;
- forming a speedy computational pipeline for optimization;
- construction of metrics for evaluation of performance all along with the training and inference.

While analysing medical images, maximum constituents have specific ranges; details of specific idiosyncrasies, in Sect. 4, like healthcare images have its formats that will take care of the massive 3D graphics images with anisotropic types and conceal supplementary spatial data and information of the patient, requiring various information that are loaded on pipelines. High-resolution images motivate with the sample of customer data that the domain-specific memory-efficient networks processing need high computerized power. Physical characteristics motivating domain-related information augmentation and model priors can be represented in an image, as the captured images are in standard anatomical form. Custom evaluation metrics can be warranted by some errors in clinical implications. Large duplication occurs when independently all of the custom architecture results are implemented again; it poses a barrier to spread the research tools which ruins the fair comparisons between the computing methods.

The recent development of research has been seen in the field of medical image analysis with emphasis on brain tumours detection [1], skin lesion [2], monitoring of heart ventricles [3] and diagnosis of the liver [4]. Many o fthe automatic brain tumours segmentation processing algorithms make use of various handcrafted theories and features which include: corners, edges, local binary pattern, a histogram of gradient and many more [5]. The main focuses on these methods have been given to the use of classical machine learning algorithms. Bahadure et al. [6] discussed MRI-based brain scanning which helped in the identification and attribute selection methodology. To enhance the prediction performance and subsequently limit the complex nature of image segmentation, the authors had introduced the use of Berkeley wavelet transformation-based tumour segmentation process. An SVM model classification was demonstrated to increase the rate of accuracy. Gebru et al. [7] provided a weighted data-based Gaussian mixture model. This framework helped in the proposal of new mixture model which associated with a weighted for the observed point. It also introduced the weighted data Gaussian mixture for the derivation of two EM algorithms. In [8], Sushruta et al. presented a new enhanced attribute optimization technique that helped in optimization of dengue fever prediction with optimal accuracy.

Akkus et al. [9] had implemented the use of a deep learning mechanism which is used for brain MRI segmentation. The technique is used for the provision of a succinct analysis of the present deep learning-based segmentation for MTI scans. The efficiency o fdeep learning approach is dependent on the use of several vital steps like preprocessing of data, initialization and the post-processing of output. Anupurba Nandi [10] has discussed a process of morphological operator and segmentation which would help in the detection of brain tumours. The process made use of k-means algorithm where the abnormality is depicted by the tumour which is later corrected by using various operators. Alongside this, the use of basic image processing is used to segregate the tumour-affected cells from the normal cells of the brain. In [11], authors proposed and implemented an efficient methodology involving biologically motivated technique for tumour classification. Katiyar et al. [12] analysed the use of unsupervised segmentation process for tumour tissue population. This process is used multiparametric MRI. The procedural approach followed in this work was Multiparametric MRI.

2 A Typical Deep Learning Framework

Generally, a machine learning framework has three steps which are followed by deep learning: selecting a model (choosing and adapting a model according to its training set), evaluating a model (model effectiveness is measured on its testing samples) and distributing a model (models are shared for use of the wide range of population). Ample complications retain in the interior of those simple phrases which are shown in Fig. 1. Learning about the implementation of the network is the complex part. Simple functions are used in deep learning, but their hierarchies are complex. Researchers have implemented the previous network (which is incomplete specified),

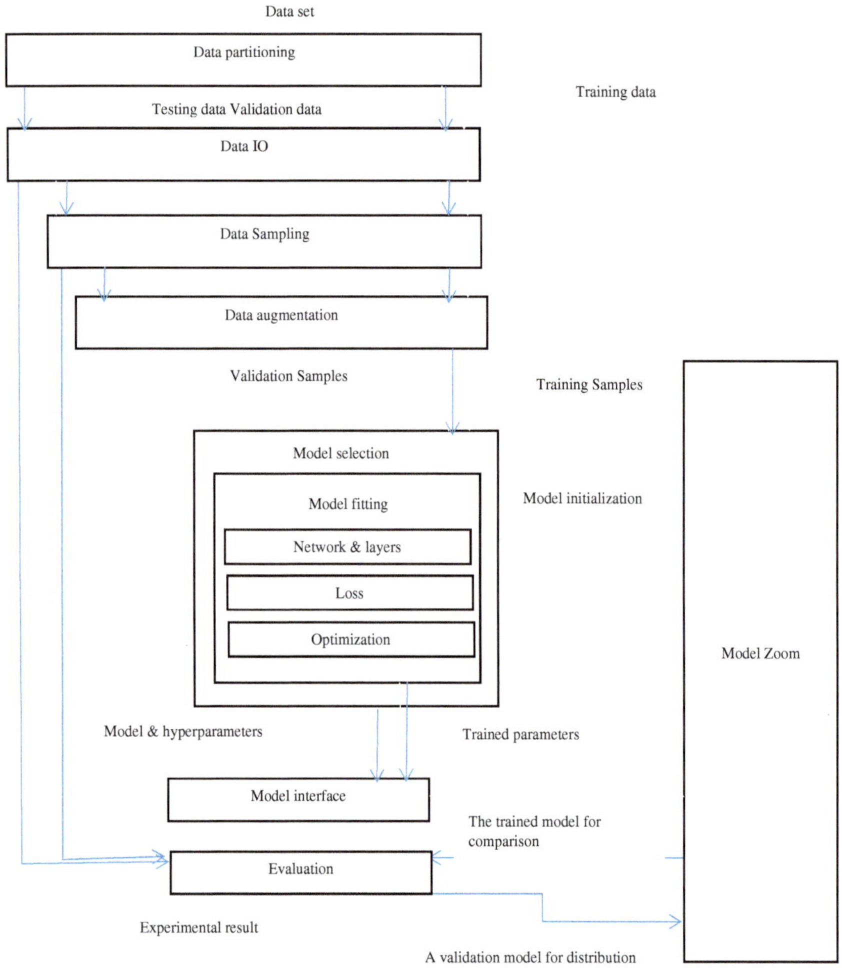

Fig. 1 General data flow illustration in deep learning framework

and the new network is also implemented. New infrastructure is needed to train, evaluate and distribute this network. Evaluation of data sets correctly is much important; otherwise, sometimes data collection occurs (e.g. graphical illustration of the same medical centre may be more identical to each other than from other centres). Data samples are a vital part of any experimental evaluation. After successful loading of these datasets is done then it is passed to the internetwork using distinct methods depending on pipelining phases. Methodologies used to tune hyperparameters in a classic set of methods and optimizing model metrics on the training samples are required. At the time of training and debugging, dissection of the developed model are required and hence logging and visualization are used for this purpose. Applications consisting of restricted data and data sets must be integrated by perturbing; by this, overfitting can be prevented at the time of training data practically. Adapting architecture of the previous network, trained or untrained, similar or different work in part or full, all these are common practices in deep learning; community repository is needed (also known as a model zoo); adaptable format is storing model and parameters. A deep learning project that has a framework is regenerated by the researcher or research group, and most of its part depends on the addressed application domain.

3 Domain-Specific Criteria in Medical Image Analysis Using Deep Learning

Deep learning has different applications for different medical images as images have various characteristics of data and every application uses different types of images. This section would discuss the need for domain-specific considerations driving the system design.

3.1 Data Availability

Collection, annotating and the proper distribution of the medical image data sets have been found to have a higher cost than many of the computer's vision-based tasks. Many of the medical imaging models have a costly image generation process. Annotation of various medical images requires the knowledge from highly expertise clinicians who can provide with limited time. Apart from these, due to the privacy concerns, the sharing of the data collected among institutions as well as over international connections is legally challenging for the organizations. The recent development of various tools are DeepIGeoS which works for semi-automation of annotations and GIFT-Cloud which is used for reducing the size of the data sets for easy sharing. The use of smaller data sets helps in the increase in the importance in the field of data augmentation, data regularization as well as cross-validation of the

writing efficient processing and also in the competition of objective-based evaluation. Some review works have been from the main focus which has been placed on the use of traditional methods for the process the MRI image of brain tumour segmentation. Apart from the other papers in the field, this study is on the basis of the recent learning trend methodology in the field. The succinct analytical design of deep learning is depicted in Fig. 2.

The definition of cancer is the uncontrolled growth of abnormal cells in the body. Having these similar mass of strange and uncontrolled development of cell and dissection in the brain is known as a brain tumour. Some occurrences of brain tumour are abnormal and so they have been considered to be one of the most lethal forms of cancer.

Based on the initial origin of brain tumours, it can be categorized into two forms—primary brain tumours and metastatic brain tumours. In the case of primary brain tumours, the cells originate in the brain tissue itself, whereas in metastatic brain tumours, cancer originates from another portion of the body and spreads to the brain.

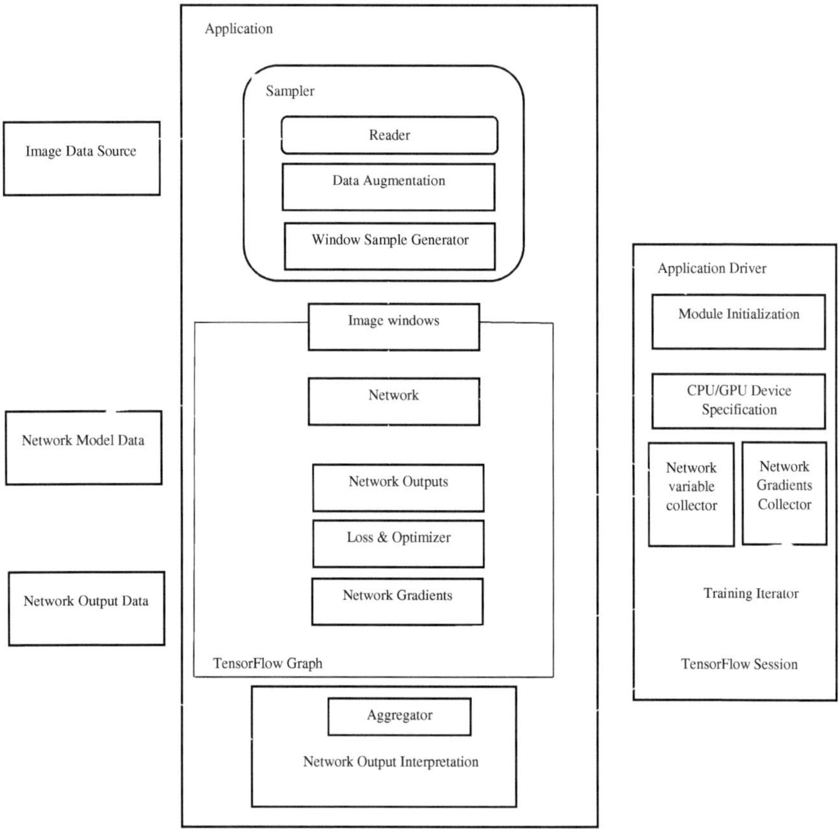

Fig. 2 Design analysis of deep learning

Glioma is one type of brain tumour that originates from glial cells. These are the most important types of brain tumours which are being focused on in tumour image segmentation research field. Glioma is the general term that is often used to describe various types of glioma. The glioma range from low-graded ones such as oligodendroglioma and astrocytoma to the higher graded (IV) glioblastoma multiform (GBM), which is considered to be one of the most aggressive and common types of primary malignant brain tumour. Chemotherapy surgery and radiotherapy are used in different combinations to treat glioma.

Glioma diagnosis in the early stages of development is considered to be of high importance. To understand the tumour properly, there needs to be an analysis of different variables of the tumour. These variables include size, shape, metabolism and location. These factors are determined with the help of methods like computed tomography (CT), magnetic resonance spectroscopy (MRS), single-photon emission computed tomography (SPECT), magnetic resonance imaging (MRI) and positron emission tomography (PET). With the help of these different combinations of techniques, we can calculate the highest detailed data about the tumours. As the tumour is a soft tissue mass, the use of MRI has been considered to be one of the most standard forms of tests. The MRI makes use of imaging techniques where radio frequency signals are used to excite the target tissue to generate internal images of the tumour under the influence of the radio frequency. With the help of altering the radio signals, various MRI image sequences are generated. The use of different MRI processes helps in the generation of different types of images of tissue contrast which in turn provide valuable information about the tumour. This collection of information can be used to provide better diagnosis and also helps to understand various sub-regions of the tumour easily. There are typically four techniques of the MRI process for the glioma analysis. These include T1-weighted MRI and T2-weighted MRI; T1-weighted MRI includes gadolinium contrast enhancement (T1-GD) and fluid-attenuated inversion recovery (FLAIR). These four processes are shown in the image below. MRI techniques may differ from one another, but approximately, one hundred and fifty slices of 2D plane images are required for the generation of a complete volume 3D image of a brain. These data when combined create very high populated and complicated data.

On a general diagnosis, the use of T1 is made for the determination of healthy tissues, and T2 is used for the delineation of the oedema region of the brain that helps to produce bright signal feedback on the images. The T1-GD is used to determine the tumour border with the help of the contrast agent known as gadolinium ions which accumulate the tissue to activate a region of the tissues in the tumour. Due to the non-interaction of the necrotic cells with the contrast agent, these cells can be easily observed with the help of the hypo-intensive part of the core in the tumour. This makes it possible to make segmented from the non-passive cell sites of the tumour. With the help of FLAIR imaging, molecules of the water are suppressed, and these helps in the differentiation of the cerebrospinal fluid from the oedema region. MRI models with high-grade glioma with enhanced sub-regions of tumour are shown in Fig. 3.

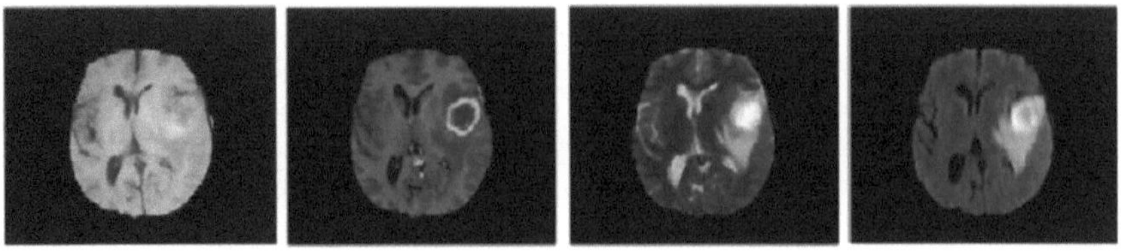

Fig. 3 Four different MRI models with high-grade glioma each improving distinct tumour region. T1, T1-GD, T2 and FLAIR (from left)

It is important to understand the tumour and differentiate the healthy tissues from the tumour cells. The healthy tissues need to be kept intact, while the tumour tissues need to be destroyed completely. Understanding the brain tumour requires the segmentation of the brain tumour. The segmentation includes diagnosis, delineated and separation of the tumour cells from the active brain cells. The different segmentation has been shown below. The current scenario includes the study and manual annotation of a large number of MRI images. Any type of manual process will take lots of time-consuming work, and then, developing a robust system that will able to automate the whole annotation and segmentation process is becoming an interesting and popular research field. Currently, the use of deep learning methods is of the highest segmentation performance and thus makes it an important candidate in the study.

4.1 Brain Tumour Image Segmentation Methodology

Based on the user interaction levels, the brain tumour segmentation process can be categorized as conventional segmentation, semi-automatic segmentation and completely automated segmentation. A sample brain tumour segmentation is shown in Fig. 4.

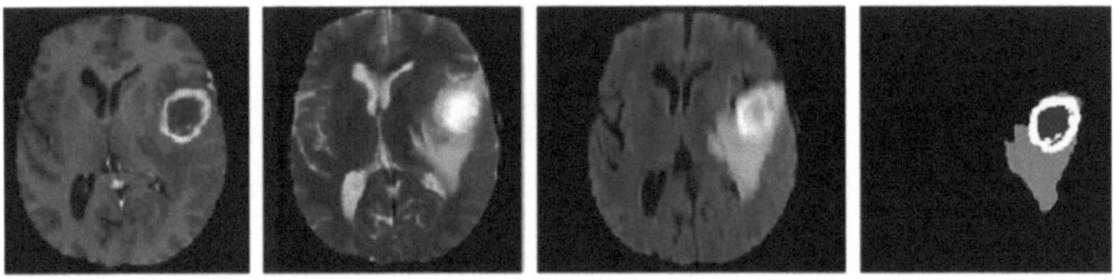

Fig. 4 Brain tumour segmentation. From left T1-Gd, T2, FLAIR and segmented tumour

4.1.1 Methods of Manual Segmentation

The radiologist is required in this process to use the multimodal information which we get from the MRI images and with the knowledge that is gained from training the segment images. The process includes the radiologist to study the portion of the image by portion, understand the tumour and physically draw the areas of the tumour. Alongside, this process consumes a lot of time, and the process is also dependent on radiologist, and the outputs are subjective to various intra- and inter-rating variables. It has been found that the process of manual segmentation is used alongside the process of semi-automatic as well as fully automated processes.

4.1.2 Semi-automatic Segmentation Methods

A partially automated segmentation process requires the communication among human from certain specific purposes: initialization of the process, feedback response and evaluation of the result. Initialization of the process is done by defining the interest domain of images which approximately constitutes tumour regions. This region of interest is then processed by the automated algorithm. Adjust the preprocessing parameters according to the provided images as input. In addition to the process of initialization, the automated algorithm can be used to steer the process to provide the desired result with the help of responses and feedbacks. In the end, the user can estimate the output obtained and adjust the result or replication of the whole process again if the user is not fulfilled.

Hamachi et al. had proposed the use of the 'tumour-cut' method for the semi-automated segmentation process. The process needs a drawing of a maximum possible tumour diameter on the MRI image of the user. After the initialization of the cellular automata-based tumour seed segmentation, the process is executed two times; one tumour seeds given by the user and another for the background seeds to obtain a probability map of the tumour. This process requires the use of all four MRI models (T1, T2, T1-GD and FLAIR), and then, results are merged to note the ultimate volume of the tumour.

Even if the semi-automatic segmentation process is faster than the manual segmentation process, there lies the involvement of inter- and intra-variables. Thus, the research in this field is mainly connected to the fully automated segmentation process.

4.1.3 Fully Automatic Segmentation Methods

In this method, no user interaction is required for the completion of the process. The use of artificial intelligence and prior tested knowledge of tumours are used for the segmentation process.

One of the most challenging problems is the automated segmentation of glioma. The brain carrying tumour is studied as a 3D model. The location, shape and size of the tumour vary from patient to patient. Tumour boundaries are irregular and have

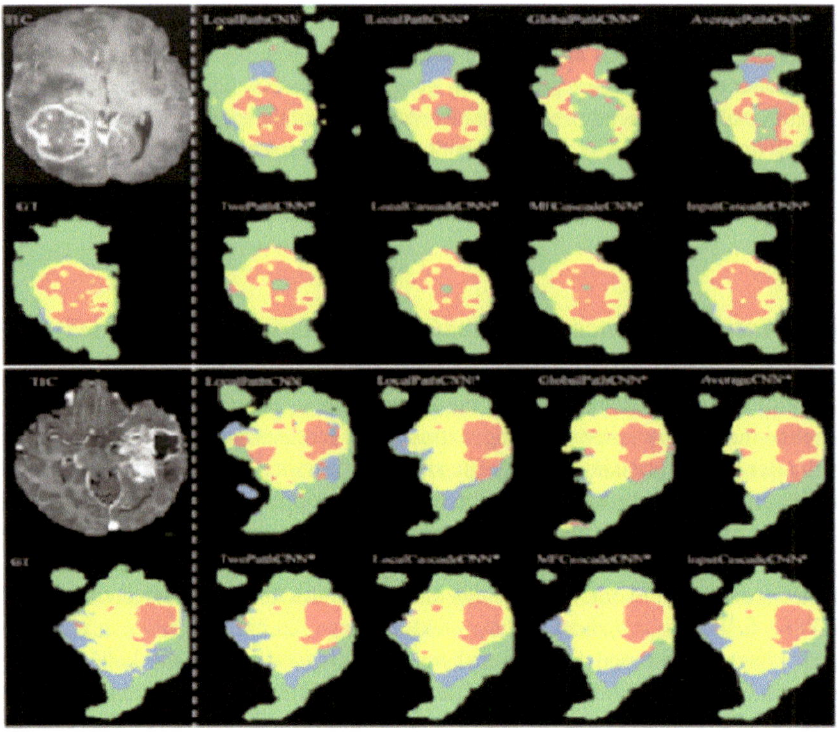

Fig. 6 Visual axial view analysis of CNN architecture

5 Summary

The use of an automated segmentation process for cancer diagnosis has been considered to be a challenging task. With the rise of public data sets and widely accepted BRATS benchmark, it helps the researchers for the development of new methods and thereby compare them with the existing ones. The study discusses different deep learning methods with a little discussion of the traditional methods. Due to the generation of high throughput, the use of deep learning methodology has been considered as the best method for glioma segmentation. The discussion focuses on both the objective of the introduction of deep learning in the image segmentation method and the provision of a general outline for future research as well as the current scenario in the field. The safety and proper methodical understanding of the deep learning method are still not known, but research is being conducted to find these values. It is believed that deep learning will be there in the active research field for many years to come.

Table 1 Brain tumour segmentation approaches with BRATS data analysis

Author	Method	Level of user interaction	Performance (dice scores)		
			Whole tumour	Core tumour	Active tumour
Human rater	Experience in clinical training	Manual	0.88	0.93	0.74
Kwon et al.	A model able to perform registration and joint segmentation	Semi-automated	0.88	0.83	0.72
Tustison et al.	A model trained with the help of first-order mathematical features concatenated RFs	Fully automated	0.87	0.78	0.74
Havaei et al.	With the help of SVM implementation of training and segmentation	Semi-automated	0.86	0.77	0.73
Davy et al.	Two-path CNN able to process global and local processes	Fully automated	0.85	0.74	0.68
Hamamci et al.	A incremental model makes use of automata of cells to generate the probability map of tumour	Semi-automated	0.72	0.57	0.59
Pereira et al.	A CNN with a 3 × 3 filter for deeper framework model	Fully automated	0.88	0.83	0.77
Havaei et al.	Cascaded two-path CNN able to process global and local processes	Fully automated	0.88	0.79	0.73
Urban et al.	3D convolution filters used for 3D-CNN architecture	Fully automated	0.87	0.77	0.73

(continued)

Artificial Intelligence for Internet of Things and Enhanced Medical Systems

Salome Oniani, Gonçalo Marques, Sophio Barnovi, Ivan Miguel Pires, and Akash Kumar Bhoi

Abstract Internet of things (IoT), Big Data, and artificial intelligence (AI) are related research fields that have a relevant impact factor on the design and development of enhanced personalized healthcare systems. This paper discussed the review of AI for IoT and medical systems, which include the usage and practice of AI methodology in different fields of healthcare. The literature review shows that four main areas use AI methodology in medicine, such as heart disease diagnosis, predictive methods, robotic surgery, and personalized treatment. The results confirm that k-nearest neighbors, support vector machine, support vector regression, Naive Bayes, linear regression, regression tree, classification tree, and random forest are the leading AI methods. These methods are mainly used for patient's data analysis to improve health conditions. Robotic surgery systems such as Transoral Robotic Surgery and Automated Endoscopic System for Optimal Positioning lead to several advantages as these methods provide less aggressive treatments and provide better results in terms of blood loss and faster recovery. Furthermore, Internet of medical things addresses numerous health conditions such a vital biophysical parameters supervision, diabetes, and medical decision-making support methods.

S. Oniani · S. Barnovi
Faculty of Informatics and Control Systems, Georgian Technical University, Tbilisi, Georgia
e-mail: s.oniani@gtu.ge

S. Barnovi
e-mail: s.barnovi@gtu.ge

G. Marques (✉) · I. M. Pires
Instituto de Telecomunicações, Universidade Da Beira Interior, 6201-001, Covilhã, Portugal
e-mail: goncalosantosmarques@gmail.com

I. M. Pires
Computer Science Department, Polytechnic Institute of Viseu, Viseu, Portugal
e-mail: impires@it.ubi.pt

A. K. Bhoi
Department of Electrical & Electronics Engineering, Sikkim Manipal Institute of Technology (SMIT) Sikkim Manipal University (SMU), Sikkim, India
e-mail: akash730@gmail.com

A. K. Bhoi et al. (eds.), *Bio-inspired Neurocomputing*, Studies in Computational Intelligence 903, https://doi.org/10.1007/978-981-15-5495-7_3

Keywords Artificial intelligence · Internet of things · Internet of medical things · Medical systems

1 Introduction

IoT can be defined as the ubiquitous and pervasive appearance of cyber-physical systems with communication and sensing capabilities. IoT has been used in several domains such as agriculture [1–4], noise pollution supervision [5–8], indoor quality monitoring [9–14], and other numerous applications for enhanced living environments and promote health. Most relevant, IoT has transformed hospital settings and created a new paradigm named the Internet of medical things (IoMT). IoMT provides several opportunities, as most people use wearable sensors for enhanced health and well-being closely related to eHealth and mHealth [15–20]. Its low cost, availability, and accessibility can explain the increase in the adoption of these mobile sensors. Moreover, these personalized healthcare systems collect relevant biophysical data to support medical decisions and diagnostics [21, 22].

IoT, Big Data, and AI are related research fields that influence the design and development of enhanced personalized healthcare systems [23]. Wearable medical systems with Big Data can provide continuous monitoring features that can collect a high amount of medical data, on which doctors can predict patient's future condition [24]. This data analysis and knowledge extraction are a complex process that must ensure improved security methods [25].

Big Data and AI can offer several opportunities for healthcare systems based on IoT [26]. The Big Data processing methods based on AI can significantly improve global public health [27–30]. IoMT technologies allow to decrease global cost for chronic illnesses prevention. The real-time health data collected by these systems can be used to support patients during self-administration therapies. Mobile devices with mobile applications are often used and integrated with telemedicine and mHealth via the IoMT [31]. The results of medical data analytics from those platforms increase the relevance of data interpretations and decrease time to analyze data outputs [32]. Furthermore, a new system, "Personalized Preventative Health Coaches" has been developed. It maintains experiences and can be used to explain and understand health and well-being data [33, 34].

Networked sensors enable the monitoring of people who do not have access to an efficient health monitoring system [35]. Moreover, using machine learning (ML) and wireless communication together is possible to analyze medical data which helps doctors make suitable recommendations. This paper proposes a comprehensive review of application IoMT into the medical fields. It discusses what kind of IoMT devices was used in different therapeutic areas and which methodology was used to collect clinical data and to support analysis and diagnostics. The second aim of the manuscript is to create a review of AI methodologies in healthcare IoT.

2 Applications of Artificial Intelligence on the Internet of Medical Things

IoMT aims to support healthcare systems and consists of acquisition data from medical devices and applications. It uses various sensors such as visual sensor, accelerometer sensor, carbon dioxide sensor, temperature sensor, electrocardiogram (ECG)/electroencephalogram/electromyogram sensors, gyroscope sensor, pressure sensor, humidity sensor, respiration sensor, blood oxygen saturation sensor, and blood-pressure sensor to monitor patient's health in real-time [36]. These medical devices monitor the patients' health conditions collect clinical data and send it to doctors through remote cloud data centers [37, 38]. However, how to control clinical applications which generate a large amount of medical data from connected devices is the main challenge for IoMT. This review discusses the IoMT tools and AI methodology used in heart disease diagnosis (Sect. 2.1), the use of AI and IoMT for disease prediction (Sect. 2.2), the influences of robotic surgery in healthcare (Sect. 2.3), and the use of IoMT combined with AI for personalized treatment and its benefits (Sect. 2.4).

2.1 Heart Disease Diagnosis

Numerous researchers explore machine learning algorithms (MLA) in heart disease diagnosis systems. MLA has the most important place in healthcare systems, especially heart disease diagnosis filed [39]. The support vector machine (SVM) and Naive Bayes (NB) ML methods are used in the heart disease diagnosis systems. The SVM develops a set of hyperplanes in the infinite-dimensional area. It estimates the linear separation surface with a maximum margin in a provided training dataset. Also, SVM has been widely used in pattern analysis puzzles and nonlinear regression. However, it does not make any strong hypotheses about the data [40]. The NB AI algorithm is applied in text classification, document categorization, spam filtering, and news classification. This machine learning method functions well if the input data are classified into predefined groups. Also, it requires less data than logistic regression [41].

Furthermore, mobile cardiac telemetry devices are essential for atrial fibrillation detection. These devices perform real-time monitoring for an extended period of heart rhythm and contribute to arrhythmia detection. Besides, medical wearables and AI can be used for developing efficient methods for heart disease diagnosis through remote patient monitoring [42]. A machine learning system (MLS) can identify symptoms of heart disease by examining scans data from previous patients, and based on the analysis of prior data. Moreover, MLS can predict the possibility of an occurrence of a heart attack in the future. At the Mayo Clinic, researchers examined using AI and ECG. MLS has been used to analyze the heart's electrical activity and identify asymptomatic left ventricular dysfunction (ALVD). These researchers studied by Schurr et al. at Eberhard Karls University developed MIS, who propose a master-slave manipulator system called ARTEMIS [55].

Furthermore, in MiTech laboratory, a miniature robotic system prototype has been designed. The miniature robots can improve computer-enhanced colonoscopy methods. It consists of a voice-activated robotic endoscope, Automated Endoscopic System for Optimal Positioning (AESOP) system, and master–slave systems [72]. The miniature robotic system has tree hands, one of them is AESOP. In this system, AESOP is controlled via voice and provides surgery visualization features. Using Zeus system, the operations become more safe and flexible. It has a distance monitoring system and its cameras can create full images of the surgeon room. Therefore, the surgeons can stay another room, make directives according to the monitoring system and move robotics hands (left and right) with millimeters accuracy.

2.4 *Personalized Treatment*

In modern medicine, personalized treatment is a popular field. It means collecting and applying Big Data, using ML and data science methodology, creating statistical analysis and provides reliable results. Especially in recent years, when Big Data and ML technologies development stage has almost reached its last point and medical devices have started integrating them. Personalized medicine is going to use ML and Big Data into clinical practice. Using Big Data and ML algorithms for personalized treatment means more complex model predictions, more reliable predictive models, and validated clinical trials [73]. Recently, research results show that personalized medicine has the potential to become more reliable and sustainable field in the medicine world.

Nowadays, the therapeutic area is separated in five main terms. The first is "personalized medicine" with its special IoMT devices, second is "individualized medicine" with genius technologies, third is "stratified medicine" with treatment methodology, fourth is "segmented medicine" with Big Data, and the last is "targeted therapies" with classification of the type of cancer. Thus, scientists have to work and analyze large amounts of data which requires high-degree accuracy. It means the improvement of analyzing strategies, integration and combination of new high-quality technology, generation and interpretation of Big Data. AI methodology offers several possibilities for personalized treatment, especially in data science and statistical analysis [74]. Using AI technics in personalized medicines creates new ways to advance this medical field. IoMT, Al, and ML together enable the doctors can access, collect, store, and improve statistical analyses of patient's condition. Based on deep neural network (DNN) has created DeepSurv system [75] which unites state-of-the-art survival method and Cox proportional hazards analyses. DeepSurv services personalized treatment recommendations and creates a statistical model for patients' treatments effectiveness.

Developing medical technology has posed new challenges for patient's treatment, particularly in patients who require more accuracy during diagnosis and treatment.

Therefore, intelligent health monitoring systems have been created, which provides high performance in real-time. There is an Intelligent IoMT-based health monitoring system, such as the BioSenHealth 1.0 [76]. This system is created using Things-peak.com and uses could flatform to send patient live data to doctors. BioSenHealth 1.0 is a prototype. It can monitor pulse rate/heart rate, body temperature and body oxygen level, make statistical analysis of real-time data, and create visual graphs.

The fact is that Alzheimer's disease is a difficult illness, which needs everyday examination. IoMT made this process more easy and comfortable. Khan and his companions [77] created an algorithm to monitor Alzheimer's disease patients. This algorithm uses a hybrid feature vector combining methodology to create statistical analysis of IoMT data. Its three-dimensional view creates a complete picture of the patients' condition. The experimental results show an average of 99.2% and 99.02% for binary and multi-class classifications via using this new algorithm.

Based on Diabetes 1.0 and Diabetes 2.0, methodology has been created was 5G-Smart Diabetes personalized treatment system [78]. It combines Big Data methodology, wearable 2.0 technology, and ML algorithm. 5G-Smart Diabetes system works for patients who have diabetes and make analysis sensors data to predict diabetes progress.

3 Discussion and Results

The literature review on heart disease diagnosis shows that there are different artificial algorithms, frameworks, and methodologies. Table 1 presents the distribution of the studies which apply AI methodology for heart disease diagnosis. Based on the results of our review, the most used methods are NB and deep learning, and the less-used methods used in studies analyzed are multilayer perceptron, RFRS, cluster analysis, and EDPDCS. Furthermore, studies proposed by [35] incorporate different AI methodologies such as SVM and NB.

Table 2 presents a comparison summary of the predictive methods used for IoMT applications. Based on Table 2, it is possible to identify that most of the studies analyzed in this paper incorporate NNs for prediction propose in several medical fields such as clinical visits, chemical composition, risk, diagnosis, privacy, and cancer recognition. Moreover, it is possible to conclude that AI methods have accurate and useful results in the medical field.

Robotic surgery is playing an essential role in the development of medicine, but professionals are required too. Table 3 presents a summary of the analyzed studies on robot surgery. The robotic surgery systems are great assistant competent professionals. We found that robot surgery applications lead to numerous advantages. Studies analyzed present promising results for the medical field. On the one hand, the non-autonomous technologies such as TORS and AESOP lead to several benefits as these methods provide less aggressive treatments and provide better results in terms of blood loss and a faster recovery. On the other hand, MIS provides hardware and software solutions to address an important topic in these kinds of applications, which

Table 1 AL methodology for heart disease diagnosis

AI method	Description	Function	Studies
Support Vector Machine	Creation of classification and regression analysis	Develop a hyperplane. Use in pattern analysis puzzles and nonlinear regression	[35, 40]
Naive Bayes	Probabilistic classifiers	Creation of classification, sentiment analysis, spam filtering, and news classification	[23, 35, 38]
Cluster analysis and efficient differentially private data clustering scheme	Classify a sample of subjects (or objects) based on a set of measured variables in different groups	Interaction using the K-means algorithm by comparing the NICV	[46]
Convolutional neural network	Class of DNNs, most commonly applied to analyze visual imagery. Known as shift invariant or space invariant artificial neural networks (SIANN)	Classify patients with ventricular dysfunction	[43, 60]
Deep learning	DNN learning and a prediction model	Enable machines to process data with a nonlinear approach	[40, 47, 50]
Recurrent fuzzy neural network	Neural Classifier	Using DPSS to help pretention healthcare services and data security	[55, 56]
Grasshopper optimization	Gravity force on it GI and the wind advection AI	MLP NN for tackling optimization with flexible and adaptive searching methodology	[79]
ReliefF and RS	Approach as an integrated feature selection system for heart disease diagnosis	To data analysis and data mining that has been applied successfully to many real-life problems in medicine, pharmacology	[45]
Multilayer perceptron	SLPs and ANNs technology together	Provide various continuous functions	[79]

is the human–robot interaction and all the relevant related issues such as safety and accuracy.

The personalized healthcare applications analyzed in this review are summarized in Table 4. These applications address several health complications, such as vital biophysical parameters supervision, diabetes, and medical decision-making support methods. Several medical systems for personalized healthcare applications incorporate several wearables and mobile sensors for oxygen, heart rate, and temperature monitoring. Some of these systems provide real-time data consulting methods to promote medical intervention in healthcare facilities.

Table 2 Summary of predictive methods assisted by IoMT

Study	Application	IA methods	Results
Jin et al. [51]	Predict respiratory the clinic visits	Backpropagation NN	The proposed model fits the nonlinear relationship between the respiratory outpatient visits
Yao et al. [52]	Deep learning model for predicting chemical composition	CNN	Developed CNN, like stacked auto-encoders, deep belief networks, and RNN
Fki et al. [53]	ML with IoT data for risk prediction	K-nearest neighbors, NB, SVM, LR, support vector regression, classification trees, regression trees, and random forests	The model is a set of hypotheses about dialysis biomarkers proved in a probabilistic format
Kumar et al. [55]	Fuzzy neural classifier for data prediction and diagnosis	LR and NN	Fuzzy rule-based neural classifier for diagnosing the disease and the severity
Devarajan et al. [56]	Privacy-Aware Disease Prediction Support System	K-nearest neighbor and case-based reasoning	The combinatorial advantage of Fuzzy set theory
Masood et al. [60]	Cancer detection and stage classification on CT images	Deep FCNN and CNN	Potential for the proposed technique in helping radiologists in improving nodule detection accuracy with efficiency

Table 3 Summary of AI methods used in robot surgery

Robot surgery	Description	Advantages	Number of studies
TORS	Transoral Robotic Surgery	Fewer blood losses, faster postoperative recovery, and fewer adhesions	[58, 62, 64]
MIS	Management Information Systems	Improve consistency, safety and accuracy, advances in imaging and human-robot interaction	[65–69]
STAR	Smart Tissue Autonomous Robot	Nascent clinical viability of a self-governing soft-tissue surgical robot	[70, 71]
AESOP	Automatic Endoscopic System for Optimal Positioning	Robotic endoscope and surgical robotic systems	[72]

Table 4 Personalized healthcare applications summary

Project	Application	Sensors	Results	References
BioSenHealth 1.0	Real-time monitoring of vital statistics of patients	Body oxygen, temperature level, and pulse rate/heart rate	Live data access using thingspeak.com cloud platform	[76]
DeepSurv	DNN and state-of-the-art survival method	Pulse rate/heart rate	Provide individual treatment recommendations	[75]
5G-Smart Diabetes	Personalized diabetes diagnosis	Temperature, electrocardiography, and blood oxygen	Real-time system to analysis diabetes suffering	[78]

The results presented state that using AI methodology improves the main medicine areas such as heart disease, predicting methods, surgery, and personalized treatment. AI gives a chance to realize patients' privies conditions and get a result in a short time. Its methodology can see the deeper reason for illnesses and predict a patient's health in the future.

4 Conclusions

IoMT and personalized healthcare systems can be seen as the global support of the data collected from medical devices. It is a large area that includes the blockchain for IoMT in medicine, wearable sensors, mHealth Things, IoT networks for health care, Big Data, ambient sensors, MEMS devices, robotics in the Healthcare Industry, mobile health systems, health informatics security, privacy methods and AI. As a consequence, this paper reviews AI for IoT and medical systems, which include the usage and practice of multiple methodologies in different fields of medicine. The study results show that k-nearest neighbors, NB, SVM, LR, support vector regression, classification tree, regression tree, and random forest are the most used AI methods. These methods are applied to analyze and measure the patient's data to improve their health condition. Regarding the heart disease diagnosis, the primary usage AI algorithms are NB, SVM, and deep learning multilayer perceptron, RFRS, cluster analysis, and EDPDCS. In the predictive methods, the most used methods are CNN and k-nearest neighbors.

Furthermore, robotic surgery systems are relevant to support health professionals. Nevertheless, the non-autonomous technologies such as TORS and AESOP lead to several advantages as these methods provide less aggressive treatments and provide better results in terms of blood loss and a faster recovery. IoMT and its personalized healthcare applications address numerous health symptoms such as vital biophysical parameters supervision, diabetes, and decision-making support methods.

References

1. Marques, G., Pitarma, R.: Environmental quality monitoring system based on internet of things for laboratory conditions supervision. In: Rocha, Á., Adeli, H., Reis, L.P., Costanzo, S. (eds.) New Knowledge in Information Systems and Technologies, pp. 34–44. Springer International Publishing, Cham (2019). https://doi.org/10.1007/978-3-030-16187-3_4
2. Mehra, M., Saxena, S., Sankaranarayanan, S., Tom, R.J., Veeramanikandan, M.: IoT based hydroponics system using deep neural networks. Comput. Electron. Agric. **155**, 473–486 (2018). https://doi.org/10.1016/j.compag.2018.10.015
3. Marques, G., Aleixo, D., Pitarma, R.: Enhanced hydroponic agriculture environmental monitoring: an internet of things approach. In: Rodrigues, J.M.F., Cardoso, P.J.S., Monteiro, J., Lam, R., Krzhizhanovskaya, V.V., Lees, M.H., Dongarra, J.J., Sloot, P.M.A. (eds.) Computational Science—ICCS 2019, pp. 658–669. Springer International Publishing, Cham (2019). https://doi.org/10.1007/978-3-030-22744-9_51
4. Marques, G., Pitarma, R.: Agricultural environment monitoring system using wireless sensor networks and IoT. In: 2018 13th Iberian Conference on Information Systems and Technologies (CISTI), pp. 1–6. IEEE, Caceres (2018). https://doi.org/10.23919/CISTI.2018.8399320
5. Marques, G., Pitarma, R.: Noise mapping through mobile crowdsourcing for enhanced living environments. In: Rodrigues, J.M.F., Cardoso, P.J.S., Monteiro, J., Lam, R., Krzhizhanovskaya, V.V., Lees, M.H., Dongarra, J.J., Sloot, P.M.A. (eds.) Computational Science—ICCS 2019, pp. 670–679. Springer International Publishing, Cham (2019). https://doi.org/10.1007/978-3-030-22744-9_52
6. Skouby, K.E., Lynggaard, P.: Smart home and smart city solutions enabled by 5G, IoT, AAI and CoT services. In: 2014 International Conference on Contemporary Computing and Informatics (IC3I), pp. 874–878. IEEE, Mysore, India (2014). https://doi.org/10.1109/IC3I.2014.7019822
7. Marques, G., Pitarma, R.: Noise Monitoring for Enhanced Living Environments Based on Internet of Things. In: Rocha, Á., Adeli, H., Reis, L.P., Costanzo, S. (eds.) New Knowledge in Information Systems and Technologies, pp. 45–54. Springer International Publishing, Cham (2019). https://doi.org/10.1007/978-3-030-16187-3_5
8. Dutta, J., Roy, S.: IoT-fog-cloud based architecture for smart city: prototype of a smart building. In: 2017 7th International Conference on Cloud Computing, Data Science & Engineering—Confluence, pp. 237–242. IEEE, Noida, India (2017). https://doi.org/10.1109/CONFLUENCE.2017.7943156
9. Marques, G., Pitarma, R.: An internet of things-based environmental quality management system to supervise the indoor laboratory conditions. Appl. Sci. **9**, 438 (2019). https://doi.org/10.3390/app9030438
10. Lohani, D., Acharya, D.: SmartVent: a context aware IoT system to measure indoor air quality and ventilation rate. In: 2016 17th IEEE International Conference on Mobile Data Management (MDM), pp. 64–69. IEEE, Porto (2016). https://doi.org/10.1109/MDM.2016.91
11. Marques, G., Pitarma, R.: Monitoring and control of the indoor environment. In: 2017 12th Iberian Conference on Information Systems and Technologies (CISTI), pp. 1–6. IEEE, Lisbon, Portugal (2017). https://doi.org/10.23919/CISTI.2017.7975737
12. Marques, G., Pitarma, R.: A cost-effective air quality supervision solution for enhanced living environments through the internet of things. Electronics **8**, 170 (2019). https://doi.org/10.3390/electronics8020170
13. Wei, S., Ning, F., Simon, F., Kyungeun, C.: A deep belief network for electricity utilisation feature analysis of air conditioners using a smart IoT platform. J. Inf. Process. Syst. **14**, 162–175 (2018). https://doi.org/10.3745/JIPS.04.0056
14. Marques, G., Pires, I., Miranda, N., Pitarma, R.: Air quality monitoring using assistive robots for ambient assisted living and enhanced living environments through internet of things. Electronics **8**, 1375 (2019). https://doi.org/10.3390/electronics8121375
15. Marques, G., Pitarma, R.: mHealth: indoor environmental quality measuring system for enhanced health and well-being based on internet of things. JSAN **8**, 43 (2019). https://doi.org/10.3390/jsan8030043

16. Buckingham, S.A., Williams, A.J., Morrissey, K., Price, L., Harrison, J.: Mobile health interventions to promote physical activity and reduce sedentary behaviour in the workplace: a systematic review. Digital Health **5**, 205520761983988 (2019). https://doi.org/10.1177/2055207619839883
17. Marques, G.: Ambient Assisted Living and Internet of Things. In: Cardoso, P.J.S., Monteiro, J., Semião, J., Rodrigues, J.M.F. (eds.) Harnessing the Internet of Everything (IoE) for Accelerated Innovation Opportunities, pp. 100–115. IGI Global, Hershey, PA, USA (2019). https://doi.org/10.4018/978-1-5225-7332-6.ch005
18. Silva, B.M.C., Rodrigues, J.J.P.C., de la Torre Díez, I., López-Coronado, M., Saleem, K.: Mobile-health: A review of current state in 2015. J. Biomed. Inf. **56**, 265–272 (2015). https://doi.org/10.1016/j.jbi.2015.06.003
19. Marques, G., Pitarma, R., M. Garcia, N., Pombo, N.: Internet of things architectures, technologies, applications, challenges, and future directions for enhanced living environments and healthcare systems: a review. Electronics. **8**, 1081 (2019). https://doi.org/10.3390/electronics8101081
20. Lake, D., Milito, R.M.R., Morrow, M., Vargheese, R.: Internet of things: architectural framework for ehealth security. J. ICT Stand. **1**, 301–328 (2014)
21. Firouzi, F., Rahmani, A.M., Mankodiya, K., Badaroglu, M., Merrett, G.V., Wong, P., Farahani, B.: Internet-of-Things and big data for smarter healthcare: from device to architecture, applications and analytics. Future Gener. Comput. Syst. **78**, 583–586 (2018). https://doi.org/10.1016/j.future.2017.09.016
22. Marques, G., Garcia, N., Pombo, N.: A survey on IoT: architectures, elements, applications, QoS, platforms and security concepts. In: Mavromoustakis, C.X., Mastorakis, G., Dobre, C. (eds.) Advances in Mobile Cloud Computing and Big Data in the 5G Era, pp. 115–130. Springer International Publishing, Cham (2017). https://doi.org/10.1007/978-3-319-45145-9_5
23. Martis, R.J., Gurupur, V.P., Lin, H., Islam, A., Fernandes, S.L.: Recent advances in big data analytics, internet of things and machine learning. Future Gener. Comput. Syst. **88**, 696–698 (2018). https://doi.org/10.1016/j.future.2018.07.057
24. Marques, G., Pitarma, R.: Smartwatch-based application for enhanced healthy lifestyle in indoor environments. In: Omar, S., Haji Suhaili, W.S., Phon-Amnuaisuk, S. (eds.) Computational Intelligence in Information Systems, pp. 168–177. Springer International Publishing, Cham (2019). https://doi.org/10.1007/978-3-030-03302-6_15
25. Manogaran, G., Varatharajan, R., Lopez, D., Kumar, P.M., Sundarasekar, R., Thota, C.: A new architecture of Internet of Things and big data ecosystem for secured smart healthcare monitoring and alerting system. Future Gener. Comput. Syst. **82**, 375–387 (2018). https://doi.org/10.1016/j.future.2017.10.045
26. Özdemir, V., Hekim, N.: Birth of industry 5.0: making sense of big data with artificial intelligence, "The Internet of Things" and Next-Generation Technology Policy. OMICS J. Integr. Biol. **22**, 65–76 (2018). https://doi.org/10.1089/omi.2017.0194
27. Allam, Z., Dhunny, Z.A.: On big data, artificial intelligence and smart cities. Cities **89**, 80–91 (2019). https://doi.org/10.1016/j.cities.2019.01.032
28. Marques, G., Roque Ferreira, C., Pitarma, R.: A system based on the internet of things for real-time particle monitoring in buildings. Int. J. Environ. Res. Public Health. **15**, 821 (2018). https://doi.org/10.3390/ijerph15040821
29. Pitarma, R., Marques, G., Ferreira, B.R.: Monitoring indoor air quality for enhanced occupational health. J. Med. Syst. **41**, (2017). https://doi.org/10.1007/s10916-016-0667-2
30. Marques, G., Pitarma, R.: An indoor monitoring system for ambient assisted living based on internet of things architecture. Int. J. Environ. Res. Public Health. **13**, 1152 (2016). https://doi.org/10.3390/ijerph13111152
31. Marques, G.M.S., Pitarma, R.: Smartphone application for enhanced indoor health environments. J. Inf. Syst. Eng. Manag. **1**, (2016). https://doi.org/10.20897/lectito.201649
32. Dimitrov, D.V.: Medical internet of things and big data in healthcare. Health Inform Res. **22**, 156 (2016). https://doi.org/10.4258/hir.2016.22.3.156

33. Marques, G., Pitarma, R.: IAQ Evaluation using an IoT CO_2 monitoring system for enhanced living environments. In: Rocha, Á., Adeli, H., Reis, L.P., Costanzo, S. (eds.) Trends and Advances in Information Systems and Technologies, pp. 1169–1177. Springer International Publishing, Cham (2018). https://doi.org/10.1007/978-3-319-77712-2_112
34. Marques, G., Ferreira, C.R., Pitarma, R.: Indoor air quality assessment using a CO_2 monitoring system based on internet of things. J. Med. Syst. **43**, (2019). https://doi.org/10.1007/s10916-019-1184-x
35. Kaur, P., Kumar, R., Kumar, M.: A healthcare monitoring system using random forest and internet of things (IoT). Multimed Tools Appl. **78**, 19905–19916 (2019). https://doi.org/10.1007/s11042-019-7327-8
36. Manogaran, G., Chilamkurti, N., Hsu, C.-H.: Emerging trends, issues, and challenges in internet of medical things and wireless networks. Pers. Ubiquit. Comput. (2018). https://doi.org/10.1007/s00779-018-1178-6
37. Kaur, P., Sharma, N., Singh, A., Gill, B.: CI-DPF: a cloud IoT based framework for diabetes prediction. In: 2018 IEEE 9th Annual Information Technology, Electronics and Mobile Communication Conference (IEMCON), pp. 654–660. IEEE, Vancouver, BC (2018). https://doi.org/10.1109/IEMCON.2018.8614775
38. Dankwa-Mullan, I., Rivo, M., Sepulveda, M., Park, Y., Snowdon, J., Rhee, K.: Transforming diabetes care through artificial intelligence: the future is here. Popul. Health Manag. **22**, 229–242 (2019). https://doi.org/10.1089/pop.2018.0129
39. Dua, S., Acharya, U.R., Dua, P. (eds.): Machine Learning in Healthcare Informatics. Springer, Berlin (2014)
40. Brink, H., Richards, J.W., Fetherolf, M.: Real-world Machine Learning. Manning, Shelter Island (2017)
41. Cielen, D., Meysman, A., Ali, M.: Introducing Data Science: Big Data, Machine Learning, and More, Using Python Tools. Manning Publications, Shelter Island, NY (2016)
42. Kumar, P.M., Devi Gandhi, U.: A novel three-tier internet of things architecture with machine learning algorithm for early detection of heart diseases. Comput. Electr. Eng. **65**, 222–235 (2018). https://doi.org/10.1016/j.compeleceng.2017.09.001
43. Attia, Z.I., Kapa, S., Lopez-Jimenez, F., McKie, P.M., Ladewig, D.J., Satam, G., Pellikka, P.A., Enriquez-Sarano, M., Noseworthy, P.A., Munger, T.M., Asirvatham, S.J., Scott, C.G., Carter, R.E., Friedman, P.A.: Screening for cardiac contractile dysfunction using an artificial intelligence–enabled electrocardiogram. Nat. Med. **25**, 70–74 (2019). https://doi.org/10.1038/s41591-018-0240-2
44. Krittanawong, C., Zhang, H., Wang, Z., Aydar, M., Kitai, T.: Artificial intelligence in precision cardiovascular medicine. J. Am. Coll. Cardiol. **69**, 2657–2664 (2017). https://doi.org/10.1016/j.jacc.2017.03.571
45. Li, B., Wen, T., Hu, C., Zhou, B.: Power System Transient Stability Prediction Algorithm Based on ReliefF and LSTM. In: Sun, X., Pan, Z., Bertino, E. (eds.) Artificial Intelligence and Security, pp. 74–84. Springer International Publishing, Cham (2019). https://doi.org/10.1007/978-3-030-24274-9_7
46. Guan, Z., Lv, Z., Du, X., Wu, L., Guizani, M.: Achieving data utility-privacy tradeoff in internet of medical things: a machine learning approach. Future Gener. Comput. Syst. **98**, 60–68 (2019). https://doi.org/10.1016/j.future.2019.01.058
47. Allouzi, M.A., Khan, J.I.: Soter: trust discovery framework for internet of medical things (IoMT). In: 2019 IEEE 20th International Symposium on "A World of Wireless, Mobile and Multimedia Networks" (WoWMoM), pp. 1–9. IEEE, Washington, DC, USA (2019). https://doi.org/10.1109/WoWMoM.2019.8792971
48. Yanambaka, V.P., Mohanty, S.P., Kougianos, E., Puthal, D.: PMsec: physical unclonable function-based robust and lightweight authentication in the internet of medical things. IEEE Trans. Consumer Electron. **65**, 388–397 (2019). https://doi.org/10.1109/TCE.2019.2926192
49. Manogaran, G., Varatharajan, R., Priyan, M.K.: Hybrid recommendation system for heart disease diagnosis based on multiple kernel learning with adaptive neuro-fuzzy inference system. Multimed Tools Appl. **77**, 4379–4399 (2018). https://doi.org/10.1007/s11042-017-5515-y

A Brief Review on Brain Tumour Detection and Classifications

K. Sri Sabarimani and R. Arthi

Abstract Recognition of unusual pixels in a source brain Magnetic Resonance Image (MRI) remains a difficult activity owing to various similarities with that of surrounding pixels. The presence of brain tumour pixel detection process has been simplified using preprocessing steps before the proposed method starts. The preprocessing steps earns an attempt to enhance the internal pixel regions for improving the brain tumor pixel detection rate. The preprocessing stage may include noise reduction, pixel resolution enhancement, image registration, edge detection methods and artifact detection and reduction. The available techniques in preprocessing stage has different methods for improving the clarity of the source brain MRI image that leads to further processing such as segmentation and classification of tumor images. The proposed work discusses various conventional methods for brain tumour detection and classifications with the limited number of available information.

Keywords Detection · Abnormal pixel · Tumour · Preprocessing · Segmentation

1 Introduction

Preprocessing stage is an important stage for the identification of tumors in brain Magnetic Resonance Images (MRI). It enhances the internal pixel regions for improving the brain tumour pixel detection rate. The preprocessing stage may include noise reduction, pixel resolution enhancement, image registration, edge detection methods and artifact detection and reduction. Each method in preprocessing stage permit different methods for improving the clarity of the source brain MRI image for further processing such as segmentation and classification of tumor images. Some of the conventional methods for preprocessing stage use transformation techniques

K. Sri Sabarimani (✉) · R. Arthi
Electronics and Communication Engineering, SRM Institute of Science and Technology, Ramapuram Campus, Chennai 600089, India
e-mail: srisabak@srmist.edu.in

R. Arthi
e-mail: arthir2@srmist.edu.in

A. K. Bhoi et al. (eds.), *Bio-inspired Neurocomputing*, Studies in Computational Intelligence 903, https://doi.org/10.1007/978-981-15-5495-7_4

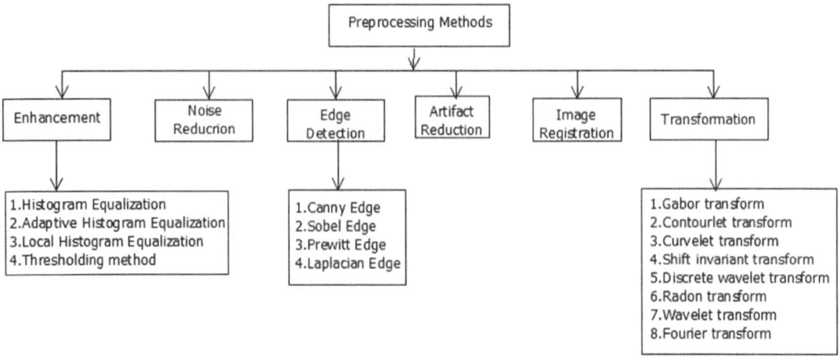

Fig. 1 Preprocessing methods on brain MRI images

that relate each pixel in an image into other domain pixels which is suitable for further feature extraction process. This transformation of one domain pixels into other domain pixels consumes precise amount of time period for their conversion of pixels. Many conventional methods use the combination of two or more preprocessing stages to directly develop the clarity in pixels of brain MRI images. This paper focuses on the main methodologies that has been used in every preprocessing stages for analyzing their performance.

2 Classifications on Preprocessing Stages in Brain Tumor Detection

The classifications on preprocessing stages for brain tumour detection are shown in Fig. 1 with remarks.

3 Edge Detection Method

Edge detection method uses edges in the brain MRI images to detect the presence of tumor pixels. The abrupt variation of intensity of pixels in an image is known as edge. Many edge detection methods are available for the brain tumour detection process has been stated as Canny edge detection, Sobel edge detection, Prewitt edge detection, Laplacian edge detection. Various remarks on the classification of Edge Detection are listed in Table 1.

Canny edge detection has been used as operator [1] for detecting edge pixels in brain MRI images. The Canny edge detection process has eliminated the problem of threshold in brain images for the detection of tumor pixels. This work has analyzed the impact of edge detection algorithms on the detection process of brain growths in both

Table 1 Various remarks on the classification of edge detection

	Classification	Author	Remarks
Edge detection	Canny edge detection	Alexander Zotin et al. (2018)	• The Figure of Merit and maintains its accuracy of the values and does not change with the increase in noise level
			• Sensitivity index decreased within acceptable range
		Manasa et al. (2016)	• The performance of the canny edge detector was superior
			• Higher the Peak signal to Noise Ratio (PSNR), better the reconstruction
		Ismail Yaqub Maolood et al. (2018)	• This method obtains MSE and PSNR for a limit of the high-level description of images compared with previous methods
	Sobel edge detection	Nagabushanam et al. (2018)	• Tumor detected in this method was closer to the tumor in the original image
			• Improved closed contour algorithms was needed that increases the area of the region and decrease their thickness of boundaries
		Asra Aslam et al. (2015)	• Lesser false edges and contain Closed contours. Closed contour algorithm enhances the region and diminishes the width of boundary lines of regions
	Prewitt edge detection	Esmail Hassan et al. (2015)	• Fast and efficient detection of tumors
	Laplacian edge detection	Xu Yan et al. (2011)	• The details of examined image are improved without the evidence of an artifact

brain MRI and CT images. The evaluation parameter considered are performance analysis of Peak Signal to Noise Ratio (PSNR) and the edge detection rate.

Sobel edge detection method has been applied [2] for recognizing the unusual edges in brain MRI images. The Gradient mask was created to use in the design of Sobel kernel and this mask was convolved with the source brain image to obtain abnormal edges in the source images. Then, morphological function along with watershed segmentation method was applied on the edge detected brain images for improving the edges detection rate. An improved edge detection method has been developed for using the Prewitt Edge detection technique [3] by applying this edge detection operator on basis of brain MRI images in demand to detect the edges around brain images.

Edge detection centered [4] has been used on Fuzzy C-Mean Clustering (FCM) for detecting the fine edges in medical image processing. The noise of the picture has been lowered with a non-linear digital filtering technique called the median filter to remove noise and improved by BCET (Balanced Contrast Enhancement Technique). The depiction has been processed by two methods such as FCM clustering and the Canny edge detector, that has been employed to build the border plot of a brain tumour. The sharp border has been resolved by the Canny Edge detection technique that was centered atop by the inclined rate of a pixel intensity. Initially, the Gaussian filter was exercised to eliminate primary representations, each pixel was used to find the magnitude and direction of the inclined values in the image. Along the direction of the gradient, the magnitude of the pixel should be greater than two neighbouring pixels.

An advanced edge observation technique [5] has been used for MR image division. This work blends the Sobel edge detection with the image threshold method and determines distinct sections using closed contour algorithm. Lastly, malignant growth was separated from the image using intensity information within the closed contours.

A procedure has been suggested that uses the original image [6] for extracting resistance from noise edging, then unites the denoised image and High Frequency Components from Laplacian Pyramid technique for enhancement. This procedure was verified through artificially manufactured data as well as real data, that has resulted in smoothening of the image in homogeneous regions, augmented facts and limits of functional shape lacking the initiation of obvious objects. The author has used NLM (Non-Local Mean) method for denoising and SUSAN operator that provides good noise resistance. The edge detection procedure has stated the normal way of the attractive picture by means of a fixed frame that was positioned on the individual pixel, then using nearby usual procedures to provide an edge response. This reaction was processed to contribute the output as a set of edges.

Segmented Cluster edge detection limit has been used [7] in brain MRI using Fuzzy C-means threshold with a Gaussian filter based on edge-detection method. With an MRI as an input, Fuzzy C-Means (FCM) clustering created image clusters, showing white and gray matter with an addition of 5% Gaussian blur noise.

4 Noise Reduction Methods

The source of brain MRI images has been affected by the noise of brain tumour detection process. The noisy pixel intensity was similar to the intensity of the tumour abnormal pixel in acquired brain images. There are a number of noises generated on the source brain MRI images under different criteria. These types of noise contents should be detected and removed before applying the proposed method on brain tumor detection. In this regard, various filters are used to spot and reduce the noise contents in brain images. They are classified into Mean filter, the Median filter and weighted median filter. These filters have been employed as the mask kernel, and it was applied over the initial part of the pixels to ending of the pixels in source brain images as depicted in [8].

Figure 2a shows the noise included in the source brain MRI image and Fig. 2b shows the noise content reduced brain MRI image using the mean filtering method. An adaptive threshold filter based non-linear evaluation method [9] has been used to detect and reduce the noise contents in images of the brain. The author has analyzed its effectiveness with respect to various performance evaluation factors. The noise reduction result of the proposed adaptive threshold filter was compared with linear filter noise reduction results.

A pre-model procedure for detection and reduction of noise [10] has shown improvement in brain MRI images. The author has reviewed a number of noise reduction algorithms and compared with various filters. The performance of each filtering algorithm was compared with many stated of art methods. The author has introduced a new, confide to a limited part filter [11] for the noise decline in Magnetic Resonance Image Sets. Being a multispectral extension of confide to the limited part

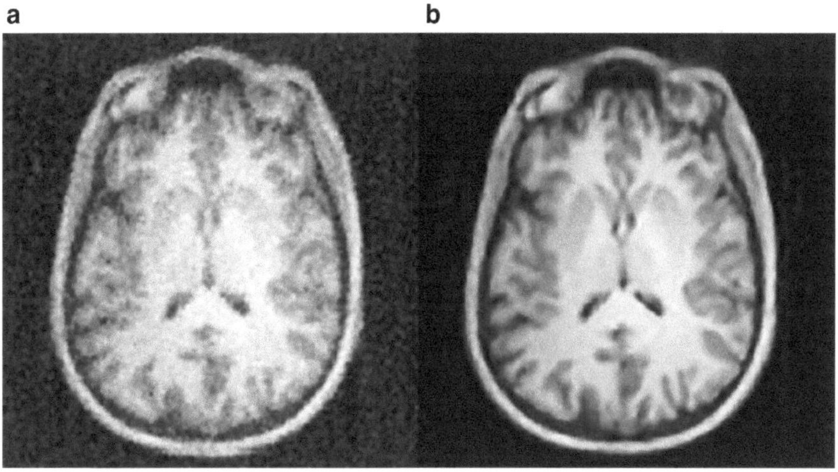

Fig. 2 **a** Noise included in the source brain MRI image **b** Noise content reduced brain MRI image using mean filtering method

of maximum likelihood filter, it has combined both spatial and spectral information namely, signals intensities across various image emphasis to perform the efficient reduction of the noise. Noise estimation using multispectral NLML filters for the assessment of noise Standard deviation was more effective.

5 Transformation Methods

Transforms has been used in medical imaging in order to improve the abnormal pixel detection rate. In general, the transforms convert each pixel in an image into various domain pixels. These transforms play an essential role in the brain tumour detection process. Many transformation methods or techniques are accessible for brain tumour detection process and they are listed as Gabor transform, Contourlet transform, Curvelet transform, Shift invariant transform, Discrete wavelet transforms, Radon transform, Wavelet transform, Fourier transform.

The detection rate parameter varies with respect to different transform methods. Many brain tumor detection methods used Gabor and Contourlet transform due to its effectiveness in multi domain pixel conversion. These transforms convert the pixel belonging to the spatial mode into the pixel belonging to frequency and amplitude mode.

The Contourlet transform has been divided into non-sub sampled Contourlet transform and sub sampled Contourlet transform based on the size of each sub bands with respect to different levels. In case of non-sub sampled Contourlet transform, the size of each sub band was equal to the size of the entire source image. It does not lead to any mis-detection of the pixel during the transformation process. In case of sub sampled Contourlet transform, the size of each sub band is not equal to the size of the entire source image. It leads to mis-detection of the pixel during the transformation process. Hence, non-sub sampled Contourlet transform was used in various brain tumour detection process, as depicted in Fig. 3.

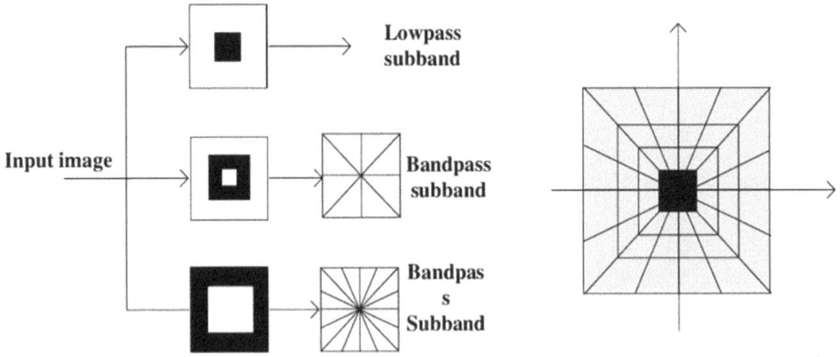

Fig. 3 The non-sub sampled Contourlet transform

The author has applied Gabor transform on the source brain MRI image [12] to detect tumour affected areas in brain images. The author has combined both the wavelet features along with Gabor transform for improving the tumor pixel detection rate. The Gabor scales for Gabor transform was set between 1 and 5 and the angle of orientation was set to 180° for each pixel in a source image.

The author has used a method that compresses and combines the image [13] with the help of a reduction property of auto encoder with the inclusion of a property that decomposes the image of wavelet transform. This combination has a remarkable effect that reduces the size of the feature set results in the lasting of the effect that can be classified using Deep Neural Network (DNN). The proposed deep wavelet auto encoder-DNN image classifier was considered for a brain image dataset, the results of classifier were compared with existing classifiers DNN, and it was observed that the new method offers better results.

A Combined Fourier Transform (CFT) algorithm has been implemented which dynamically aggregated the k-space computation [14] after preceding tested edges creating perfect prognostic picture for the present structure. Besides it combined the CFT procedure by a 3D path-based dictionary learning procedure called DLCFT aimed at fast real time dMRI reestablishment. The suggested procedure was judged with the four-state-of-the-art online and offline methods on two real and complex perfusion MR structures and a real functional brain MR sequence. The trial outcome showed the algorithms that were suggested was better than the methods by swifter that come together and greater rebuilding accuracy.

Non-sub sampled Contourlet transform [15] has been used for transforming each pixel in an image into multi resolution pixels. The author has applied the proposed method on several MRI brain images and their outcomes remained contrasted through many conventional methods in terms of sensitivity and tumor pixel classification rate.

6 Enhancement and Image Registration Methods

Enhancement is the process of increasing the intensity level of each pixel in an image for improving the clarity of the image. The situation develops the feature of each pixel in an image for better detection of abnormal pixels in an image. The enhancement algorithms are classified as Histogram Equalization, Adaptive Histogram Equalization, Local Histogram Equalization, Threshold method.

Brightness Preserving Bi-level Fuzzy Histogram Equalization (BPBFHE) has been applied that combines Fuzzy logic [16] with Adaptive Histogram Equalization (AHE) for improving MRI brain images. This technique has displayed improved contrast while preserving the natural look of an image. The author has analyzed efficiency of the proposed method with respect to Entropy, Feature Similarity Index (FSIM), and Contrast Improvement Index (CII).

A medical image contrast enhancement algorithm based on quad weighted histogram equalization has been proposed [17] using adaptive (QWAGC-FIL) gamma correction and homomorphic filtering. This technique was aimed to progress the images of low contrast including high entropy preservation along with control on over enhancement. QWAGC-FIL technique provides the utmost entropy that was approximately equal to entropy of original input images.

Histogram equalization technique has been applied [18] on every detail using a source brain MRI to enrich the pixel intensity to greater extent. The pixel clarity was improved for detecting and reducing the abnormal tumor pixels in an image. This method has used 3 * 3 sub window mask and was applied over the entire region of the image. Adaptive Gamma Correction and Homomorphic Filtering (AGC-FIL) has been proposed for preserving brightness [19] and contrast of the images. This technique yields images including sufficient contrast and effective interpretation of local details. This method reduces the error in brightness in the processed image.

The author has used local histogram equalization method [20] for enhancing the pixel intensity of every grain in an image. It has computed histogram count for each pixel throughout the image and it was used for pixel enhancement. The enhanced image was compared with the source image for analysing the performance of the proposed system. Table 2 discusses the various remarks on the classification of Image Enhancement.

Histogram equalization technique has been used [21] for enhancing the pixel intensity to a higher level that was most suitable for brain tumor detection. The authors analyzed their functioning of the suggested approach with respect to PSNR and Mean Square Error (MSE).

Figure 4 illustrates the image registration process which registers the misaligned brain image $u(x)$ into 0–90° orientation with respect to reference image $v0(x)$.

The source brain images may be misaligned due to the fault in the image capturing process. These images should be aligned with 0–90° orientation before the abnormal pixel detection process starts. Hence, there is a need for registering the captured image with respect to the reference image to make the captured image to 0–90° orientation. This process is called as image registration.

Fast adaptive bases algorithm has been used [22] for registering the misaligned image into the aligned with respect to the reference image. This method has worked based on the bases functionality of the kernel on the source brain MRI image. This non-rigid transformation method fully operated on the source image pixel by the pixel.

7 Feature Extractions and Classifications

The Features are used to describe the internal differentiation between every detail for a brain image. In general, the ensuing features has been used by conventional methods for the finding and the classification of brain tumours such as Grey Level Co Occurrence Matrix (GLCM), Local Binary Pattern, Local Ternary Pattern, Local Tetra Pattern, Grey level features, The Wavelet features and The Texture features GLCM features has been used [23] for the recognition process of abnormal tumour pixels in source brain images. In that work, the authors computed correlation, energy

Table 2 Various remarks on the classification of image enhancement

	Author name and year	Method	Performance measurement parameters	Remarks
Image enhancement	Subramani (2018)	Brightness Preserving Bi-level Fuzzy Histogram Equalization (BPBFHE)	• The entropy • Feature Similarity Index (FSIM) • The Contrast Improvement Index (CII)	• The natural image was shown and improves the image details
	Monika Agarwal (2018)	Quad Weighted Histogram Equalization with Adaptive Gama Correction and Homomorphic Filtering	• Entropy • Visual quality	• Maximum entropy preserves the image and also saves the natural details of an image
	Hardeep kaur (2016)	• Histogram Equalization • Local Histogram Equalization • Adaptive Histogram Equalization • Contrast Limited Adaptive Histogram Equalization	–	• LRE technique consumes more time • CLARE improves the quality of an image and reduces the noise
	N. Senthilkumaran (2014)	• Global Histogram Equalization • Local Histogram Equalization • Brightness Preserving Dynamic Histogram Equalization • Adaptive Histogram Equalization	• Weber contrast • Michelson contrast • The contrast • Absolute Mean Brightness Error (AMBE)	–

(continued)

Table 2 (continued)

	Author name and year	Method	Performance measurement parameters	Remarks
	Mayank Tiwari (2016)	• Adaptive Gamma Correction and Homomorphic Filtering (AGC-FIL)	• Peak Signal to Noise Ratio (PSNR) • Absolute Mean Brightness Error (AMBE)	• It has sufficient dissimilarities with good interpretation of pixel details

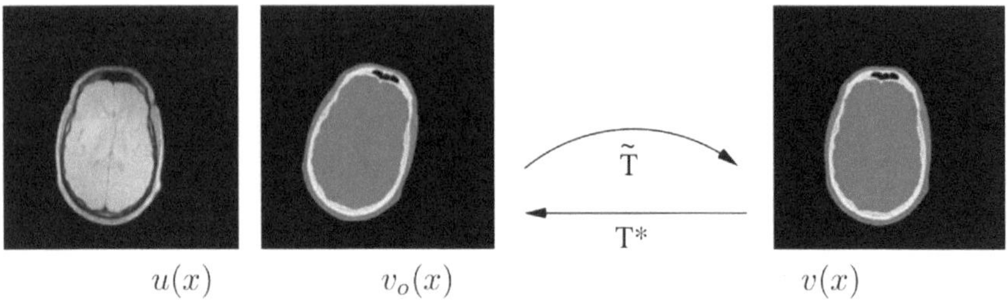

Fig. 4 Image registration process

and entropy features from GLCM matrix. These features correlate every detail in an image with its adjacent details. Further, these extracted features were classified using various soft computing approaches. The soft computing approaches are classified into Machine Learning (ML) and Deep Learning (DL) algorithms that requires Large number of trained images has been used by ML that includes Support Vector Machine (SVM), Neural Networks (NN), Decision Tree algorithms. Convolutional Neural Networks (CNN) are known as DL algorithms that require less number of images for training the network.

Soft computing approaches has been used [24] for detecting tumour pixels in an image. The authors extracted GLCM features incorporated with pixel brightness and digitization depth features and wavelet features. The SVM classification method was one of the extract methods used to categorize the source brain image into a normal or tumour affected image. ML approach to enhance the orthogonal gamma distribution has been used [25] that helped the author to understand under and over segments of malignant regions of the brain to find anomaly. This method uses spontaneous Region of Interest (ROI) detection which is unique amongst the other techniques.

CNN has been used to classify [26] diverse types of brain tumours and grade it. Machine Learning algorithm was classified into two major categories mainly supervised and unsupervised. In supervised learning technique, the machine uses the data that has been labelled whereas in case of unsupervised learning, the model was not supervised by the machine. Supervised learning allows to use the processed data and hence it can be used to train the model. In case of unsupervised learning, the data was distributed in order. They use the real-time data analysis that has resulted in less complexity of the data. The author has used 16 layer CNN structure starting from the input layer and accuracy was improved to 96.13%.

The author has employed a method for feature extraction and classification [27] aimed at precise Magnetic Resonance brain image classification with Regularized Extreme Learning Machine (RELM). The min–max normalization rule enhances the brain edges and hence produces an improved the image. The Principal Component Analysis (PCA) method has been employed in favour of feature extraction and the RELM method classifies the brain tumor that has resulted in improved classification accuracy to 94.233%.

8 Conclusion

The proposed work has discussed various conventional methods for brain tumour detection and classifications. This work has detailed about various preprocessing techniques that are mainly used before the tumor pixel segmentation process starts. It has also discussed various feature extraction methods and soft computing techniques used in the conventional brain tumour detection system.

References

1. Manasa, N., Mounica, G., Divya Tejaswi, B.: Brain tumor detection based on Canny edge detection algorithm and its area calculation. Int. J. Comput. Math. Sci. **5**(3) (2016)
2. Nagabushanam, P., Satheesh Kumar, S., Sunitha Kumar, J., Thomas George, S.: Brain MRI image segmentation using improved Sobel method. Int. J. Recent Technol. Eng. **7**(4) (2018)
3. Hassan, E., Aboshgifa, A.: Detecting brain tumor from MRI image using MATLAB GUI programme. Int. J. Comput. Sci. Eng. Surv. (IJCSES) **6**(6) (2015)
4. Zotin, A., Simonov, K., Kurako, M., Hamad, Y., Kirillova, S.: Edge detection in MRI brain tumor images based on fuzzy C-means clustering. Procedia Comput. Sci. **126**, 1261–1270 (2018)
5. Aslam, A., Khan, E., Sufyan Beg, M.M.: Improved edge detection algorithm for brain tumor segmentation. Procedia Comput. Sci. **58**, 430–437 (2015)
6. Yan, X., Zhou, M-X., Xu, L., Liu, W., Yang, G.: Noise removal of MRI data with edge enhancing. In: 2011 5th IEEE International Conference on Bioinformatics and Biomedical Engineering, pp. 1–4 (2011)
7. Maolood, I.Y., Al-Salhi, Y.E.A., ALresheedi, S., Ince, M., Li, T., Lu, S.F.: Fuzzy C-means thresholding for a brain MRI image based on edge detection. In: 2018 IEEE 4th International Conference on Computer and Communications (ICCC), pp. 1562–1566 (2018)
8. Balafar, M.A.: Review of noise reducing algorithms for brain MRI images. Int. J. Tech. Phys. Prob. Eng. (Published by International Organization of IOTPE) **4**(13), 54–59 (2012)
9. Kazmi, M., Aziz, A., Akhtar, P., Maftun, A., Bin Afaq, W.: Medical image denoising based on adaptive thresholding in Contourlet domain. In: 5th International Conference on Biomedical Engineering and Informatics (2012)

10. Vaishali, S., Rao, K.K., Rao, G.V.S.: A review on noise reduction methods for brain MRI images. In: International Conference on Signal Processing and Communication Engineering Systems, Guntur, 2015, pp. 363–365 (2015)
11. Bouhrara, M., Bonny, J.-M., Ashinsky, B.G., Maring, M.C., Spencer, R.G.: Noise estimation and reduction in magnetic resonance imaging using a new multispectral nonlocal maximum-likelihood filter. IEEE Trans. Med. Imaging **36**(1), 181–193 (2016)
12. Liu, Y., Muftah, M., Das, T., Bai, L., Robson, K., Aue, D.: Classification of MR tumour images based on Gabor wavelet analysis. J. Med. Biol. Eng. **32**(1), 22–28 (2011)
13. Mallick, P.K., Ryu, S.H., Satapathy, S.K., Mishra, S., Nguyen, G.N.,Tiwari, P.: Brain MRI image classification for cancer detection using deep wavelet auto encoder-based deep neural network. IEEE Access **7**, 46278–46287 (2019)
14. Wang, Y., Cao, N., Cui, Y., Zhou, Q.: Real time dynamic magnetic resonance imaging via dictionary learning and combined Fourier transform. IEEE Access **7**, 150924–150935 (2019)
15. Verma, K., Vijay, R., Urooj, S.: Extraction of brain tumor attributes using Contourlet transform with geometric transform invariant technique. Int. J. Pure Appl. Math. **114**(12), 119–125 (2017)
16. Subramani, B., Veluchamy, M.: MRI brain image enhancement using brightness preserving adaptive fuzzy histogram equalization. Int. J. Imaging Syst. Technol. **28**(3), 217–222 (2018)
17. Agarwal, M., Mahajan, R.: Medical images contrast enhancement using quad weighted histogram equalization with adaptive gama correction and homomorphic filtering. Procedia Comput. Sci. **115**, 509–517 (2017)
18. Kaur, H., Rani, J.: MRI brain image enhancement using Histogram equalization techniques. In: IEEE International Conference on Wireless Communications, Signal Processing and Networking (WiSPNET), pp. 770–773 (2016)
19. Tiwari, M., Gupta, B.: Brightness preserving contrast enhancement of medical images using adaptive gamma correction and homomorphic filtering. In: IEEE Students Conference on Electrical, Electronics and Computer Science (SCEECS) (2016)
20. Senthilkumaran, N., Thimmiaraja, J.: Histogram equalization for image enhancement using MRI brain images. World Congress on Computing and Communication Technologies, pp. 80–83 (2014)
21. Oak, P.V., Kamathe, R.S.: Contrast enhancement of brain MRI images using histogram based techniques. Int. J. Innovative Res. Electr Electron. Instrum. Control Eng. (2013)
22. Cheung, K.W., Siu, Y.T., Shen, T.W.: Fast adaptive bases algorithm for non-rigid image registration. J. Imaging Sci. Technol. **63**(1), 10505-1-10505 (2019)
23. Jain, S.: Brain cancer classification using GLCM based feature extraction in artificial neural network. Int. J. Comput. Sci. Eng. Technol. **4**(07) (2013)
24. Shantha Kumar, P., Ganesh Kumar, P.: Performance analysis of brain tumor diagnosis based on soft computing techniques. Am. J. Appl. Sci. **11**(2), 329–336 (2014)
25. Manogaran, G., Shakeel, P.M., Hassanein, A.S., Kumar, P.M., Babu, G.C.: Machine learning approach-based gamma distribution for brain tumor detection and data sample imbalance analysis. IEEE Access **9**(7), 12–19 (2018)
26. Sultan, H.H., Salem, N.M., Al-Atabany, W.: Multi-classification of brain tumor images using deep neural network. IEEE Access (2019)
27. Gumaei, A., et al.: A hybrid feature extraction method with regularized extreme learning machine for brain tumor classification. IEEE Access **7**, 36266–36273 (2019)

Deep Learning Techniques for Electronic Health Record (EHR) Analysis

T. Poongodi, D. Sumathi, P. Suresh, and Balamurugan Balusamy

Abstract Data is gathered from millions of patients across various healthcare organizations could be framed as electronic health record (EHR). It comprises of data in various forms that include laboratory results, diagnosis reports, medical images, demographic information about patient, and clinical notes. Transformation in the health care could be done with the help of deep learning techniques since these techniques outperform than the conventional machine learning methods. In addition, a huge and complex dataset gets generated day by day, and thus, it enables the need for the deployment of deep learning models. It is examined that the clinical decision in certain fields that vary from analysis of medical research issues and to suggest as well as prioritize the treatments in order to detect abnormalities, and in addition, the identification of genomic markers in tissue samples is done. Deep learning is applied in EHR system to derive patient representation to provide enhanced clinical predictions, and augments clinical decision systems. Moreover, it is applied in order to detect the disease in the earlier stages, determining the clinical risk, forecasting the future need of regular checkups and in addition prediction of hospitalization in the near future if required. According to statistical report, the deep learning software market is estimated to reach 180 million US dollars in size by 2020. The availability of a huge amount of clinical information particularly EHR has stimulated the growth of deep learning techniques which assist in rapid analysis of patient's data. Nowadays, EHR is being incorporated with deep learning techniques and tools into EHR systems which provide deep insights for health outcomes. The market for EHR deep learning

T. Poongodi (✉) · B. Balusamy
School of Computing Science and Engineering, Galgotias University, Uttar Pradesh, India
e-mail: tpoongodi2730@gmail.com

B. Balusamy
e-mail: kadavulai@gmail.com

D. Sumathi
SCOPE, VIT-AP, Amaravati, India
e-mail: sumathi.research28@gmail.com

P. Suresh
School of Mechanical Engineering, Galgotias University, Uttar Pradesh, India
e-mail: psuresh2730@gmail.com

A. K. Bhoi et al. (eds.), *Bio-inspired Neurocomputing*, Studies in Computational Intelligence 903, https://doi.org/10.1007/978-981-15-5495-7_5

tools is predicted to exceed $34 billion by the mid of 2020s, motivated massively by an emerging desire to automate tasks and provide deeper insights into clinical issues. The global health IT market is anticipated to be worth a surprising $223.16 billion by 2023, driven in part by deep learning. EHR is a thriving research domain for deep learning researchers since it helps to achieve higher accuracies in medical conditions. Clinical information could be extracted by applying deep learning techniques. The extraction is of various types, namely retrieval of single idea, extracting the association, time-based extraction, and advancement of abbreviation. In EHR representation learning process, concept and patient representations are included to obtain detailed analysis and decisive predictive functionalities. Outcome prediction in deep EHR is categorized as static and temporal prediction to predict patient outcomes. The clinical deep learning is more interpretable under the category of maximum activation, constraints, qualitative clustering, and mimic learning in EHR analysis. The objective of this chapter is to provide an insight into the digital transformation of EHR through deep learning techniques. It discusses the deep learning framework and challenges that occur during the development of deep learning models for EHR. Moreover, it also focuses on the deep learning prediction techniques for various diseases and recent advancements on deep learning techniques in the aspect of providing precise medicine and future generation health care.

Keywords Deep learning · EHR · Feature learning · Clinical data · Unsupervised learning

1 Review on Deep Learning and Objectives of Electronic Health Record Data

Health care is a hot emerging topic where massive amount of biomedical data plays many significant roles. For instance, precision medicine assures that "precise treatment should be provided to the correct patient at the exact time" and it requires patient's data in several aspects which includes environment details, lifestyle, difference in molecular traits, and electronic health record (EHR). The availability of immense amount of biomedical data paves a way for significant challenges and tremendous opportunities in healthcare research. Particularly, the link between different units of information in the dataset should be explored to create biomedical tools focusing on machine learning and data-driven approaches. Predictive analysis and discovery require a well-connected knowledge base which could be built from various data sources [1], while the predictive tools need more attention to be used in the medical domain. Certainly, there are several challenges that exist in using biomedical data due to its heterogeneity, sparsity, high-dimensionality, irregularity, and temporal dependency [2]. It is further complicated because of the influence of medical ontologies utilized to speculate the data (International Classification of Disease-9th version (ICD-9), Unified Medical Language System (UMLS) [3], Systematized Nomenclature of Medicine–Clinical Terms (SNOMED-CT) which results in confliction and inconsistency [4]. For example, in the EHR system, the values in the laboratory report state that hemoglobin A1C > 7.0 in the existence of 250.00 ICD-9 code for the patient diagnosed with "type 2 diabetes mellitus." Consequently, it is not required to understand the correlations and synthesize all medical concepts to construct an abstract view of semantic structure. However, a domain expert is essential in biomedical research to specify the phenotypes. The feature space is scaled poorly, and discovering novel patterns are highly missed in supervised learning algorithm. The learning methodologies are in demand to discover the patterns automatically for making prediction from the training dataset [5].

Multiple levels of representation in deep learning algorithms are composed as nonlinear modules which transform the representation from low-level to next high-level abstract. Deep learning models play a major role in various domains like NLP, computer vision, and speech recognition [6, 7]. Deep learning paradigms demonstrated a rapid progression with exciting opportunities in biomedical informatics. Applying deep learning methods in health care is extensively preferable for a wide range of medical problems. It has been illustrated that the deep learning mechanisms have been deployed in the healthcare industries and on computed tomography (CT) scans and X-rays with the help of deep mind and enlitic, respectively [8, 9]. Various characteristics of deep learning techniques would assist in health care include end-to-end learning along with feature learning, capable of managing multimodality and complex data, superior performance, etc.

The deep learning domain addresses various challenges related to the features of healthcare data (i.e. heterogeneous, noisy, sparse, time-dependent) demand promising tools and methods which enable deep learning to incorporate clinical decision support and healthcare information workflow. Recently, deep learning techniques are employed to process the aggregated EHR which includes structured (e.g., medications, laboratory tests, diagnosis) and unstructured data (clinical notes). The processing of EHR based on deep architecture is specifically done for supervised clinical task. Particularly, deep learning achieves better results when compared to conventional machine learning techniques. Unsupervised models are proposed to obtain the latent patient representations and evaluated using classifiers such as logistic regression, random forests. The subcategories of EHR deep learning functionalities along with the input data are mentioned in Table 1.

The likelihood of diseases could be predicted with the application of deep learning mechanisms. A four-layered CNN approach has been devised by Liu et al. [10] so that the enduring disruptive respiratory illness and myocardial infarction. RNN along with the long short-term memory (LSTM) was applied in DeepCare [11], deep dynamic network infers the illness of the patient, and anticipated the medical outcomes. LSTM is moderated with a decay effect in order to leverage irregular events and medical interventions that are incorporated to figure out the predictions dynamically. DeepCare was examined to model the disease progression, risk prediction on patient's mental health and diabetes. RNN with gated recurrent unit (GRU) used, and this end-to-end model utilizes patient's history for predicting the medications of various problems. The evaluation shows the significant results by incorporating the model without losing accuracy. In this method [12], three-layer stacked denoising

Table 1 EHR deep learning functionalities

Input data	Functionalities	Sub-functionalities
Clinical notes	Information extraction	• Single concept
		• Temporal event
		• Relative-based
		• Abbreviation-based
Medical codes	Representation learning	• Conceptual view of representation
		• Patient representation
Mixed information	Outcome prediction	• Static and temporal prediction
Mixed information	Phenotyping	• Novel phenotype identification
		• Improving the available phenotypes
Clinical notes	De-identification	• Clinical text de-identification

autoencoder (SDA) is proposed to understand patient representation from EHR and it predicts the disease based on random forest method as classifiers. The evaluation has been carried out on 76,214 patients who have 78 diseases from different clinical domains. The result showed the better predictions in deep representation than using conventional learning algorithms (e.g., k-means, principal component analysis (PCA)). The results are significantly improved by adding logistic regression layer to refine the complete supervised network. Liang [13] followed RBM to deep learn the representations from EHR which provides better prediction accuracy for many diseases. Deep learning techniques were employed to simulate perpetual time signals like laboratory reports for the instinctive identification of certain phenotypes. In the context, RNN with LSTM is efficiently utilized to identify patterns in different time series of clinical data.

1.1 Introduction to Deep Learning

The input features for conventional machine learning algorithm are taken from raw data that relies on domain knowledge and practitioner expertise to extract the patterns. The feature extraction involves analyze, choose and evaluate; which is time consuming and laborious in machine learning. In contrast, optimal features are learned directly from the available data in deep learning without any human intervention and permit self-discovery of dormant data relationship otherwise that might be hidden or unknown. The composition of simpler representations is expressed as intricate data representation in deep learning. In the context, to recognize a human in an image involves discovering representation of facial features from corners and contours, contours and corners from edges, and edges from pixels [14]. Recurring deep learning mechanism is the principle behind the enhancement in the complexity in the unsupervised ordered representation. The idea behind the deep learning architectures is the artificial neural network (ANN) framework. It comprises of nodes that are connected together and organized in layers. These nodes are known as neurons. A few neurons are located in the input and output layers which are known as hidden neurons. The set of weights W is stored in every hidden unit that is updated in the trained representation.

ANN weights can be optimized by considering the reduction of loss function such as negative log likelihood, and it is denoted below,

$$E(\theta, D) = -\sum_{i=0}^{D}[\log P(Y = y_i|x_i, \theta)] + \lambda||\theta||p \quad (1)$$

In Eq. 1, the first term decreases the total of log loss in the complete training dataset *D*; the second term decreases the p-norm of θ_i (learned model parameter) controlled by λ (tunable parameter). The second term is referred to as regularization, and it could be exploited to increase the capability for generalizing unseen examples. A back-propagation mechanism is used for weight optimization in the loss function that reduces loss in the final layer of the network. Some open-source tools for deep learning algorithms include Keras3, Torch4, Theano2, TensorFlow1, Deeplearning4j8, Caffe6, PyTorch5, CNTK7, etc.

Deep learning differs from the conventional machine learning approaches in which how the learning process is carried out to create representations from the raw data. A computational model which involves various layers so that the data representation could be learned. This process is performed by applying deep learning. The significant variations among deep learning and conventional ANN are about the connections, number of hidden layers, and the ability to learn various levels of substantial abstraction view of the inputs. However, the conventional ANNs are restricted to three layers and the dataset is trained to attain the supervised representations that can be optimized to do a specific task [15]. In a deep learning system, each and every layer in the deep learning system is used to denote the illustrations of the patterns which depend on the data obtained from the layer underneath. Features in deep learning are discovered and identified from the existing raw data using learning algorithm. In deep neural networks, the inputs are processed layer by layer in a nonlinear way to set the nodes in the consequently hidden layers in order to learn the representations.

The learned representations are given into the supervised layer to refine the complete network by employing the back-propagation algorithm, and then the representations are improved according to a particular end-to-end task. The novel methodologies to prevent overfitting, the unsupervised pre-training [16], the graphical processing units are used to speed up the computational process, and high-level modules are developed to construct the neural network where the deep learning paradigms provide state-of-the-art solutions for several tasks. Complex structures are identified in high-dimensional data using deep learning and achieved high-level performances for detecting an object in speech recognition, images, and NLP. The intelligent tools using deep learning procedures play a vital role in the classification of skin cancer [17], detection of diabetic retinopathy [18], anticipating the particular DNA and RNA binding proteins [19] in the real-world clinical e-health care.

1.2 Role of Electronic Health Record

The patient medical information details such as history, allergies, medications, therapy procedure are maintained as medical records. In fact, an individual's medical record is significant in case of emergencies for the clinicians to take necessary steps for the patient to provide appropriate treatment. EHR provides distinct solutions for several kinds of critical as well as normal issues. As like different applications, medical domain is also following digitized route for record maintenance and EHR is the digital form of patient's medical information. EHR offers accurate, fast access, and real-time data with the latest tools and technologies. EHR provides the solutions for various challenges in healthcare service, and it is shown in Fig. 1.

It is a real-time digital version of the patient data, and the web-oriented EHR is completely patient-centric. It furnishes data in a fast, accurate, and reliable manner in a secured way. Patients can avail medical treatment with various medical facilities, where physicians at different places can be communicated simultaneously. The digitized patient details can be circulated irrespective of location and time to meet the challenges in an efficient manner. The mentioned features in EHR assist the professional and individuals in the medical service domain,

- Medical history of every individual in complete form
- Administrative and billing details
- Patient demographics
- Allergies and medication information
- Vital signs and symptoms

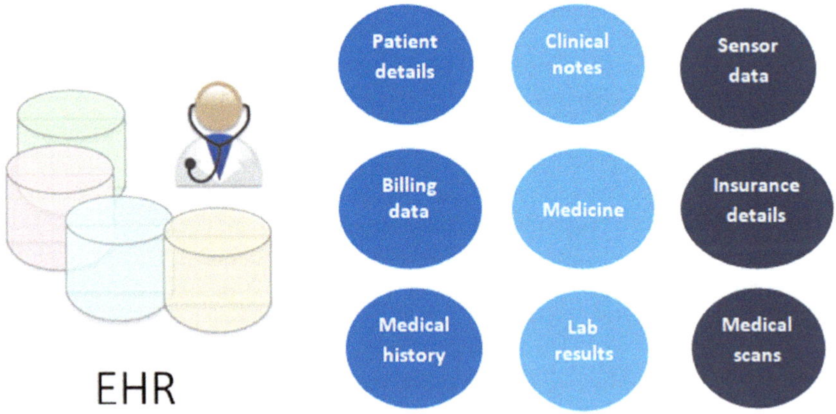

Fig. 1 Medical entities in EHR system

- Diagnoses and medications
- Laboratory tests and results
- Reports and immunization details.

EHR brings down medical expenses and improves efficiency in providing better health care. It facilitates high-quality treatment and secured health care for patients by transferring information efficiently, and it supports medical service provides. Well-being of individual patient can be improved indeed using EHR for maintaining medical records. Some of the advantages of EHR system are given below:

- Legible document without error
- Patient-centric
- Fast, complete, and accurate data access
- Diagnoses and care-taking improved significantly
- Decision making and on-time treatment
- Precise prescription and billing services
- Privacy and secured data handling
- Workflow maintenance among various departments and service providers
- Healthcare professionals work together in real time.

The usage of EHR is drastically increased in both ambulatory care center services and hospitals. It improves the patient care-taking by increasing efficiency, reducing errors, enhancing care coordination, and it acts as a medical data repository for researchers in various domains. EHR systems differ in the aspect of functionality and fundamentally classified into (i) primary EHR with clinical notes, (ii) primary EHR without clinical notes, and (iii) comprehensive systems. This system also provides patient's wealth-related information on tracking medical history, medication usage, and complications. The various EHR services are depicted in Fig. 2.

Initially, EHR was planned to handle core administrative functionalities in hospitals; controlled vocabularies and different classification schema are available for recording medical events and related information. The examples for various classification schemas such as International Statistical Classification of Diseases (ICD), Current Procedural Terminology (CPT), Logical Observation Identifiers Names and Codes (LOINC), and RxNorm are given in Table 2.

Synthesizing and analyzing medical data with massive terminologies and many institutions with the huge array of schemata are an on-going research area. EHR systems contain many types of patient information such as diagnoses, sensor measurements, demographics, physical exams, laboratory report, prescribed medications, and clinical notes. The main significant challenges to represent the heterogeneous data with its type in EHR are mentioned below:

- Datetime objects (Time of admission, DOB)
- Numerical quantities (Body mass index)
- Categorical values (ICD-10 diagnoses, CPT procedures)
- Natural language textual information (Clinical notes, discharge summaries)
- Derived time series (Multimodal patient history, perioperative vital sign signals).

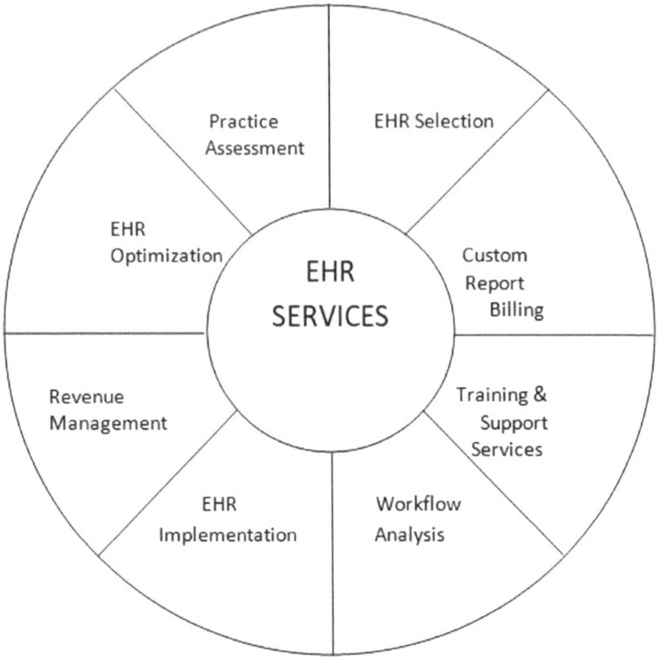

Fig. 2 EHR services

Table 2 Classification schema

Classification schema	Administrative tasks	Examples
ICD-10	Diagnosis	Heart failure, acute respiratory failure, systolic heart failure
CPT	Procedures	Eyelid skin biopsy, MRI thoracic spine, partial mastectomy
LOINC	Laboratory	Ethanol, blood, salicylate, serum, buprenorphine screen
RxNorm	Medication	Morphine, acetaminophen, buprenorphine

1.3 Challenges and Issues

Despite promising solutions achieved following deep architectures, there are various challenges faced in deep learning-based clinical application. Naylor [20] defines seven significant factors drives the adoption of deep learning in health care:

1. potential of digital imaging for further interpretation
2. digitizing biomedical records and data sharing techniques
3. incorporating deep learning techniques for analyzing diverse datasets
4. strength of deep learning for hypothetical analysis in research

5. deep learning helps in empowering patients and streamlining clinical overflows
6. rapid diffusion among proprietary and open-source deep learning programs
7. adequacy in deep learning technology provides improved performance even if the datasets become larger day by day.

The incredible advancement in the dataset size and computational power, the researcher's attention began to focus on deep learning techniques. Initially, algorithms were categorized as "supervised learning" and the training dataset is labeled first by human experts and the algorithms learned the characteristics and weights from the available data directly. In the context, retinopathy screening algorithm is a kind of supervised deep algorithm. Hinton [21] shows an interest in developing novel deep learning techniques which are absolutely unsupervised. Recently, deep learning algorithm enters the mainstream of clinical database. Gulshan [22] summarized the development of deep learning algorithm for identifying diabetic retinopathy in the photographs of retinal fundus. Wong and Bressler [23] quoted the limitations and further validation of deep learning algorithms with different populations and how to incorporate algorithms in clinical workflow by convincing patients and clinicians. Some of the other significant issues in deep EHR are also discussed and highlighted.

Data volume: The set of intensive computational models are involved in deep learning. In multilayer neural networks, huge network factors are required to be anticipated promptly with the available immense amount of data. A common rule is followed to have 10 as the minimum number of samples as network factors. The deep learning is successful in several domains like speech recognition, computer vision, NLP, where massive amount of data could be easily gathered. Understanding and predicting the diseases and its variability were highly complex. Moreover, the massive volume of medical information is required to train a powerful deep learning paradigm.

Temporality: Diseases are changing and progressing forever in a non-deterministic way. The existing deep learning models are not much efficient to handle the time factor dynamically since it is assumed to accept static, vector-based inputs. Deep learning approaches are required to handle temporal healthcare data and provide novel solutions.

Interpretability: The phenotypes drive the predictions and convincing professionals are critical regarding the actions recommended such as high risk of diagnosing a specific disease, providing prescription of a particular medication. There is a promising future for deep learning in health care by introducing several research possibilities and opportunities to face the challenges.

Data quality: In health care, data are highly ambiguous, noisy, heterogeneous, and incomplete as it is not well-structured and clean as in other domains. An efficient deep learning paradigm is required to train such kind of variant and massive datasets, and it is a challenging task in the perspective of several issues such as redundancy, missing values, and data sparsity.

Domain complexity: The problems in health care and biomedicine are more complicated and riskier as it is completely different from other domains such as speech analysis and image processing. The diseases are mostly heterogeneous, and there is no thorough knowledge about the causes, progression of it. Furthermore, the number of patients is not finite and increasing day-by-day in a practical medical scenario.

Expert knowledge: Uniting the expert knowledge along with the deep learning process directed in the right direction to make decisions for various problems with massive amount of medical data. For instance, the reliable content is extracted from PubMed or online medical encyclopedia and that could be fed into deep learning architecture to maximize the overall performance. In semi-supervised learning, leverages both labeled and unlabeled samples were from the huge number of unlabeled samples with the limited labeled samples.

Temporal modeling: The time factor is highly significant in several health-oriented problems, particularly in EHR training time-based deep learning model is obviously critical to identify the patient condition and provide on-time clinical decision. RNN will play a significant role in temporal deep learning to solve health-related problems.

Feature enrichment: The features are essential to characterize each patient, and the data sources for creating such features must be placed in EHR. The efficient integration of diversified data and the way to utilize them in deep learning is a challenging research topic. The existing solutions are inadequate to combine data from heterogeneous sources by incorporating deep learning techniques. A dormant solution could process data efficiently retrieved from heterogeneous sources and employ the hierarchical way of deep learning, and the result could be presented as a combined paradigm toward a holistic abstraction level.

2 Deep Patient Representation Based on Unsupervised Deep Learning for Processing EHR Data

The primary objective of precision medicine is to construct the quantitative models for predicting the people health status in order to prevent disability or disease. EHR is a promising technique in medical domain to accelerate predictive analysis and clinical research. In addition, EHR facilitates discovery of type 2 diabetes subgroups, data-driven prediction about the drug effects and causes, improvement in admitting patients for clinical trials, identification of comorbidity clusters in autism spectrum disorder. Yet, there are no further machine learning techniques which does not have much attention in the clinical decision-making systems. EHR data is highly complex to represent because of different characteristics such as scarcity, huge dimensions, heterogeneous, random errors, noise, dearth, and systematic biases. The identical clinical phenotype is denoted by using different terminologies and codes. Hence, it is really complicated to identify patterns using machine learning techniques that obtain clinical models for various real-world applications. Depiction of deep patient is illustrated with the deployment of high-level theoretical framework. Clinical data warehouse generates the EHR; patient vectors represent the clinically pertinent phenotypes that have been determined, controlled, and collected. The patient details can be represented using a single vector/set of vectors. The collection of vectors is given as input to the feature learning algorithm to search the abstract level descriptors. Such kind of deep representation is appropriate for various clinical tasks. The steps involved in deep patient representation via unsupervised deep learning are:

(i) Preprocessing of data to retrieve the patient data from EHR.
(ii) The raw patient data is modeled using unsupervised deep learning which in turn gives the robust features.
(iii) The deep feature learning is employed to the hospital database for obtaining the patient representation that could be utilized for numerous clinical tasks.

The patient representation in deep learning architecture is represented as multi-layer neural network. All layers are trained to obtain the abstract view of patterns by optimizing the unsupervised criteria. The last layer in the network produces the ultimate patient deep representation. In Fig. 3, the process of obtaining deep patient representation from EHR data warehouse via unsupervised deep learning is illustrated and the various components are described below.

2.1 Hospital Data Warehouse

Predictive algorithms are highly dependent on feature selection and data representation. In supervised learning, the feature space does not generalize well, scales poorly and fails in discovering novel features and patterns. To resolve this issue, data-driven approaches are introduced for feature selection process in EHR. The problem in this approach is that the patients are viewed as two-dimensional vector in a clinical data warehouse. It is again noisy, repetitive, and sparse; hence, the model is not suitable for organizing hierarchical information which is available in EHR.

2.2 EHR Processing

Few general demographic details such as gender and age are retained for every patient in the dataset, and some clinical descriptors are also organized in a structured format such as medications, laboratory tests, diagnoses, procedures, and clinical notes. Preprocessing is done for all clinical records using biomedical annotator to normalize medications, obtaining codes for laboratory tests, and to obtain biomedical conceptual information from clinical notes. The normalized records are varied according to the datatype, and it facilitates modeling of clinical events include diagnoses, procedures, medications, and laboratory tests. The tool discussed in Pendu et al. [24] identifies the negated tags related to historical clinical events and such tags can be discarded which are considered as irrelevant. The negated tags were found out exploiting a regular expression algorithm called NegEx, that filters sentences which contain negation phrases, and it restricts the scope of it. Similarities would be

Table 3 Types of autoencoder and its characteristics

Type of encoder	Characteristics
Vanilla autoencoder	It comprises of three layers. Here there is a one concealed layer. In this, the input and the output are the same
Multilayer autoencoder	More hidden layers are present
Convolutional autoencoder	Instead of compressed 1D vectors, images are used. The input is downsampled so that the concealed information that is in small dimensions is seen, and thus, it is a mandatory task to be aware of the compressed images
Regularized autoencoder	Two kinds of regularized autoencoders are available. They are the sparse autoencoder and the denoising autoencoder
	Sparse autoencoder: Sparse autoencoder shall be implemented in order to reduce the reconstruction error because the single-layer network has been implemented to learn a code dictionary. This minimization is done during the restriction of the number of code words needed for the reconstruction. Naturally, the classification task could be considered as a type of scarifying algorithm since it diminishes the input size to a single class value which results in less prediction error
	Denoising autoencoder: The loss function is modified with the cost of implementing noise function for the input image. The objective in this encoder is to eradicate the noise so that the significant attributes could be extracted

due to the complications in representing the human physiology, and in addition, it comprises of several non-obvious dependencies between observed measurements. Time series data is required to build a model that develops clinical information, but this data is found to have missing values, irregular sampled data, and varying length. Autoencoders are normally used for physiological time series signal reconstruction. The low-dimensional embedding is performed in which the target values look similar to the input values and the hidden layers compress the inputs. This type of embedding process is done in order to restructure the input which in turn acquires the fundamental features about input time series, and thus, it could be used as a measure of patient perspicacity.

3.2.3 Convolutional Neural Networks

In mathematical terms, convolution refers to the ultimate measuring of the way of the two functions that get overlapped. The objective of convolutional networks is to recognize images as dimensions, i.e., three-dimensional objects. Usually, digital color images could be encoded with the colors of red-blue-green (RGB). Thus, the combination of these three colors is done so that the humans could perceive the color spectrum. A convolutional network absorbs the images which are arranged

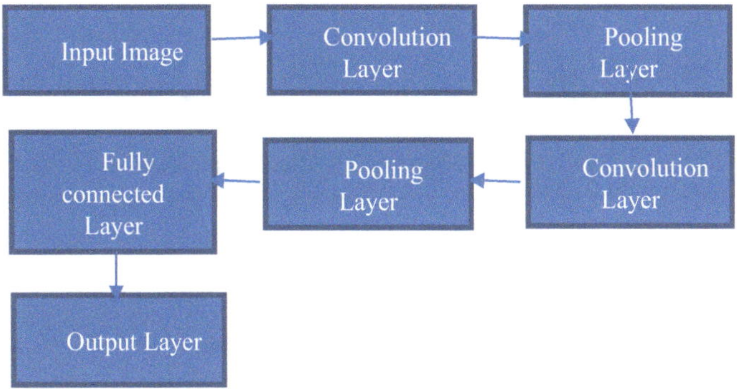

Fig. 6 Architecture of CNN

in three different colors arranged on one on another. The convolutional network takes the input as the image which is surrounded by the rectangular box in terms of dimensions. These depth layers are denoted as channels. The architecture of CNN is shown in Fig. 6.

CNN comprises of several nodes or artificial neurons located in a tiered network of consecutive convolutional, max pooling, and fully connected layers as shown in Fig. 6. Due to the ordered arrangement, the model approximates the complex functions and maximal discrimination among the classes is done through the nonlinear feature combinations. For example, instead of considering 50 * 50 image as 2500 unrelated pixels, the collection of local pixel patches could be considered by extracting the important features by viewing the image. Correspondingly, the collection of the local elements is the single dimension time series data. The equation for 1D convolution is shown in Eq. 2.

$$C_{1d} = \sum_{a=-\infty}^{\infty} x(a) \cdot w(t-a) \tag{2}$$

Here the input signal is "x" and the "w" is used to denote the convolutional filter or the weighing function. The two-dimensional convolutional equation is given in Eq. 3.

$$C_{2d} = \sum_{m}\sum_{n} X(m,n)K(i-m, j-n) \tag{3}$$

In this equation, a two-dimensional grid is denoted as "X" and "K" denote the kernel. The objective of the filter is used to segment the weights that are denoted in terms of the matrix with the help of the input so that the feature maps are extracted. Parameter sharing is encouraged in this network since each and each filter is related with the entire input.

A CNN classifier has been developed in [38] so that the heart failure has been detected from sub-images that have been sampled from whole slide images (WSI) of cardiac tissue, and in addition, it is also shown that cardiac histopathology is adequate enough toward the prediction of the patients who suffer from cardiac arrest precisely. In addition, it is also observed that these algorithms were found to be perceptive to sense the tissue pathology and also helps in detecting the disease that precedes the definitive clinical diagnosis. A two end-to-end frameworks to combine the learning of EHR exemplifications and a comparison are done without the feature collections. It has been proposed in [39]. Through this framework, accuracy obtained seems to be high. The significant characteristics adopted in this approach are

- Sequential structure is used to study and learn about the local important information of EHR.
- Certain examples from the same class and as well as from the several dissimilar classes are distinguished due to the triplet loss, and thus, it makes certain of large margin.
- Pairwise labels are classified properly based on the softmax cross-entropy loss.

From the experimental results, it has been proven that the learning framework could show better depictions in terms of vectors for clear understanding of patients' past information, and thus, the precision obtained is high when compared with the other baseline models.

Deepr (Deep record) a system has been devised so that the features from the medical records are extracted so that the future risk could be forecasted. This is approached in [40]. The critical task of Deepr is to convert an EMR into several sentences that are separated by distinct words. It describes about the patient's health right from the beginning stage to the advanced stage. Irregular timing, transfers, and time gaps are represented as special words. With these depictions, EMR is converted into a sentence of adaptable extent that preserves all substantial actions. The first layer comprises of words that are implanted into an incessant vector space. The words are then given to the convolutional operation which in turn discovers the local themes. The local themes thus obtained are combined to generate a comprehensive feature vector. This is given as the input to the classifier for predicting the future risk. From the given data, all mechanisms are well-read at the similar time. The input signals carry the data, and they are transmitted to the output, whereas the training signals have been dispersed to the design indicators. Henceforth, Deepr is considered to work from one end to the other end. Various architectures of CNNs act as the key to developing algorithms. They are

- LeNet
- AlexNet
- VGGNet
- GoogLeNet
- ResNet
- ZFNet.

3.2.4 Unsupervised Embedding

Representations for the medical codes could be denoted with the word2vec algorithms. A dual-level illustration for therapeutic codes and scientific visits is denoted by joining together with the help of word2vec algorithm. This algorithm has two alternatives:

- Prediction of target (codes) through the continuous bag of words (CBOW)
- Skip-gram is used to forecast the predicts that surround contexts given target (codes).

3.2.5 Generative Adversarial Networks (GANs)

This comprises of three parts, namely

- Generative: This is to describe the way of generating the data in terms of a probabilistic model.
- Adversarial: Model is trained in a confrontational setting.
- Networks: Deep neural networks are used as the artificial intelligence algorithms for the training.

In GAN, there are two parts, namely discriminator and generator, as shown in Fig. 7.

Fake samples such as audio and image are given as input to the discriminator by the generator. The duty of the discriminator is to segregate the real and fake samples. Both the parts participate in competition to prove themselves in the training phase. Several iterations are carried out in order to make the both parts perform in a better manner. Training a GAN is done in two divisions:

1. During the training of discriminator, generator remains idle. No back-propagation is performed. Discriminator takes the real data for training and check whether

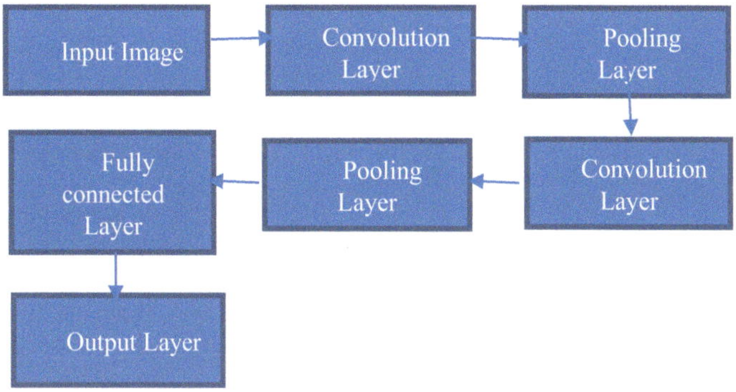

Fig. 7 Generative adversarial network

the data is real or not. In addition, it is also trained on the fake data that has been generated by the generator to check whether it could predict them as fake.
2. Discriminator remains idle during the training of generator. Results obtained during the training of generator could be used to obtain better results when compared with the previous state and fool the discriminator.

Missing data imputation is done in the first stage, and disease prediction is performed in the two-stage framework. This resolves the issues that occur due to the missing data in EHR. In this process, the raw data is partitioned into records with sparsity of greater than or equal to a specific threshold and records below this threshold. Then the records below the threshold value are taken, and it is partitioned into training dataset and validation dataset. The autoencoder is trained with the training data, and patterns of the training data are stored in the parameters used in autoencoder. Validation error is minimized with the help of validation data by identifying an epoch. A trained autoencoder is used to fill the most suitable values for the missing values of EHR data. The AC-GAN model is trained in order to predict whether the record is patient's record or a normal person using the data attributed by the autoencoder. Therefore, the framework does the prediction efficiently. The missing values are imputed in two ways.

1. Substitute the missing values with the mean values of the attributes
2. Autoencoder is used to replace the missing values.

Different types of GANs are shown in Table 4.

Table 4 Different kinds of GANs

Type of GANs	Characteristics
Vanilla GAN	It is one of the modest types of GAN
	Generator and the discriminator are multilayer perceptron
	Stochastic gradient descent is used for the optimization
Conditional GAN (CGAN)	It is a deep learning method
	'*y*' a supplementary parameter is supplemented to the generator so that the corresponding data is generated
	Discriminator takes the input in addition to labels since these labels aid in differentiating the real data from the fake data
Deep convolutional GAN (DCGAN)	It implements ConvNets without max pooling and replacement is done by convolutional stride. The layers are not connected properly
Laplacian pyramid GAN (LAPGAN)	Generator and discriminator networks are used more in number along with the several levels of the Laplacian pyramid. High-quality images are produced
Super-resolution GAN (SRGAN)	High-resolution images are generated. Optimal up-scaling native low-resolution images are done

3.3 Distinctive Issues and Promising Solutions

Exceptional challenges that arise from EHR data could be identified as temporality, indiscretion, modalities, dearth of label, and model features as interpretability.

3.3.1 Temporality and Irregularity

EHRs are intrinsically chronological in nature as information about patient's health state is collected and stored over time. Accurate analysis of the medical state of the patient is done with the collected information. Analyzing the temporal progression in patient's health over time is done in the first phase. Collecting information and analyzing it becomes more pertinent as exposures over time matter. Sequence is important since predication of certain outcomes as events that appear close to the outcome are significant than the events that occur in the distant past. This type of context affects the hidden relations with the clinical variables and outcomes of the future health of patients. Conversely, it is identified that the factual signal's identification from the continuing context among the clinical events is a perplexing task.

To solve the irregularity, numerous techniques have been recommended. A dynamic time warping and an algorithm determining relationship have been devised. Among the two, varying speed temporal sequences have been proposed so that the gate parameters of 2D-GRU are modeled which results in sequence generation of EHR. A subspace decomposition of the long short-term memory (LSTM) memory is suggested in the direction of guaranteeing the consequence of memory with respect to the elapsed time.

3.3.2 Multimodality

The complex task is to produce the EHR data when a patient visits a hospital and to store in various formats is also a challenging task. There are chances of providing the unbalanced information about the patient which reflects in the status of the patient's health. It does not mean that the incorporated visualization of the EHR data is the combination of various visualization techniques. Therefore, the relationships among the data must be considered for identifying the hidden information conceded by the data. It is advisable to represent the multimodal EHR data with a structure so that the visualization could be achieved. Several kinds of EHR data are represented with a unified structure. Then, it is organized accordingly based on the relationships. The integrated viewer-based method which has been proposed illustrated the overall state of the patient.

A strategy must be used so that it must be shared among all tasks. In [41], a multitask learning is done in order to combine the model based on two data modalities for prediction. They are medical codes and natural language text.

3.3.3 Data Interoperability

Many healthcare organizations get benefits by using EHR. The benefits would be still much better if the interoperability of EHRs among the heterogeneous health systems could be achieved. It is observed that the lack of interoperability standards and solutions is found to be a major hindrance during the data exchange. It could be categorized into three levels as mentioned below.

Structural: Exchange of data could take place among information technology systems, and interpretation could be done at the data field level.

Foundational: There is no necessity of interpretation during the receiving of data from one HER system.

Semantic: During the exchange of data, necessity might arise to interpret the data.

3.3.4 Interpretability

Clinicians do not accept the recommendations generated by the machine since they feel that it lacks clarity and support factor for the basic reasoning. Several methods like attention mechanisms, knowledge distillation, and knowledge injection through attention have been proposed to augment the interpretability. Attention mechanism-based learning is used to study the contribution of historical information toward the prediction of disease onset or future events. A dormant representation of medical codes (medication codes and diagnosis codes) is obtained through the attention mechanism. With the help of biomedical ontology, interpretability could be enhanced. The clinical code embeddings are obtained by the amalgamation of the code embeddings and ancestors on the ontology graph. Through the process of knowledge distillation, the complex model is compressed into a simpler model.

4 Review of Clinical Databases

EHR data typically encompasses a massive number of variables that facilitate high tenacity of data and more efficient in handling complex clinical interactions. Several datasets are integrated, and the information available in the system is completely accessible. Healthcare research using huge datasets would become readily accessible as like other data sources in an open-source environment. Initially, EHR was designed to archive and organize the medical data. In the next step, it has been concentrated on billing and quality enhancement processes. EHR-driven databases become very dynamic, comprehensive, and interconnected. In fact, medical industry is little behind in big data usage when compared to other industries. Research with huge datasets is slowed down due to the poor organization of datasets and the quality of gathered data. Contemporary medical data is continued to be evolved that allows the researchers to analyze the data in deep. Traditionally, chronic disease management systems (CDMS) are acted as repositories for the gathered data and often specific for a particular disease. Such systems are failed in decision making, and the data is pertained to patients in one geographical location.

4.1 The Medical Information Mart for Intensive Care (MIMIC) Database

The MIMIC database was found to be established in 2003 and is sponsored by the National Institute of Biomedical Imaging and Bioengineering. It is designed for patients in intensive care units (ICU) who would like to undergo surgery in an urban tertiary care hospital. The most important proclamation is announced as MIMIC-III which presently embraces more than 40,000 patients with huge number of variables. Then the database is available to be accessed openly for the researchers' as it contains detailed clinical and physiological data [42]. It is evolved to adopt advanced patient observing and policy framing systems that automatically improves accuracy, competence, and taking timely decision in critical clinical care. The researchers who are competent to handle delicate information of patients can utilize the open-access database MIMIC-III. It is maintained by a group of physicists, scientists from the domain of computer science, biomedicine, mathematicians, and educationalists all-round the globe.

There are two primary types of data in MIMIC-III database:

(i) Medical information retrieved from EHR such as patients' history, laboratory test reports, vivacious signs, imaging reports, and diagnoses.
(ii) The monitor located near the bed of patients displays the waveforms with relevant parameters, and it is stored in binary files. It includes high-resolution data such as patients' electrocardiograms (ECGs or EKGs), electroencephalograms (EEGs), real-time recordings of vital signs of patients in ICU.

4.2 PCORnet

The National Patient-Centered Clinical Research Network (PCORnet) is a powerful starting process of the Patient-Centered Outcomes Research Institute (PCORI). It was started in 2013 by integrating data from Patient-Powered Research Networks (PPRNs) and Clinical Data Research Networks (CDRNs) [43]. It enables feasible access to massive amount of medical and healthcare data. It acquires health information and clinical data which are created during patient visits. It utilizes data shared by individual users with community networks or personal health records with other patients to manage day-to-day life. Patients and healthcare researchers can work together and share their experiences for the betterment.

4.3 Open NHS

The National Health Service (NHS), a government entity, contains huge repository of medical data about people's health around the world. Open NHS was established in Oct 2011, and it is actively migrated to open new large repositories across many departments and agencies. Most prioritized information about the health information is needed in the healthcare sector so that it meets the necessity of the people. NHS allows clinicians, patients to associate the quality of care delivery in different regions and effectively identifies in which region the quality is less or perfect. It is open-source database which contains information released by government or public bodies [44].

4.4 Open Data Platform

Open Data Platform is implemented by the National Institute of Health Research (NIHR) Clinical Research Network (CRN). The value of the data obtained from electronic records is compared with the data gathered manually. CRN facilitates the accessibility for the staff members in the open data platform, and the count of patients involved in the investigation studies pertained to a particular hospital is determined. Moreover, the hospital that is admitting patients rapidly under which specialty field is also identified.

4.5 VistA

The Veterans Health Information Systems and Technology Architecture (VistA) is an information system that is constructed about EHR and it is used in Veterans Affairs (VA) healthcare medical system. VA medical system activates over 800 ambulatory clinics, 125 hospitals, and 135 nursing homes. VistA interface is utilized since 1997, and it amalgamates pharmacy, ambulatory, hospital, and ancillary services among 8 million US veterans. The database is being involved in the number of medical research that has been performed in the past 25 years.

4.6 NSQUIP

The main objective of the National Surgical Quality Improvement Project (NSQUIP) is to improve the quality surgical care delivery worldwide [45]. It gathers massive amount of data from EHR related to surgical procedures, baseline demographics, complications, adverse events, hospital information, and outcomes. NSQUIP database is an open source that is available for the members belonging to affiliated institutions. It is an efficient database specifically for quality care and surgical care delivery; thus, it is highly generalizable to several real-world surgical practices across the world.

5 Obstacles, Ethical Issues, and Limitations of Deep Learning with EHR

- Rapidly growing medical data in EHR are complex to monitor, poorly annotated, unstructured, and generally heterogeneous.
- Modeling and learning of heterogeneous data sources are the challenging tasks in the medical domain in which the principle of deep learning has to be deployed.
- For effective processing in clinical domain, the contemporary deep learning techniques have to be enriched with features like intractability, data integration, temporal modeling, and security.
- Deep learning supports the future generation predictive healthcare system in which scalability in terms of patient records and provides effective support for physicians in their day-to-day activities.
- Deep learning can act as a regulatory support for organizing research based on several postulates and meticulous exploration on clinical databases.
- Deployment of deep learning techniques for treating the patients is in the starting phase.
- Identification of drug targets and communications or the prediction of drug response.
- Drug interfaces and prediction of bioactivity could be done with the help of deep learning algorithms.
- Drug repositioning through deep learning is also considered to be an exhilarating area in research.
- Relation between the patient and physician, patient caring, patient sovereignty, concealment, and secrecy needs more focus.

References

1. Xu, R., Li, L., Wang, Q.: dRiskKB: a large-scale disease-disease risk relationship knowledge base constructed from biomedical text. BMC Bioinform. **15**(105) (2014)
2. Jensen, P.B., Jensen, L.J., Brunak, S.: Mining electronic health records: towards better research applications and clinical care. Nat. Rev. Genet. **13**, 395–405 (2012)
3. Unified Medical Language System (UMLS). https://www.nlm.nih.gov/research/umls/
4. Mohan, A., Blough, D.M., Kurc, T, Post, A., Saltz, J. Detection of conflicts and inconsistencies in taxonomy-based authorization policies. In: Proceedings IEEE International Conference on Bioinformatics and Biomedicine, pp. 590–594 (2011)

5. Bengio, Y., Courville, A., Vincent, P.: Representation learning: a review and new perspectives. IEEE Trans. Pattern Anal. Mach. Intell. **35**, 1798–1828 (2013)
6. Abdel-Hamid, O., Mohamed, A.-R., Jiang, H., Deng, L., Penn, G., Yu, D.: Convolutional neural networks for speech recognition. IEEE/ACM Trans. Audio Speech Lang. Process. **22**, 1533–1545 (2014)
7. Deng, L., Li, X.: Machine learning paradigms for speech recognition: an overview. IEEE Trans. Audio Speech Lang. Process. **21**, 1060–1089 (2013)
8. Clark, J.: Deep mind forms health unit to build medical software. https://www.bloomberg.com/news/articles/2016-0224/google-s-deepmind-forms-health-unit-to-build-medical-software
9. Enlitic uses deep learning to make doctors faster and more accurate. http://www.enlitic.com/index.html
10. Liu, C., Wang, F., Hu, J.: Risk prediction with electronic health records: a deep learning approach. In: ACM International Conference on Knowledge Discovery and Data Mining, Sydney, NSW, Australia, pp. 705–714 (2015)
11. Pham, T., Tran, T., Phung, D., Venkatesh, S.: DeepCare: a deep dynamic memory model for predictive medicine. https://arxiv.org/abs/1602.00357
12. Miotto, R., Li, L., Kidd, B.A., Dudley, J.T.: Deep patient: an unsupervised representation to predict the future of patients from the electronic health records. Sci. Rep. **6**, 26094 (2016)
13. Liang, Z., Zhang, G., Huang, J.X., Hu, Q.V.: Deep learning for healthcare decision making with EMRs. In: IEEE International Conference on Bioinformatics and Biomedicine, pp. 556–559 (2014)
14. Goodfellow, I., Bengio, Y., Courville, A.: Deep Learning. MIT Press (2016)
15. Lv, X., Guan, Y., Yang, J., Wu, J.: Clinical relation extraction with deep learning. Int. J. Hybrid Inf. Technol. **9**(7), 237–248 (2016)
16. Choi, E., Schuetz, A., Stewart, W.F., Sun, J.: Using recurrent neural network models for early detection of heart failure onset. J. Am. Med. Inf. Assoc. JAMIA **292**(3), 344–350 (2016)
17. Beaulieu-Jones, B.K., Greene, C.S.: Semi-supervised learning of the electronic health record with denoising autoencoders for phenotype stratification. J. Biomed. Inform. **64**, 168–178 (2016)
18. Che, Z., Purushotham, S., Cho, K., Sontag, D., Liu, Y.: Recurrent neural networks for multivariate time series with missing values. Sci. Rep. **8**, 6085 (2018)
19. Lasko, T.A., Denny, J.C., Levy, M.A.: Computational Phenotype discovery using unsupervised feature learning over noisy, sparse, and irregular clinical data. PLoS ONE **8**(6) (2013)
20. Naylor, C.D.: On the prospects for a (deep) learning health care system. JAMA. https://doi.org/10.1001/jama.2018.11103
21. Hinton, G.: Deep learning—a technology with the potential to transform healthcare. JAMA. https://doi.org/10.1001/jama.2018.11100
22. Gulshan, V., Peng, L., Coram, M., Stumpe, M.C., Wu, D., Nararayanaswamy, A., Venugopalan, S., Widner, K., Madams, T., Cuadros, J., Kim, R., Raman, R., Nelson, P.C., Mega, J.L., Webster, D.R.: Development and validation of a deep learning algorithm for detection of diabetic retinopathy in retinal fundus photographs. JAMA **316**(22), 2402–2410 (2016)
23. Wong, T.Y., Bressler, N.M.: Artificial intelligence with deep learning technology looks into diabetic retinopathy screening. JAMA **316**(22), 2366–2367 (2016)
24. Lependu, P., Iyer, S.V., Fairon, C., Shah, N.H.: Annotation analysis for testing drug safety signals using unstructured clinical notes. J. Biomed. Semantics **3**, S5 (2012)
25. Tou, H., Yao, L., Wei, L., Zhuang, X., Zhang, B.: Automatic infection detection based on electronic medical records. BMC Bioinform. **19**(Suppl. 5), 117 (2018)
26. Kaji, D.A. Zech, J.R., Kim, J.S., Cho, S.K., Dangayach, N.S., Costa, A.B., Oermann, E.K.: An attention based deep learning model of clinical events in the intensive care unit. PLoS One **14**(2), e0211057 (2019). https://doi.org/10.1371/journal.pone.0211057.eCollection
27. Yoon, H.-J., Ramanathan, A., Tourassi, G.: Multi-task deep neural networks for automated extraction of primary site and laterality information from cancer pathology reports. https://doi.org/10.1007/978-3-319-47898-2_21

28. Qiu, J., Yoon, H.-J., Fearn, P.A., Tourassi, G.D.: Deep learning for automated extraction of primary sites from cancer pathology reports. IEEE J. Biomed. Health Inform. (2017). https://doi.org/10.1109/JBHI.2017.270072223
29. Esteban, C., Schmidt, D., Krompa, D., Tresp, V.: Predicting sequences of clinical events by using a personalized temporal latent embedding model. In: International Conference on Healthcare Informatics, ICHI 2015, Dallas, TX, USA, October 21–23, 2015
30. Edward, C., Mohammad, T.B., Andy, S., Walter, F.S., Jimeng, S.: Doctor AI: predicting clinical events via recurrent neural networks. In: Proceedings of Machine Learning for Healthcare, pp. 301–318 (2016)
31. Rajkumar, A., Oren, E., Chen, K., et al.: Scalable and accurate deep learning for electronic health records (2018). http://arxiv.org/abs/1801.07860
32. Choi, E., Bahadori, M.T., Searles, E., Coffey, C., Thompson, M., Bost, J., Tejedor-Sojo, J., Sun, J.: Multi-layer representation learning for medical concepts. In: Proceedings of the 22Nd ACM SIGKDD International Conference on Knowledge Discovery and Data Mining, pp. 1495–1504. ACM, New York, NY (2016)
33. Beam, A.L., Kompa, B., Fried, I., Palmer, N.P., Shi, X., Cai, T., Kohane, I.S.: Clinical Concept Embeddings Learned from Massive Sources of Multimodal Medical Data, pp. 1–27 (2018). http://arxiv.org/abs/1804.01486
34. Mikolov, T., Chen, K., Corrado, G., Dean, J.: Efficient Estimation of Word Representations in Vector Space, pp. 1–12 (2013). https://doi.org/10.1162/153244303322533223
35. Pennington, J., Socher, R., Manning, C.G.: Global vectors for word representation. In: Proceedings of 2014 Conference in Empirical Methods and National Language Processes, 1532–1543 (2014). https://doi.org/10.3115/v1/D14-1162
36. Zhou, J., Liu, J., Naryan, V.A., Je, J.: Modeling disease progression via multi-task learning. Neuroimage **78**, 233–248 (2013)
37. Riccardo, M., Li, L., Brian A.K., Joel, T.D.: Deep Patient: An Unsupervised Representation to Predict the Future of Patients from the Electronic Health Records. https://www.nature.com/articles/srep26094.pdf
38. Jeffrey, J.N., Andrew Janowczyk Eliot, G.P., Renee, F., Kenneth, B.M., Michael, D.F., Anant, M.: A deep-learning classifier identifies patients with clinical heart failure using whole-slide images of H&E tissue. PLoS ONE **13**(4), e0192726. https://doi.org/10.1371/journal.pone.0192726
39. Qiuling, S., Fenglong, M., Ye, Y., Mengdi, H., Weida, Z., Jing, G., Aidong, Z.: Deep patient similarity learning for personalized healthcare. IEEE Trans. Nanobiosci. **17**(3), 219–227 (2018)
40. Nguyen, P., Tran, T., Wickramasinghe, N., Venkatesh, S.: Deepr: a convolutional net for medical records. IEEE J. Biomed. Health Inform. **21**, 22–30 (2017)
41. Nagpal, C.: Deep Multimodal Fusion of Health Records and Notes for Multitask Clinical Event Prediction. http://www.cs.cmu.edu/chiragn/papers/ml4hnips17.pdf
42. Scott, D.J., Lee, J., Silva, I, Park, S., Moody, G.B., Celi, L.A., Mark, R.G.: Accessing the public MIMIC-II intensive care relational database for clinical research. BMC Med. Inform Decis Mak **13**(9) (2013)
43. Fleurence, R.L., Curtis, L.H., Califf, R.M., Platt, R., Selby, J.V., Brown, J.S.: Launching PCORnet, a national patient-centered clinical research network. J. Am Med. Inform. Assoc. JAMIA **21**(4), 578–582 (2014)
44. Open data at the NHS [Internet]. Available from: http://www.england.nhs.uk/ourwork/tsd/datainfo/open-data/
45. NSQUIP at the American College of Surgeons [Internet]. Available from: https://www.facks.org/ quality-programs/acs-nsqip

A Review on Psychological Brainwaves Behavior During Sleep: Causes and Diagnosis

Santosh Kumar Satapathy, Akash Kumar Bhoi, and D. Loganathan

Abstract There are many factors responsible for improper sleep quality and sleep quantity in day-to-day life manner. Out of these factors, some are more important impact on human for the disorder of sleep patterns that is psychological, behavioral, physiological, social, cultural, and environmental. Therefore, sleep quantification is also one basic requirement with subject to analyze the brainwaves of subject during sleep time and investigate the disorder patterns. Most of the sleep experts were hypothesized that there could be a certain relationship between sleep disturbances and its consequence diseases. In this chapter, we are more focused about analysis of the brain patterns of different stages of sleep of a subject during sleep and its behaviors. This chapter introduces the background of sleep, electrophysiological parameters of sleep, individual sleep stages behavior, different types of sleep-related disorders, and its treatment solutions.

1 Introduction

Sleep is a basic requirement for human life and it directly interlinks with a proper balance in terms of our health aspects both physical and mental balance. A proper sleep quality also maintains the smooth functioning of internal parts of our body such as proper control in immunity system and cognitive regulations. It has been found that from different surveys and studies, the improper sleep tends so many neurological disorders happened, out of that sleep disorder is the most challenging for medical science nowadays. We understand that everything does due to improper

S. K. Satapathy (✉) · D. Loganathan
Pondicherry Engineering College Puducherry, Pondicherry, India
e-mail: santosh.satapathy@pec.edu

D. Loganathan
e-mail: drloganathan@pec.edu

A. K. Bhoi
Department of Electrical & Electronics Engineering, Sikkim Manipal Institute of Technology (SMIT), Sikkim Manipal University (SMU), Sikkim, India
e-mail: akash730@gmail.com

A. K. Bhoi et al. (eds.), *Bio-inspired Neurocomputing*, Studies in Computational Intelligence 903, https://doi.org/10.1007/978-981-15-5495-7_6

sleep, however, first of all, it is necessary that we need to understand what sleep is and its physiology changes behavior in different moments. This chapter provides about the fundamental information on sleep, sleep stages, sleep disorders, and polysomnography [1].

1.1 Background of Sleep

The basic behavioral of sleep stages are segmented with a period of rest associated with subdivision-specific posture, reduced responsiveness toward internal and external environment activities. Additionally getting proper sleep during the night can help to protect our mental health and physical health, and it indirectly interlinks with the day-to-day basis quality of life activities such as dynamic behaviors happened with conscious and unconscious biological activities during awake or sleep [2]. The voluntary decisions are carried by the subject itself through such activities like switch off lights, reducing sound, and laying down in bed, but the outcome is involuntary decisions like hormone secretion. The whole sleep contains a series of activity patterns happened in the brain throughout the night. Sleep is completely in the balance between its regular sleep pattern behaviors and its biological systems, and a deficit happened in any of the section will disturb the sleep.

1.2 Sleep and Health Behavior

Conceptualizing sleep as health behavior is mainly focused on how sleep behavior and neurobiology interact and how one can maintain his/her health with proper sleep. With this conceptual thought, sleep can be defined into three ways with subject to its sleep requirement, ability, and opportunity.

Sleep requirement. The sleep requirement is one of the biological requirements and proper sleep requires individuals to perform his/her day-to-day activities. Sleep factor also be sometimes influenced by society and by changes in the lifestyle of societies. It has been seen that now in the job scenario, the working hours are longer and shift-work trends also one reason to not get proper sleep and once a night of sleep loss, it cannot be managed with oversleeping in weekends. Unfortunately, there is no standard procedure to measure the actual need for sleep, but as per epidemiological studies, one can need to rest a minimum of 6–8 h per day. The requirement of sleep at night varies from individuals to individuals with different points such as age groups, health conditions, and psychiatric conditions as well. In general, the average sleep duration for a subject can take rest of 7–8 h, but it may also sometimes be different as it depends upon individual behavioral characterization like short sleepers and long sleepers. To ideally behave the normal activities in day to day, quality of sleep essential for each individual, for short sleepers require minimum of sleep hours must be five to six hours and for long sleepers, it may be continue for minimum of eight hours. The sleep requirements are different from each age bar subjects, and this parameter is highly important for balancing between physical health and mental health and consistently fails in this part that may lead to the creation of sleep-related disorders.

Sleep ability. This factor is totally depends on individuals total amount of sleep can achieve is called sleep ability. It completely depends on individuals lifestyle and in some cases also environmental impact can be responsible with subject to sleep ability of a person. Sometimes its reduced level may lead to sleep deprivation and one-night sleep deprivation may cause drowsiness and tiredness to the next immediate day.

Sleep opportunity. It has directly correlated with sleep ability. Sleep opportunity is the total amount of time one individual can spend in sleeping. It is the total time one individual stays in bed.

All these three processes are highly dependent and interrelated to each other in the real world; they all worked in combine nature to control the sleep. The sleep requirement demands for sleep opportunity and sleep opportunity provides context for sleep ability to produce sleep [2].

1.3 Electrophysiological Parameters of Sleep

EEG. The structure of the brain composed of two hemispheres, the brain stem and cerebellum. Generally, structure of the brain stem is complicated; for better observation brain activities, again it includes midbrain, pons medulla, and reticular formation. It is one of the non-invasive measurements of electrical activities of the brain. These recorded electrical activities shown the cortical behavior. These activities recorded by placed sensor named electrodes positioned on brain. It is one of the most important tools in the case of neurology disorder such as sleep, epilepsy, and coma. Currently, both in medical and technical aspects, EEG signal is considered as one of the effective inputs for the diagnosis of any type of neurological-type disorders. Neurologists generally inspect and visualize the brain behaviors of a patient who is under treatment through polysomnography study. The sleep experts go for analysis each and every epoch during sleep through a computer interface for a specific time window and classify the brain activities into different sleep stages. EEG is measured using multiple electrodes positioned in the scalp according to the international 10–20 placement standards. During brain activity recordings, the positioned electrodes fixed according to 10–20. When we have considered multi-channel EEG data recordings at that time, we have followed international 10–10 placement standards. EEG shows continuous recordings of electrical activities. The EEG waveform can be analyzed through different frequency ranges, and as per frequency band, the whole EEG signal can be segmented into alpha band, delta band, theta band, beta band, and gamma band [3–5].

Delta band. This band frequency ranges from 0 to 4 (Hz) and it is one of the slowest wave patterns. Generally, it has observed that delta waves have shown normal for children with age limit to less than or equal to three years. It reached to maximum to 100 μV amplitude during sleep. Generally, delta waveforms are associated with deep sleep and sometimes it may happen with wake state also.

Theta band. It is reflected in the range of frequency from 4 to 8 (Hz). It mostly appears toward beginning of the sleep stages and drowsiness stages. Basically, the theta waves are used in the clinical diagnosis part if its amplitude range must be greater twice when compared to alpha. It is considered as normal when its wave patterns' amplitude is within the ranges of 15 μV. For healthy subjects, it has appeared in central, temporal, and parietal parts [6, 7].

Alpha band. Basically, this frequency band ranges from 8 to 13 (Hz). Generally, its waveform representation looks like sinusoidal shapes. It has been shown in the occipital region of the brain. Alpha waves in general are reduced with opening the eyes, with more anxiety and hearing unfamiliar sounds. The amplitude of alpha waves is in the range of 20–100 μV.

Beta band. The frequency range of beta band ranges from 14 to 25 (Hz). It is one of the fastest wave patterns and in general, it is localized in the frontal part. It is associated with an active and busy environment. This type of band appears mostly in such cases in which the subject most of the time is busy with anxious thinking, active concentration and focused on outside activities or solving the concentrated task. The amplitude is up to a range of 30 μV.

Gamma band. Generally, it appears when the frequency range is greater than 30 Hz and its amplitude ranges are less when compared to other frequency bands and its occurrence is very less.

Besides all fact, the EEG signal is dependent on age and consciousness of subjects. With the proceed of activity, the EEG shifts to higher frequency ranges and lower amplitude. When the subject's eyes are closed in that scenario, alpha waves of EEG are dominated. In the case of the subject falls asleep, the dominant frequency range decreases.

EOG. The EOG signal is generally collected from eye movements of subjects and it is also utilized in the sleep research field. It is a non-invasive procedure and the required EOG signals recorded through fixed electrodes are placed near eye holes. Different expressions of eye movements are recorded during sleep study like slow, waking, and eye movements (REMs). The eye movements continuously change during sleep time from awake state to deep sleep state [8].

EMG. It is obtained for the purpose of recording the muscle behavior and its movements during sleep. EMG signals are utilized mostly during treatment of various types of neurological disorder diagnosis [9].

ECG. It is basically obtained to keep tracking the behavior of hearts according to different instants. A single channel ECG is also sufficient for identifying the abnormalities of sleep patterns. The ECG data collected from fixed electrodes fixed in the region of sternum and another to be fixed on the location lateral part of chest. Generally, ECG is frequently utilized for the detection of arrhythmias and sleep apnea syndrome diseases.

1.4 Polysomnography and Sleep Stages

Human sleep is not a single entity, but it is a series of structured continuous events that follows a regular, cyclic program. The whole sleep duration contains different stages and these individual stages characterized by a specific behavior of tonic and physiological activity and its behaviors extracted by considering the electrophysiological signals and all these activities recorded the electrical activity of human body. In general for sleep pattern analysis, the central electrophysiological signals are considered and for this acquisition of signals, we have obtained electroencephalogram (EEG) signals, which measure the electrical activity from fixed electrodes on human scalp. In Fig. 1, we mention the 21 electrode positions and its naming convention followed according to the international electrode placement rule [10, 11].

EEG electrodes record voltage changes occurred in the surface of the brain and those changes reflected during the exchanges of electrical and chemical signals of brain neurons with each other. The measuring unit of EEG is volts, more specifically sometimes represented as also in micro-volts (μV) units. In general, the EEG signals are values and each EEG signal from a particular electrode is the difference in recordings of voltage and it completely depends on the placement of the individual electrode and its reference electrode in different locations [12]. These EEG derivations (electrode-reference pairs) followed according to AASM manuals during sleep stage scoring. According to AASM rule, we record the signals from different regions of the human brain through fixed electrodes, the brain behavior recorded from different parts from the brain and such regions are frontal, central and occipital locations. To measure sleep, the EEG signal is also associated with other electrophysiological signals like electrooculogram (EOG), electromyogram (EMG), and electrocardiogram(ECG). With consideration of electrooculogram, we have recorded the behavior of eye movements through the placed electrodes on the forehead. The

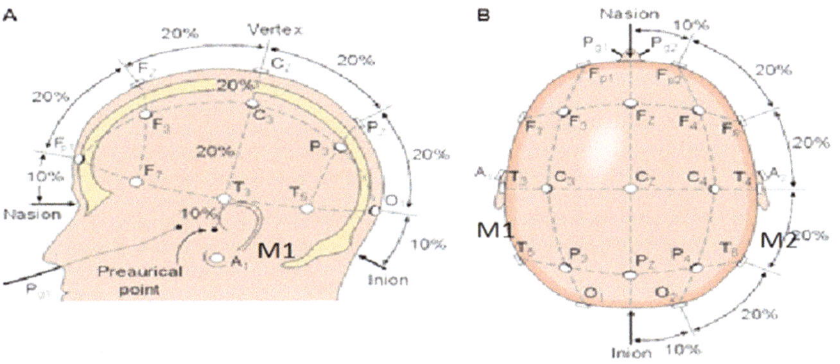

Fig. 1 Placed electrode locations according to the standards of international 10–20 system of electroencephalography (EEG) [10, 11] (http://www.diytdcs.com/tag/1020-positioning/, and edited to include the mastoid electrodes M1 and M2)

electromyogram (EMG) signal is used to measure the muscle movement's behaviors from fixed electrodes on the chin and it can also be used to measure the leg movements, respiratory variables, and body movements. These three signals are the basic sensing modalities in polysomnography. Polysomnography test conducted majorly for diagnosis the sleep irregularties occurred during sleep time. To obtain the PSG signal, the subject must be enrolled in the sleep laboratory for complete one or two nights and sometimes also it has collected from some portable devices fixed on subjects at their own home as well.

Sleep stages. The sleep stages of subjects are segmented based on different behaviors occurred during sleep and the sleep stages are recognized through the behavior reflected in brain, in eyes, and the muscles during sleep, and all these characteristics are recorded through different physiological signals such as characteristic patterns electroencephalogram (EEG), electrooculogram (EOG), and electromyogram (EMG). The continuous recording of this array of electrophysiologic parameters is to define polysomnography. The whole sleep is divided into two main types and it generally occurs in the cyclic process of sleep that is rapid eye movement (REM) and non-rapid eye movement (NREM) stages. Throughout the night, these two phases are repeated cyclically with an average duration of one cycle typically around 45–90 min. A committee headed for analyzing the sleep stages and this process carried through two eminent researchers named as Rechtschaffen and Anthony Kales (R&K), who have designed a set of rules of different sleep standards and scoring of sleep stages in the year 1968. According to R&K, non-REM sleep stages are segmented further into 4 subsleep stages. According to the R&K standards, the sleep cycle of adults consists of six sleep stages: awake, rapid eye movement stage, and four non-rapid eye movement stages: NREM1, NREM2, NREM3, and NREM4 [13]. This standard became widely accepted during polysomnography test. According to R&K, sleep stage scoring is done through sleep stages epochs. The duration of each epoch considered is either a 20 s or 30 s period. The duration of the epoch cannot be changed till the completion of whole recordings. When two stages occurred at one epoch, in this case, the one which takes the largest portion in the epoch, that sleep stages considered for that epoch in the sleep staging scoring system. Till 2007, the R&k manuals followed by sleep experts for analyzing the irregularities happened during sleep. In 2007, small changes happened in sleep manuals and redefined again the new sleep standards and the whole process carried by American academy of sleep medicine (AASM) [14]. With reference to AASM standards, the non-REM sleep stages further divided into three sleep substages N1, N2, and N3. According to new sleep rule by AASM, both N3 and N4 stage combined as one stage called as N3. The whole sleep characterized into five stages according to new sleep manuals which are published by AASM. As per the new rule of AASM manuals, the whole sleep is considered into five sleep stages, awake, three non-REM sleep stages: NREM1, NREM2, and NREM3 and REM [15]. Each sleep stage is recognized through different activities and those behaviors have mostly well reflected with the EEG signals. The EEG activity is typically separated with different ranges of frequency bands such as delta band, theta band, alpha band, beta band, and gamma band, and the same information is represented in Table 1 [16].

Table 1 Sleep frequency bands defined according to AASM manuals [16]

Different patterns of EEG signal	Recommended frequency range (Hz)
Delta	0–4
Theta	4–7
Alpha	8–13
Beta	13–30
Lower gamma	30–80
Upper gamma	80–150

As non-REM sleep that progresses the brain becomes less responsive to the outside world, and this time the brain activities are very slow when compared to wake stage and in the same time more arousals are occurred, thus it is difficult to individual to wake from sleep. Most non-REM sleep occurs in the first third of the night whereas REM sleep occurred the last third part of the night.

Wake stage. It corresponds to a waking state and ranges from full alertness to the outer environment through the early stages of drowsiness. In this stage, maximum events are from alpha bands and those are occurred with low voltage and mixed frequency. In this stage, the EOG signal may be present and its range of frequency found in the range is from 0.5 to 2 Hz and its movements reduced as the subject enters into drowsiness state of sleep. As per AASM manual, if the most of an epoch behaves in alpha nature then it should be considered as wake. The chin EMG signal has high variation with regard to amplitude parameter during the wake [17]. Typical behavior of EOG, EEG, and EMG signals of a single wake epoch is shown in Table 2.

Stage N1. This stage incorporates the transition from the wake stage to the beginning of actual starts of sleep. During this alpha, rhythm is reduced. Stage N1 is scored when 50% of the epoch with low amplitude-frequency. The chin EMG behavior is reflected but it is lower comparing to wake state of sleep. Additionally, in this stage, vertex shape wave patterns may be presented with duration of 0.5 s and the EOG signal will often show slow eye movements during this stage [18]. Table 3 demonstrates an example of the activity observed in the EOG, EEG, and EMG signals during N1.

Stage N2. Here, the major portions contained the theta activity. In this stage, it is very hard to awake the sleeper. Two main characteristics that occurred in these stages are K-complex and sleep spindles. K-complexes are EEG waveforms that occurred with immediately followed by a negative sharp wave, duration of longer than 0.5 s and the maximal amplitude usually shown in the frontal region. On the other hand, sleep spindles are trains of distinct waves that occurred with a range of –14 Hz with duration of more than 0.5 s and the maximal amplitude occurred at central derivations. Epochs with mixed frequency and the high amplitude without k-complex or sleep spindles should be also considered as N2 but the epochs must be preceded to either k-complex or sleep spindles. If the time interval between the k-complex and sleep spindles occurs less than 3 min, in that condition the concerned epochs are considered as stage 2, if there is no indication of movement arousal. If

Table 2 Typical activity of the EEG (O1-A2) recordings, EOG recordings (right and left), and the EMG (chin) recordings during wake

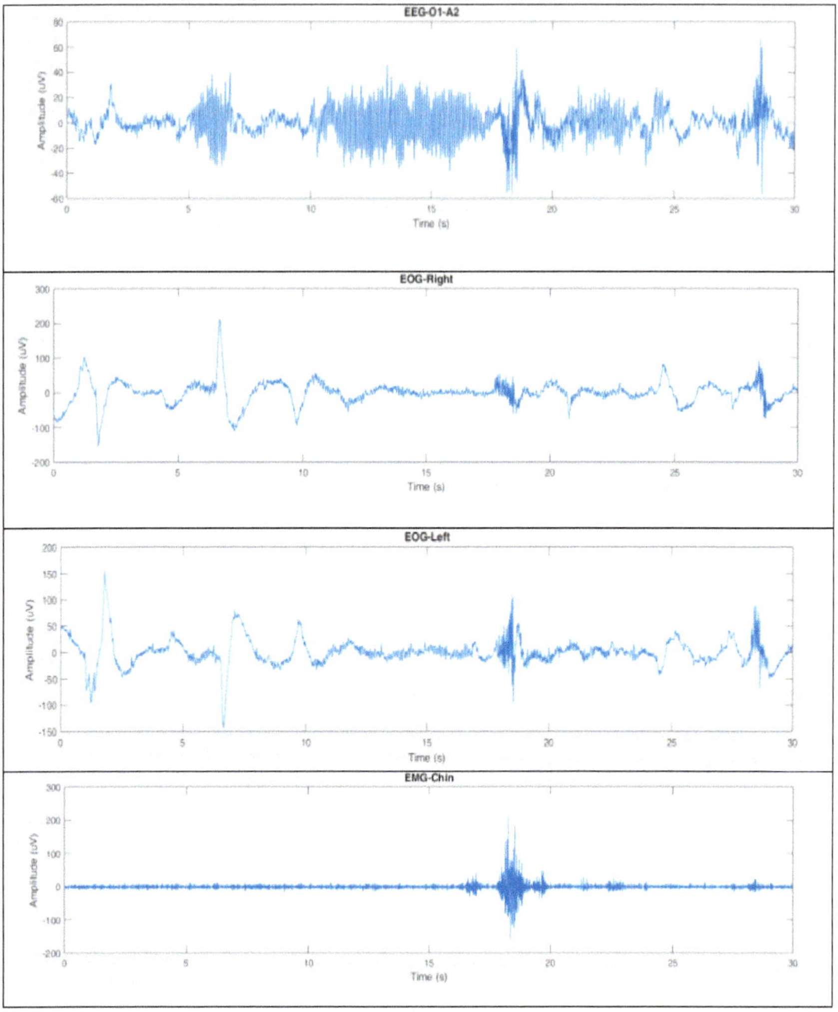

the time limit of interval has crossed more than 3 min, then it is scored as stage 1 even they have no movement arousal [18]. Table 4 demonstrates an example of the activity observed in the EOG, EEG, and EMG signals during N2.

Stage N3. In this stage, the subject went through deep sleep and maximum EEG patterns are of delta types and are characterized by slow wave with frequency range of 0.5–2 Hz. Here, in general, no eye movements are seen during N3 sleep. The arousal threshold is highest during this phase of sleep, therefore this phase is named as deep sleep [18]. Table 5 demonstrates an example of the activity observed in the EOG, EEG, and EMG signals during N3.

Table 3 Typical activity of the EEG (O1-A2) recordings, EOG recordings (right and left), and the EMG (chin) recordings during N1

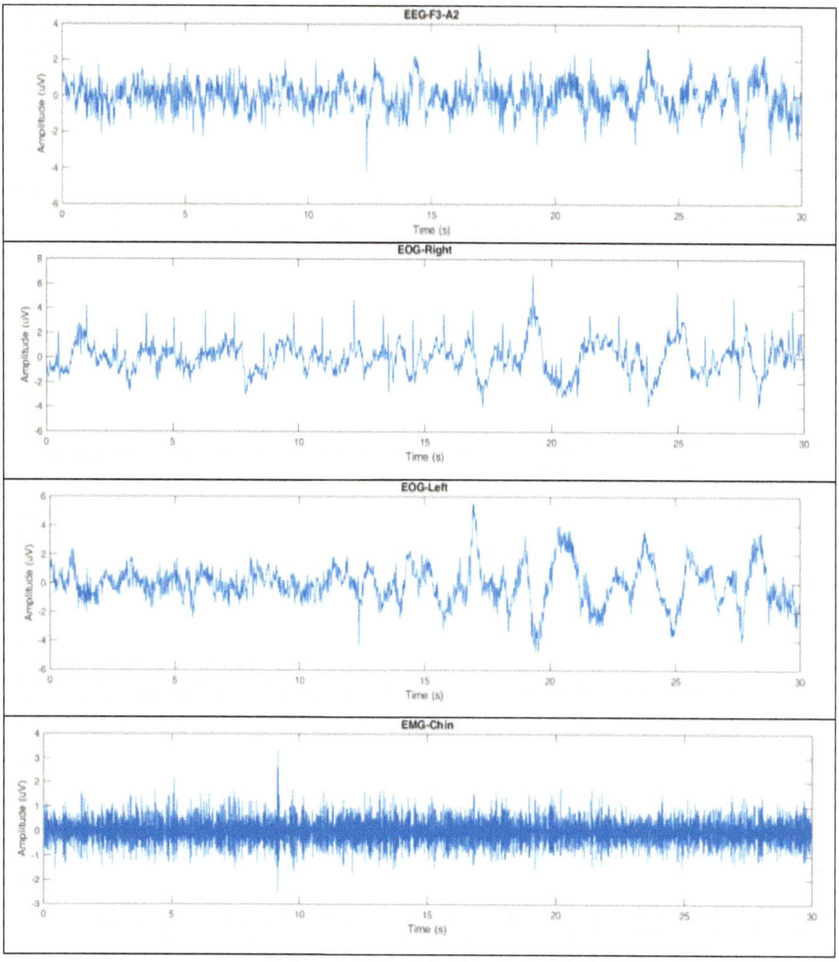

Stage REM. It typically contains 20–25% of total sleep for healthy adults. The main characteristics of the REM stage are the simultaneous occurrence of mixed frequency with low voltage and it occurs in regular intervals of eye movements. The EEG pattern is closer to stage 1 of sleep, differentiated with only vertex sharp waves. In this stage, the distinctive sawtooth waves are present and in general, it has occurred with bursts of eye movements. In this stage, the maximum events are in alpha activity with a frequency range of 1–2 Hz. Here, sleep spindles and k-complexes are completely absent [18]. Table 6 demonstrates an example of the activity observed in the EOG, EEG, and EMG signals during REM.

In Table 7, we have described the different sleep stages of brain waveform behavior in terms of different electrophysiological signals during sleep. Similarly, in Table 8,

Table 4 Typical activity of the EEG (C3-A2) recordings, EOG recordings (right and left), and the EMG (chin) recordings during N2

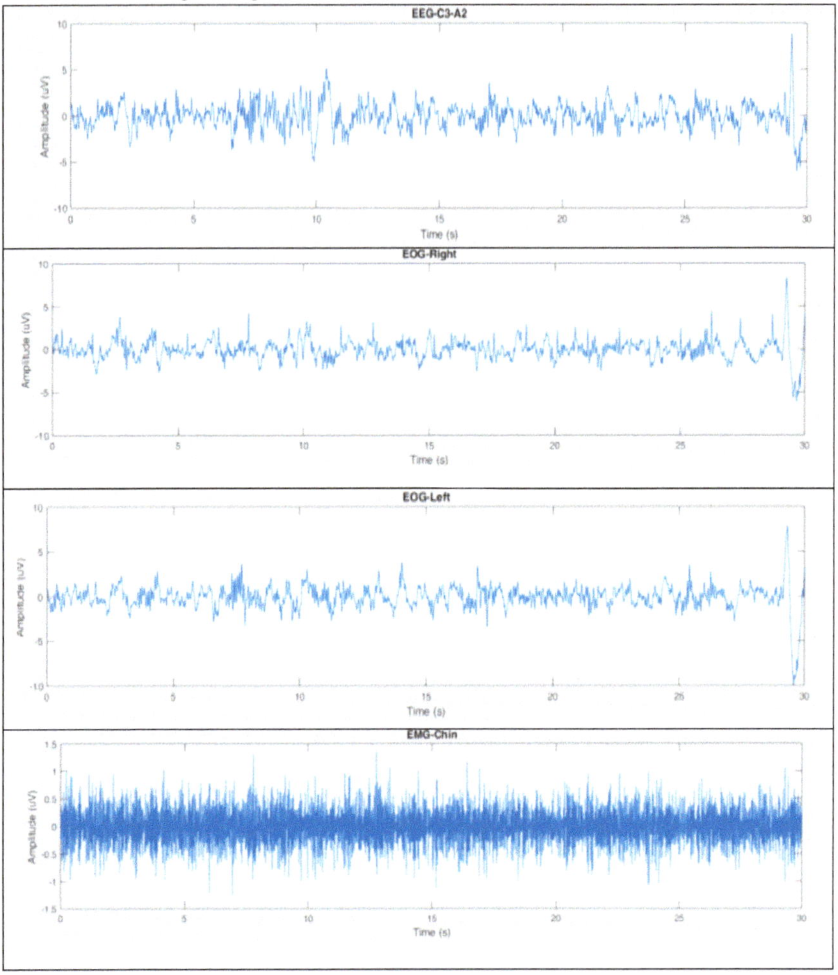

we represent different sleep stages characterization, which occurred during the sleep life cycle of a human [19].

2 Sleep Disorders Characterization

Sleep disorders happened due to improper sleep quality in day-to-day activities and these disorders very seriously affect the patient's health and quality of life. Sleep disorders are classified as one of the neurological-type disorders in which the subjects

Table 5 Typical activity of the EEG (C3-A2) recordings, EOG recordings (right and left), and the EMG (chin) recordings during N3

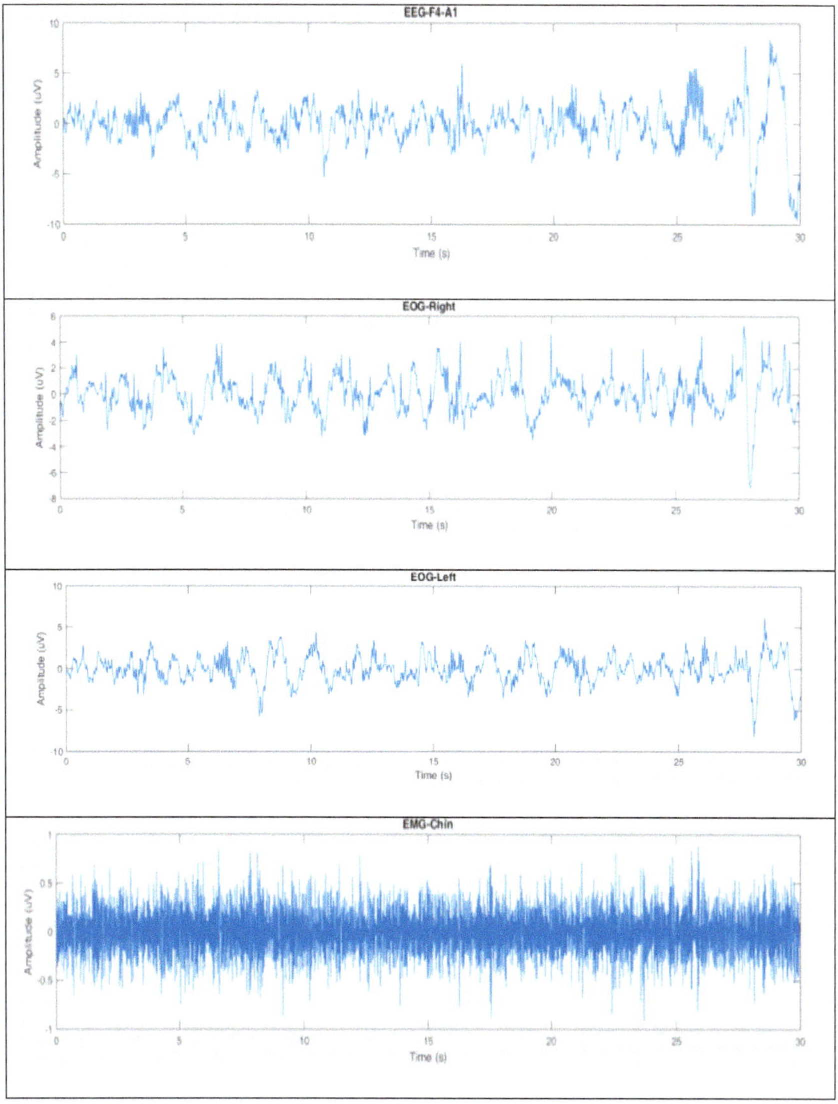

are suffered from sleep patterns and it has been shown with any category of age groups. Some of these disorders are primarily cured with a primary level of treatment through proper consultation with sleep experts, but some of the disorders may get serious for human health in a long-term basis. Here, we are majorly focused on the most occurrences of sleep-related abnormality disorders such as insomnia,

Table 6 Typical activity of the EEG (C3-A2) recordings, EOG recordings (right and left), and EMG (chin) recordings during REM

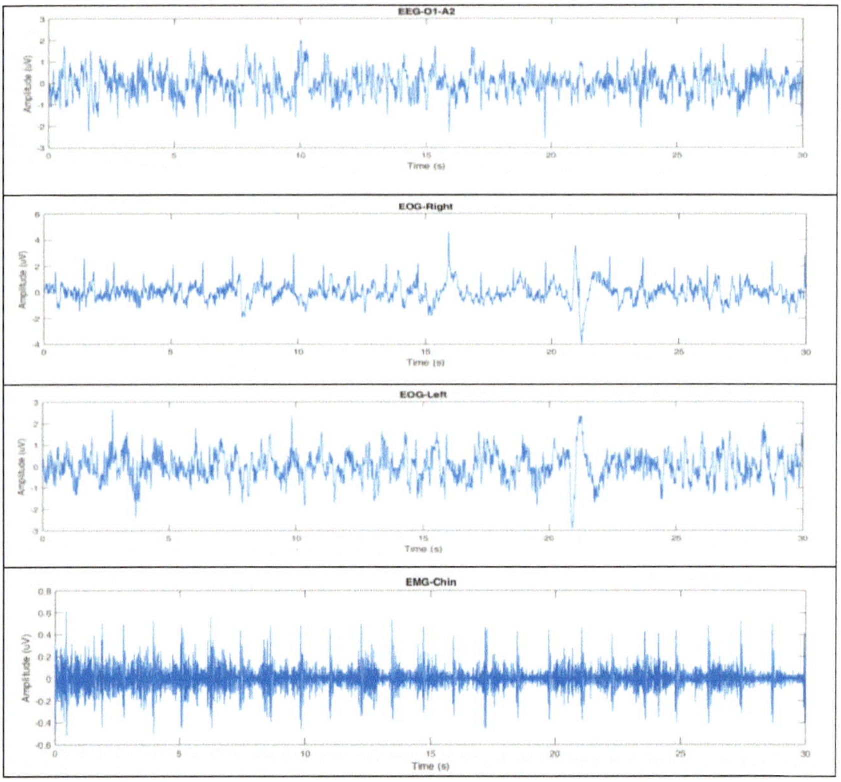

sleep apnea, narcolepsy/hypersomnia, parasomnia, circadian rhythm sleep–wake sleep disorders, and the sleep movement disorder [21].

Sleep Insomnia

It is one type of sleep disorder where one can face the difficulty to fall a proper sleep, to continue with proper sleep or both. In this type of sleep disorder, one cannot be fully refreshed even he/she wakes up after sleep, and this situation can lead to fatigue or other symptoms also reflected on body like changes of mood continuously, face difficult to concentrate for a longer period of time [22].

In general, two types of insomnia have occurred in the case of the human subject, one is primary insomnia and other one is secondary insomnia. We can tell as primary insomnia when the subject is affected with sleep disorder without any relation of other medical conditions. The secondary insomnia has happened due to some other medical situations such as asthma, depression, joint stiffness, cancer, and heart pain occurred due to chest pain. Sometimes it has also occurred due to some medications taken by the subject which was prescribed by doctors, and due to the medicine effects, some pain has occurred and it ultimately makes the subjects sleepless. This disorder

Table 7 Summary of identification characteristics of sleep stages [19]

Sleep stages	Behavior of EEG	Waveform patterns	Behavior of eye movements	Behavior of chin EMG
Wake	Most of the waveforms with alpha activity with mixed frequency and low amplitude	None	Eye blinking, slow rapid eye movements	Normal
N1	Maximum activities in nature of theta behavior with combination of low range of amplitude and mixed frequency	Vertex waves	Slow	Normal
N2	Low amplitude, maximum patterns of theta nature, some of epochs are also delta activity	K-complex sleep spindles	Very slow	Normal
N3	Maximum events in this stage with delta behavior with high voltage	None	Eye movements disappear	Normal(low)
R	In this phase, the behavior of subjects is more active with low amplitude and mixed frequency	Wave patterns are shown as sawtooth	Here the movements of eye occurred rapidly	Partially muscle movements occurred

is seen with any age group of subjects and its effect is more with the woman rather than a man. This disorder happened with the subject either in acute manner, where the subject can be sleepless for few nights, but in case of chronic situations, one person finds difficult to sleep [23].

As per report sources from National Heart, Lung, and Blood Institute (NHLBI), it has resulted that insomnia sleep disorder can create some serious situations caused in our body and to those risk factors which are more common to have insomnia such as 1. the high level of stress, 2. depression related to life events, 3. income factor, 4. traveling journey, 5. No physical activity lifestyle 6. Job professionals and their lifestyle also big impact on to get this type of sleep disorder in the subject's health. The risk performance of insomnia represents in Fig. 2.

To diagnosis this type of sleep disorder, both approaches one doctor can obtain both the ways such as pharmaceutical and non-pharmaceutical. Some experts have considered a single sleep stage or some may obtain two or three days of sleep recordings to the diagnosis of insomnia sleep disorder [24]. In some special scenarios, the subject has not gone anywhere for treatment, for them the doctor has fixed some

Table 8 Summary of characterization of sleep stages [20]

Sleep cycle stage	Duration	Characterization
Awake	16–17 h	• Completely active with all sensing units
Stage 1 light sleep	4–8 h non-REM stage	• Movements of eyes and muscle movements are slow in comparable to wake stage
Stage 2 light sleep	For adults the sleep pattern for stage 1 and 3 has 30% and stage 2 = 50%	• Here movements of eyes are and the behavior of brain become slower
Stage 3 deep sleep		• During this stage of sleep, muscles are relaxed, breathing rate and blood pressure drop down and deep sleep occurred
Stage 4 deep sleep		• This phase also called as deepest sleep or slow-wave sleep and those wave patterns are called as delta waves
		• As the subject crosses from stage 3 to stage 4, then delta waves increase and faster waves decrease
		• No eye movement and the muscles are motionless
REM stage	For adults, this phase of sleep covers 20%; in case of infants, the percentage of the REM sleep spends almost 50%	• During this phase, a subject's brain activities are energies but the body movements are motion less
		• The eye movements are rapidly fly in different directions for the time being the limbs and muscles are paralyzed

wearable smart devices into the patient body and records the behavior of the patient during sleep. Further analysis is required with subject to irregularities happened during sleep, so that it will ultimately help to find a more effective solution toward the diagnosis of insomnia sleep disorders.

2.1 Sleep Apnea

This disorder specifically occurred due to continuous stops and starts of breathing during sleep. During this apnea events, the muscle transformation during apnea events, there are no movements of the muscle of inhalation and the volume of the lungs remains unchanged. When breathing is reduced, due to some partial obstruction of the airway for a specified time frame, then it is called a hypopnea [25].

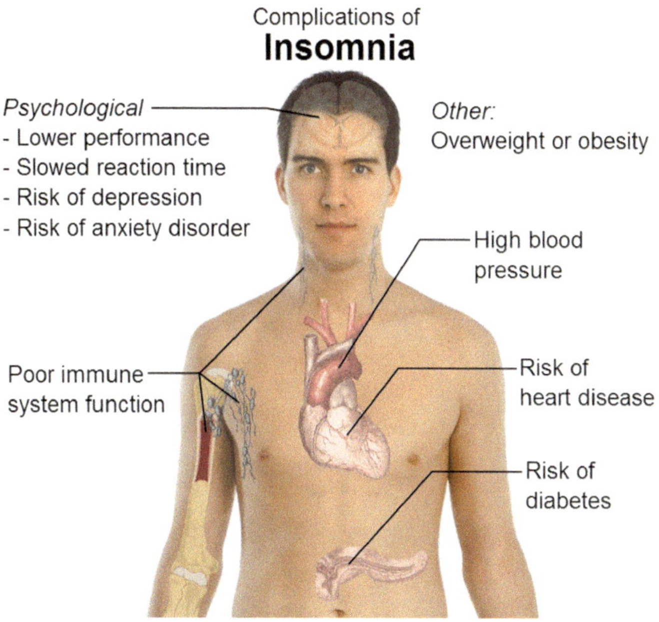

Fig. 2 Complications of insomnia sleep disorder [23]

In the same way, if one person is not breathing for 10 s and after also if he/she is not breathing for another 10 s, then this type of disorder is considered as central apnea, and it is considered as mixed apnea when the breathing is insufficient into the lungs for the duration of 10 s. The extreme situation for this type of sleep disorder is called as obstructive sleep apnea where stops breathe initially continuously stops breathing for some time of 10 s or more during sleep. This disorder occurred due to less oxygen occurred in blood and due to this one can awake easily several times in a night [26] (Fig. 3).

The general symptoms for obstructive sleep apnea and central sleep apnea are common among most of the patient cases. The general symptom for both of the obstructive and central sleep apnea includes loud snoring, gasping for air due to sleep, morning headache, excessive daytime sleepiness, and difficulty in concentrating when awake. The important reason for occurring the OSA for adults is due to excess weight, neck circumference, being as older, being as male, and the use of alcohol-related products, smoking, and obesity. It has happened when throat and tongue muscles are more relaxed and this soft tissue can block the airway. Due to sleep apnea, several other diseases are also seen in the human body like daytime fatigue, diabetes (type-2), high blood pressure and liver problems. In Fig. 4, we describe differences between normal or obstructive breathing during sleep [27].

The diagnosis solution for this type of disorder, we have obtained polysomnography test where the sleep experts analyze the irregularities happened in different stages of sleep during sleep.

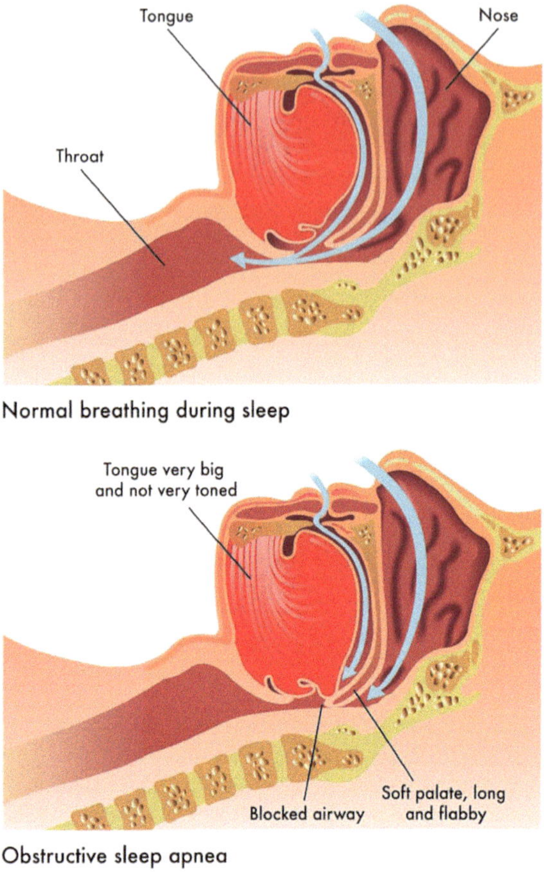

Fig. 3 Normal and obstruction breathing system [26]

To conduct the complete treatment, a patient must admit in clinics and in some exceptional cases, doctors suggest home-based treatment for diagnosing the breathing-related disorders through deploying polysomnography sleep test for diagnosing the disorder [27]. This test generally monitored breathing patterns, airflow, blood oxygen levels, and limb movements.

Narcolepsy

It is one of the neurological disorders where one patient can face difficulty regarding the ability to wake and sleep. This disorder is a lifelong disorder in which one person can feel overwhelming tiredness and it is more severe case in some situations, which may further lead to heart attack. In this disorder, one can have excessive uncontrollable daytime sleepiness and this kind of subject has uncontrollable daytime sleepiness during any time of day and any type of activity. In general, this type of disorder effective subjects spends more sleep in daytime. The other symptoms of this type of disorder are disrupted sleep, cataplexy, and hallucinations. Majorly the

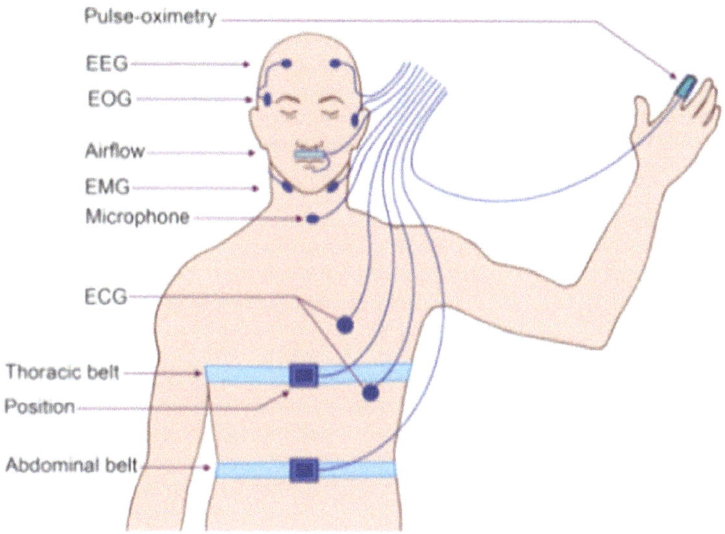

Fig. 4 Fixed location of electrodes for recording the data [32]

three types of narcolepsy are shown, the first type as narcolepsy with cataplexy, the second one is without cataplexy, and the last one is secondary narcolepsy [28].

Parasomnias

In general, this type of disorder can happen with the subject during sleep condition. In this situation, the subjects behave some unwanted activities during sleep. This disorder has affected any of the stages of the sleep cycle. In general, parasomnias effected subject may behave some abnormal behavior like sleepwalking, sleep talking, and nightmares during this type of disorder.

Generally, in some cases, it has found that this disorder was occurred due to the heredity factor and other reasons regarding this disorder are the ill-use of alcohol, medications, and depression. From maximum scenarios, it has reported by sleep experts that parasomnia occurred due to sleep apnea or sleep movement disorders.

This disorder is occurred during both phases of sleep, if it happens during non-REM sleep then most of the subject's behavior includes sleep talking, sleepwalking, and nightmares, and if it happens during REM sleep, then the subjects have faced a lot of difficulties like REM movement disorder, sleep paralysis, and catathrenia.

This disease is curable if proper consultant happened during initial level of this sleep disorders diseases because sometimes it subsists a considerable injury to other persons. One can recover from this type of disorders by getting a good quality sleep at night managing stress or depression and maintaining a sleep schedule.

There are no specific plans in concerned to treatment regarding narcolepsy, but it may be some cases controllable by putting some efforts to change in the day to day lifestyles like good sleeping habits, avoiding alcohol, nicotine and heavy meals, go for exercising regularly, proper scheduling of daytime naps and meal [29].

Sleep Movement Disorders

This type of disorder happened due to the cause of movement prior to or during sleep. In this case, the subject feels quite difficult related to go down asleep or stay asleep or restful sleep. Different types of movement irregularities occurred during sleep but some common sleep movement sleep disorders are restless leg syndrome, leg cramps, bruxism, rhythmic movement disorder, and periodic limb movement disorder.

Restless legs syndrome: This type of disorder is occurred due to unpleasant sensations happened in the legs, and that causes force to affect the legs of the subject. In this situation, one subject felt uncomfortable sensations in their legs and it raised irresistible situations and it ultimately causes an uncomfortable pain in the legs due to itchy, pins and needles and creepy crawly. Sometimes the range of severity of RLS disorder is from mild to severity. For some patients, the severity of pain increased during nighttime and it indirectly affects sleep disruption that can significantly impair the quality of sleep [30].

Periodic limb movements: This type of disorder happened due to continuously involve intolerable a series of redundant muscle movements that severely cause disrupt the sleep during the night. This disorder can cause tiredness during daytime. During sleep, the movement tends to encompass clinch or the flexing muscle can create an unpleasant environment for the illness person, and also who stayed bed near to that subject. This type disorder distinguish with regarding to the time they occurred, if it happened during awake then this type of movement disorder is called as periodic limb movement in wake condition (PLMW) and if the same condition happened with subject during sleep it is treated as periodic limb movement disorder with sleep condition (PLMS). Out of these two movement disorders, PLMS occurred most during sleep times, and generally, it happened without any notice of the subject. During this most of the time, the subject sleep was disrupted many times. One typical sign of this type of movement disorder is the extension of the big toe and the same thing can happen in the ankle, knee, and hip or even in the arms. The level o f severity of pain varies from different to different nights. PLMS usually happens in the first phase of the night during NREM sleep. When these movements are more severe, happened during wake time, then it is called PLMW [31]. Due to PLMW, some bad symptoms have occurred like the depression, bad memory, short attention span, and fatigue.

The general diagnosis of these types of disorders is physical examination and polysomnography. In a physical examination, the doctors are inquired about brief lifestyle factors of patients, medications, and family medical history. In other parts, the PSG test records sleep, breathing, movement, and other bioelectrical signals, including brainwaves and heartbeat during sleep time. This test may help to detect the cause of disrupted sleep and excessive limb movements.

Finally, in Table 9, we mentioned a brief comparison of different types of sleep disorder and made comparisons between their characteristics.

Table 9 Different types of sleep disorders according to international classification of sleep disorders (ICSD)

Category	General characteristics	Examples
Insomnia	Involves to difficult to get proper sleep and difficult to stay continue with sleep for a longer period of time	Psychophylogical insomnia, shorter sleep
Hypersomnia	Excessive sleep during daytime	Narcolepsy long sleeper
Breathing disorder	Difficulty of proper breathing during sleep	Obstructive sleep apnea hypopnea central sleep apnea
Parasomnia	Undesirable events happened during sleep like sleep walking, running, and talking	Sleep walking, talking REM sleep behavior disorder
Sleep movement disorders	Unexceptional muscle movements occurred during sleep	Leg syndrome limb movement disorder

3 Clinical Equipment and Respective Signals for Sleep Disorders Evaluation

To diagnose different types of sleep disorders, the most important step is to choose the appropriate input signals, and it ultimately helps to proper classification of sleep disorders. As we have discussed earlier, PSG test is obtained during diagnosis of different types of neurological disorders to distinguish different sleep patterns and irregularities.

From the different scenario of sleep disorder diagnosis, sleep experts recommended for acquiring different signals such as the electroencephalography signal, the electrocardiographic signal, the electrooculography signal, and the electromyography signal. Those signals' brief description is mentioned in below subsection.

3.1 Polysomnography

Polysomnography (PSG) is used basically for the physiological monitoring approaches to provide an overall assessment of the patient's quality and quantity of sleep. The word polysomnography is derived from the Greek root poly meaning many, the Latin noun somnus meaning, as sleep, and the Greek verb graphein meaning, to write. Henceforth, PSG is a term that indicates the multi-characterized physiological assessment which occurs in sleep studies.

As per the golden standard procedure to detect any kind of sleep disorders, first of all, one patient is to be admitted for a polysomnography (PSG) test. The PSG is the complex system that makes use of different electrode placement standards for acquiring signals from different parts of the body. In general, 10/20 or 10/10 electrode placement standards referred by sleep experts for recording signals from the patient during night sleep. The PSG test records several types' electrophysiological biosignals: these parameters include the electroencephalography (EEG, brainwave activity), electrooculogram (EOG, for horizontal and vertical eye movements), electromyogram (EMG, usually chin or muscle movement's activity), in addition to these parameters some parameters include while PSG test that is: airflow (nasal and/or oral), electrocardiography, pulse oximeter, and respiratory effort. In Fig. 5, we have described the placement of electrodes for acquiring different activities from different parts of the body during the overnight polysomnography [32].

The doctors have also recorded some more parameters from subjects during the PSG assessment test that is patient's age, patient's health status, body position, body temperature, and environmental temperature. Though all recordings involve instruments, during the PSG test it is carried out by sleep disorders specialists, and these tests are performed at the sleep center or hospital under surveillance of specialized clinicians.

In general, the doctors are prescribed for the PSG test if the subjects are some of sleep-related disorders symptoms to be presented in Table. For conducting a PSG test, first of all, the subjects must be admitted in the hospital or sleep laboratory, after that the sleep experts fixed electrodes in different parts of the subject body and this phase takes 1–2 h. After that, the sleep experts start recordings from the enrolled subject from different fixed channels for 7–8 h during night sleep time. During recording process, doctors or technologists will measure and record the following parameters:

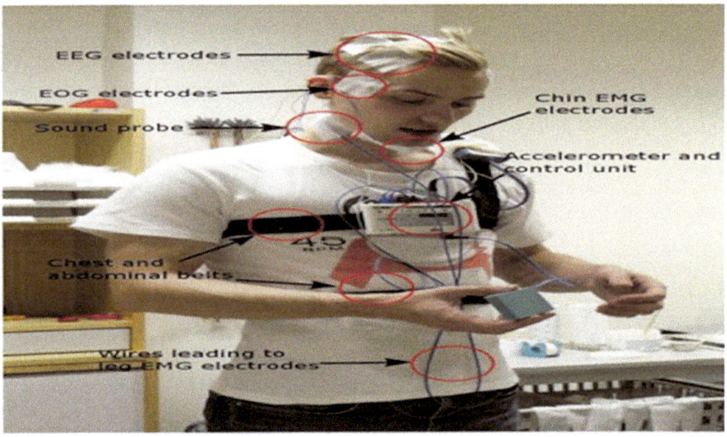

Fig. 5 Polysomnography sleep-monitoring system [34]

4 Overnight Polysomnography Test

4.1 EEG Recording

Electroencephalography records the brain activity of a subject through fixed electrodes placed on the scalp. The EEG measures the voltage fluctuations that occurred in the cerebral cortex. Through EEG, we have also observed the major differences between different electrodes placed on the scalp and also EEG provides valuable information with concern to brain behaviors and its functioning, which alternatively helpful for identifying any abnormal behavior that occurred. All these collected information from electrodes plays an important part in diagnosis and management of any type of neurological disorders. With EEG different features, sleep experts can identify the state and levels of sleep of a subject.

During the time of EEG recordings, the PSG references different leads based upon different segment-wise like C3, C4 are representative for left and right Centro cephalic, similarly for O1 and O2 are references for the region of occipital left and right leads and similarly to achieve the information from frontal portion of the brain, we have used F3 and F4 which are representing for the left and right leads. The frontal leads are used for observing the slow-wave activity of the brain. Sleep spindles the best behavior observed in central derivation. The alpha rhythm has best seen in the part of the occipitally region. Henceforth, all these electrodes help in identifying any type of irregularities that happened during sleep time. In addition, sometimes other than EEG we have considered EMG, EOG, and ECG biosignals for extracting the useful information during the sleep stage scoring evaluation process, and the EEG signal is the primary importance for inferencing any type of neurological disorder during the PSG test. In Fig. 5, we present patient under PSG test during nighttime. In Table 10, we have described how the different electrodes are fixed for acquiring the different electrophysiological signals from the patient during the overnight polysomnography test [33].

4.2 EOG Recording

In the diagnosis process of sleep irregularities, eye movements also take an important part in assessment. Basically eye movements change during different stages of sleep and these changes sometimes create disrupted sleep. To record these behaviors, two EOG recording channels are used. One electrode is fixed at the area of right outer cantus for observing the right-side eye movements, similarly for monitoring the left eye movements, the electrodes are placed at the location of left outer cantus. The EOG recordings are also an important part during the sleep disorder analysis because through EOG, firstly we know about the REM sleep and the second thing is to know about the assessment of sleep onset, so that it can be helpful for analyzing slow-rolling eye movements which are indirectly associated with sleep pattern of the subject [34].

Table 10 Placement of electrodes during PSG test

EEG
The 10/20 electrode placement is a standard for managing the electrode locations on the scalp. This placement rule recognized internationally for fixing electrodes during signal acquisitions
Letter "Z" (zero) indicates with subject of placement of electrodes in midline position of the brain
Electrodes fixed in the left side of the hemisphere indicate as odd numbers and similarly, the even numbers indicate for those electrodes which are placed to the right side of the hemispheres
During placement of electrodes in scalp, first of all we need to identify the four anatomical landmarks, first identified part is located in between the forehead and the nose; the second region of fixing electrodes is the lowest point of the skull from the back of the head, third part, the preauricular points near to the ears and the fourth space are adjusted by remaining electrodes which are placed equal interval into their adjacent locations
EOG
Basically, this signal is used for recording the eye movements of the subject during sleep. The behavior has to be recorded through fixing of the electrodes near to eye which is placed exactly 1 cm above outside corner, similarly for observing the behavior of left eye, we fixed electrode which is placed 1 cm below the outer canthus
EMG
During sleep sometimes subjects muscle movements reflected like teeth grinding, face itching, and muscle relax. All these behavior we have recorded through EMG signal by fixing the electrodes on different major parts of our body like on the legs and arms which tends the excessive amount of movements during sleep. The chin EMG electrode fixed in the parts of mental is muscle
ECG
ECG records the behavior of the heart during sleep rhythm. The electrode is placed on the chest

4.3 EMG Recording

EMG records the movements of muscle tone during the night sleep of a subject. This channel is used for measuring the latency of the actual sleep-onset duration and the REM sleep-onset occurrence. This information is extracted through the fixed electrode in the area of chin or submental. It can also be used to measure the bruxism level of a subject by putting an electrode in the part of master. For determining other sleep disorders, EMG is also used for measuring body movements, two electrodes are placed in left-side and right-side of the legs, and it ultimately helps the assessment of different types of limb movement disorders [34].

4.4 ECG Recording

During PSG recordings, one channel of ECG is sufficient to track the heart function during sleep. Two electrodes are placed for measuring the heart rhythm during sleep:

the first electrode is fixed in the rostral sternum and the second electrode is just placed in the lateral chest location. The first electrode recorded information about the heart activities during night sleep and is also monitoring any cardiorespiratory dysfunctions are occurred or not. Similarly, the other electrode only tracks the information about either any type of cardiac artifact on the EEG channels is present or not [34].

4.5 Pulse Oximetry

It is an important indicator to know about the respiratory of disturbances during sleep, and it is especially required during the diagnosis of obstructive sleep apnea disorder. This technique is used to determine the detection of blood oxygen saturation, and it has fixed on the figure tip for detecting the oxygen saturation level in blood [34].

4.6 Respiratory Movement

This recording is also important during the diagnosis of sleep-related abnormality; in general in some cases, it has found that either the partial or complete halt of breathing during sleep can cause of sleep disruption and later of times it is considered as breathing-related disorders. For monitoring the respiratory events, the piezoelectric respiratory transducer can be used and the respective electrode fixed into the region of torso. This belt specifically measures the change of behavior in thoracic and abdominal region during respiration. To track the changes in behavior in the part of thorax and abdominal part during respiration, events are managed through placing the respiratory inductance plethysmography (RIP) bands [35].

5 Conclusion

A lack of sleep is known to affect both our physical and mental health conditions. Sleep is an important impact on our body with all aspects of internal hormone secretion and proper management of our immunity system. Some sleep difficulties occurred due to the health status of different genders and it may happen during any treatment conditions or sometimes for women who are in the stage of pregnancy, they all also affect sleep disturbances. We have also found that elderly age humans also face difficult to get proper sleep. Sometimes if it continues for a longer period then it leads to some sleep-related sleep disorders. Sometimes we have seen from several studies that excessive daytime sleepiness is also one of the major symptoms regarding sleep-related diseases. There are so many strong case studies proved that improper sleep quality is one of the major reasons for any types of sleep-related disorders. Therefore, different sleep research institutes, for instance, has warned that, if one can feel difficult to stay proper sleep during nighttime, it must be consulted with physicians or sleep experts for further necessary actions and if it is neglected toward consulting with sleep experts in the primary level then it can increase the risk of cardiovascular problems, high blood pressure, and congestive heart failure. In some cases, it has found that the breathing-related disorders highly correlated with a severe risk of workplace injury or automobile accidents. The sleep disruptions can be one of the causes of hamper academic performances for the children. Therefore, sleep disturbances are one of the major challenges for the health sector for choosing the right diagnosis solutions for different types of sleep-related disorders. So that in this chapter, we have addressed sleep, different types of sleep disorders, and overnight polysomnography test.

The view expressed in this chapter is basically on different concepts of sleep and electrophysiological parameters of sleep and its changes characterization during sleep. Also, in this chapter, we have focused on several types of disorders that occur due to abnormality happened in different transition stages of sleep, and finally, we have also described the polysomnography assessment for diagnosing the sleep-related disorder. The major objective of this chapter is awareness among readers on the subject of how sleep disruption is affecting the quality of life and impact on personal and social life. Also in this chapter, we mentioned the basic common types of sleep disorders, so that it may help those who were found similar types of certain disorders, they can consult to sleep experts for diagnosis. Though sleep disturbances are one of the public health concern, with this intention we have proposed chapter on sleep and its consequences in human life. So that the content of this book chapter is therefore essential for all generations of humans and also it is essential for physicians who are associated with the diagnosis, treatment, and management of sleep disorders.

References

1. Teofilo, L., Chiong, L.I.: Sleep a comprehensive handbook. Wiley Inc. (2006)
2. Tubbs, A.S., Dollish, H.K., Fernandez, F., Grandner, M.A.: The basics of sleep physiology and behavior, Elsevier BV (2019)
3. Krauss, G.L., Fisher, R.S.: The John Hopkins Atlas of Digital EEG. The John Hopkins University Press, Baltimore (2006)
4. Chia, B.L.: Clinical Electrocardiography, 3rd edn. World Scientific Publishing Co., Pte. Ltd., Singapore (2000)
5. Niedermayer, E.: Electroencephalography: Basic Principles, Clinical Applications, and Related Fields, 4th ed. Lippincott Williams and Wilkins, Baltimore, MD (1999)
6. Rangayyan, R.M.: Biomedical signal analysis. IEEE Standards Office (2001)
7. Sanei, S., Chambers, J.A.: EEG Signal Processing. Wiley (2008)
8. Ebrahimi, F., Mikaeili, M., Estrada, E., Nazeran, H.: Automatic sleep stage classification based on eeg signals by using neural networks and wavelet packet coeffcients. In: Engineering in Medicine and Biology Society, 2008. EMBS 2008. 30th Annual International Conference of the IEEE, pp. 1151–1154. IEEE (2008)

9. Takeuchi, N., Uchimura, N., Hashizume, Y., Mukai, M., Etoh, Y., Yamamoto, K., Kotorii, T., Ohshima, H., Ohshima, M., Maeda, H.: Melatonin therapy for REM sleep behavior disorder. Psychiatry and Clinical Neurosciences **55**(3), 267–269 (2001)
10. Cohen, M.X.: Analyzing Neural Time Series Data: Theory and Practice. MIT Press, Cambridge, MA (2014)
11. Iber, C., Ancoli-Israel, S., Chesson, A., Quan, S.F.: The AASM Manual for the Scoring of Sleep and Associated Events: Rules, Terminology and Technical Specifications. American Academy of Sleep Medicine, Westchester, IL (2007)
12. Krajca, V.: Long-term electroencephalogram analysis. ERCIM news. [Online]. no. 29 (1997). Available http://www.ercim.eu/publication/Ercim_News/enw29/kraica.html. Accessed on May 2014
13. Kales, A., Rechtschaffen, A.: A manual of standardized terminology, techniques and scoring system for sleep stages of human subjects, vol. 57
14. Berry, R.B., et al.: AASM scoring manual updates for 2017 (version 2.4). J. Clin. Sleep Med. **13**(5), 665–666 (2017)
15. Moser, D., et al.: Sleep classification according to AASM and Rechtschaffen & Kales: effects on sleep scoring parameters. Sleep **32**(2), 139–149 (2009)
16. Chiu, A.W.L., Gadi, H., Moller, D.W., Valiante, T.A., Andrade, D.M.: Multistage preictal seizure analysis using Hidden Markov Model. Int. J. Biomed. Eng. Technol. **10**(2), 160 (2012)
17. Parrino, L., Ferri, R., Zucconi, M., Fanfulla, F.: Commentary from the Italian association of sleep medicine on the AASM manual for the scoring of sleep and associated events: for debate and discussion. Sleep Med. **10**(7), 799–808 (2009). https://doi.org/10.1016/j.sleep.2009.05.009
18. Khasawneh, N., Conrad, S., Fraiwan, L., Taqieddin, E., Khasawneh, B.: Combining decision trees classifiers: a case study of automatic sleep stage scoring. Int. J. Knowl. Eng. Data Mining **2**(1), 60 (2012). https://doi.org/10.1504/ijkedm.2012.044707
19. Sleep and sleep pharmacology. Clinical gate. [Online]. Available https://clinicalgate.com/sleep-and-sleep-pharmacology/. Accessed on 23 Apr 2018
20. Stages of sleep: REM and non-REM sleep cycles. [Online]. Available http://www.webmd.com/sleep-disorders/guide/sleep-101. Accessed on 20 Apr 2017
21. The internet classics archive. on sleep and sleeplessness by Aristotle. [Online]. Available: http://classics.mit.edu/Aristotle/sleep.html. Accessed on 23 Oct 2016
22. Insomnia (Chronic and Acute Insomnia) causes and symptoms. [Online]. Available http://www.webmd.com/sleep-disorders/guide/insomnia-symptoms-and-causes#1. Accessed on 25 Oct 2016
23. Chouvarda, I., Papadelis, C., Domis, N., Staner, L., Maglaveras, N.: Insomnia treatment assessment based on physiological data analysis. In: Annual International Conference on IEEE Engineering and Medical Biology—Proceedings, pp. 6694–6696 (2007)
24. Hamida, S.T., Glos, M., Penzel, T., Ahmed, B.: How many sleep stages do we need for an efficient automatic insomnia diagnosis?*, pp. 2431–2434 (2016)
25. Weng, C., Chen, J., Huang, C.: A FPGA-based wearable ultrasound device for monitoring obstructive sleep apnea syndrome. In: IEEE Ultrasonics Symposium Proceedings, pp. 2–5 (2015)
26. Health Consequences Sleep Anea. [Online] http://healthysleep.med.harvard.edu/sleep-apnea/living-with-osa/health-consequences. Accessed on 26 Oct 2016
27. Crupi, R., Faetti, T., Paradiso, R.: Preliminary evaluation of wearable wellness system for obstructive sleep apnea detection. In: Proceedings Annual International Conference on IEEE Engineering Medical Biology Society EMBS, vol. 2015, pp. 4142–4145. Novem (2015)
28. Dauvilliers, Y., Rompré, S., Gagnon, J.-F., Vendette, M., Petit, D., Montplaisir, J.: REM sleep characteristics in narcolepsy and REM sleep behavior disorder. Sleep **30**(7), 844 (2007)
29. Parasomnia: Sleep: Causes & Treatment—National Sleep Foundation. [Online]. Available https://sleepfoundation.org/ask-the-expert/sleep-and-parasomnias. Accessed on 25 Oct 2016
30. Madhushri, P., Ahmed, B., Penzel, T., Jovanov, E.: Periodic leg movement (PLM) monitoring using a distributed body sensor network. In: Proceedings of the Annual International Conference of the IEEE Engineering in Medicine and Biology Society, EMBS, pp. 1837–1840 (2015)

31. Periodic Limb Movements—Overview and Facts. [Online] Available http://www.sleepeducation.org/sleep-disorders-by-category/sleep-movementdisorders/periodic-limb-movements/overview-facts/. Accessed on 04 Dec 2017
32. Pandi-Perumal, S.R., Spence, D.W., BaHammam, A.S.: Polysomnography: an overview. In: Pagel, J., Pandi-Perumal, S. (eds.) Primary Care Sleep Medicine. Springer, New York, NY (2014)
33. Pandi-Perumal, S.R., Spence, D., Bahammam, A. (2014). Polysomnography: An Overview. https://doi.org/10.1007/978-1-4939-1185-1_4
34. Thru, X.: sensor provides quality data comparable to polysomnography medical sleep monitoring study—Xethru Blog. [Online].https://www.xethru.com/blog/posts/xethru-monitoring-study. Accessed on 04 Nov 2016
35. Ono, Y., Mohamed, D., Kobayashi, M., Jen, C.-K.: Piezoelectric membrane sensor and technique for breathing monitoring. Proceedings of the IEEE Ultrasonics Symposium (2008). https://doi.org/10.1109/ultsym.2008.0191

A Classification Model Based on an Adaptive Neuro-fuzzy Inference System for Disease Prediction

Ricky Mohanty, Sandeep Singh Solanki, Pradeep Kumar Mallick, and Subhendu Kumar Pani

Abstract Disease prediction is now prevalent in the health industry due to the need to increase the life expectancy of any human being. Diseases are of different kinds like physical, mental, environmental, human-made. Recently, machine learning is paving its path toward perfection in the field of the health industry. Machine learning (ML) with the artificial neural network (ANN) is a useful tool for solving different aspects of a complex real-time situation analysis that includes both biomedical and health-care applications. The system can help in eradicating problems faced by medical practitioners in delivering unbiased results. Patients suffering due to the unavailability of experienced as well as expensive medical help can be benefitted from this system. Machine learning has been recently one of the most active research areas with the development of computing environment in hardware and software in many application areas with highly complex computing problem definition. The medical care sector is one of them; it is capable of the automation process by saving time-consuming and subjective by nature. So, ML and ANN-based processes provide unbiased, repeatable results. The broader dimensionality nature of data in medicine reduces the sample of pathological cases made of advanced ML and ANN learning techniques to clinical interpretation and analysis. The medical understanding and disease detection mostly depend on the number of experts and their expertise in the area of the problem, which is not enough. The medical data analysis requires a human expert with the highest level of knowledge with a high degree of correctness. It is

R. Mohanty · S. S. Solanki
Department Electronics and Communication, Birla Institute of Technology, Ranchi, India
e-mail: mohantyricky@gmail.com

S. S. Solanki
e-mail: sssolanki@bitmesra.ac.in

P. K. Mallick (✉)
School of Computer Engineering, Kalinga Institute of Industrial Technology (KIIT), Bhubaneswar, India
e-mail: pradeepmallick84@gmail.com

S. K. Pani (✉)
Department of Computer Science and Engineering, Orissa Engineering College, Bhubaneswar, India
e-mail: skpani.india@gmail.com

A. K. Bhoi et al. (eds.), *Bio-inspired Neurocomputing*, Studies in Computational Intelligence 903, https://doi.org/10.1007/978-981-15-5495-7_7

prone to error, ML, and the ANN learning method can improve the accuracy with the clinical standard for computer-based decision-making models and tools with expert behavior. The traditional methods like Bayesian network, Gaussian mixture model, hidden Markov model implemented for disease recognition on humans, animals, birds, etc., applied by many researchers have failed to reach the optimum accuracy and competence. Many intelligent systems introduced for the identification of diseases like probabilistic neural network, decision tree, linear discriminant analysis, and support vector machine. Machine learning-based adaptive neuro-fuzzy inference system for disease detection and recognition is the next step of evolution in an artificial neural network. In this chapter, the usefulness of machine learning along with ANFIS utility toward a medico issue in the healthcare sector is discussed.

Keywords Adaptive neuro-fuzzy inference system (ANFIS) · Mel-frequency cepstral coefficient (MFCC) · Artificial neural network (ANN) · Disease prediction

1 Introduction

The healthcare domain is of considerable significance concerning its broad spectrum of reach to individuals and communities. For ten years from now, there was no work toward the advancement of the medical field. Unforeseen advances in artificial intelligence techniques in these recent years are the reason for application in various domains, including health care. Due to the high voluminous range of data in the medical area, this has attracted many researchers and industry experts toward application machine learning techniques in healthcare. Classical machine learning approaches were capable to hold learning-related tasks in the medical care domain where the restriction is on applications to small datasets. However, the medical care domain is growing fast, characterized by big datasets from medical management systems. These big datasets have provided a natural choice for the application of deep learning approaches on healthcare datasets, which may be mostly high-dimensional, sparse, or heterogeneous. Applications range from supervised learning tasks (recognition model), reinforcing learning task (outlier detection for trend analysis) to unsupervised tasks (clustering of data). Significant disease areas were the use of artificial intelligence (AI) tools include cancer, neurology, and cardiology [1, 2]. AI applications in these severe medical conditions like stroke that contains three major factors. The three major factors are early detection and diagnosis, treatment, as well as outcome prediction and prognosis evaluation.

The necessity for the decisional system to replace physicians is inevitable, and also, it is impossible. This decision system can be used for making suggestions and assist the physician. It only gives a recommendation based on the condition of the particular, which can aid in treatment. Humans make the ultimate decision. Nevertheless, some automated systems are beneficial for two key reasons just as

1. To avoid simple, usual things that are time-consuming and overload the medical stuff with a lot of activities that could be easily performed by a machine (for

instance, to verify if the therapeutic features of a patient recommend that person for a specific job, or to predict when is the optimal time to supply the store for a particular medicine);
2. To point out medical conditions that can hardly be detected by humans; a suggestion or an alert or indication can make the difference between life and death (for instance, to identify some cancerous cells in an image, or to predict the risk of a genetic disease).

A medical decision-making system can benefit in some significant ways just as like

1. method is not perturbed by humanly related causes that are (stress, fatigue, reduced attention),
2. it has superior speed,
3. it is efficient,
4. it can go repeating its work,
5. it can quickly store vast amounts of data being able to make complex connections between them.

These systems can select or can generate essential data, very useful for the physicians. This paper presents several mechanisms from the broad domain of artificial intelligence, focus attention on the benefits they bring to medical decision-making. There is a discussion related to some weak points of each tool; it is necessary to underline them to determine the most suitable AI method for a specific task. The work in [3] is related to mechanisms, implemented in decisional systems developed for the liver's diseases to stress on the features of these mechanisms. Some of these features (advantages or disadvantages) explained in the domain's literature and evidenced by the decisional systems presented here. Some other features got noticed once these systems got developed.

2 Background

Recently, AI techniques have fast set its foot across health care, even rendering positive areas of discussion whether AI medical practitioners will take the place of human physicians in recent years to come. Though it is impossible, machine will replace human physicians in recent years to come. AI can help as an aid for making decisions for physicians to bestow best possible and apt clinical decisions or challenge the judgment of humans in areas of a particular function in medical care such as neurology. The growing volumes of medical care data and fast growth of scientific methods based on big data that has facilitates the newly found AI applications in medical care. Supervised based on required clinical queries, strong AI methods can reveal clinically apt data hid in the vast data volumes, due to which AI can help hard medical solutions taken quickly. In this work, the study is based on AI current measures taken in the healthcare department, as well as discusses its recent years to come. The review of four essential aspects from medical practitioners:

(a) The motivations behind the application of AI in health care.
(b) The voluminous and types of data that can be investigated by AI systems.
(c) The procedure that makes possible for AI systems to speculate clinical decision with successful results.
(d) Examples of disease types that are recently being handled by the AI communities.

3 Motivation

In the medical literature, the progress of AI has been discussed in detail [3–5]. AI can use popular algorithms to fetch and learn features from a dataset of the large volume of healthcare and then utilized the findings from elements to provide aid to clinical practice. This healthcare solution is possible if learning with intelligent machines and self-evolving techniques is used to achieve an optimum result based on feedback. An AI system can help a medical individual by rendering services of updated information from clinical practices, journals, and textbooks to provide information to the patient care system [6]. Along with this, an AI system can minimize errors related to diagnoses and therapeutic measures that are the hardly avoidable case of human clinical practice [7–10]. Further, an AI system fetches information that can be benefitted to the large patient population from them to help in getting real-time results for health risk alert and resultant health prediction [11].

3.1 Types of Data

The data requires processing before applied to the AI systems deployed for healthcare applications. The system needs proper training through data that are yield decisions on screening, diagnosis, treatment, etc., from activities related to the clinic so that they can imbibe the same type of patients and establish the correlation between patient features and exciting results. These medical data present in the form of probabilistic diagnoses, notes based on medical condition, recordings were taken from electronic devices used in medicinal purposes, the examination of human body, pathology laboratory, images, and audios. In the diagnosis stage, a major part of the literature in concern with AI techniques process and deal with data from the image captured, audio signal extracted, testing of genes, and electrodiagnosis. Such exemplar is as follows as Jha and Topol requested radiologists to support AI methods for processing images captured during diagnoses that contain voluminous data. Li et al. worked to diagnose gastric cancer by the utilization of abnormal gene expression in long non-coding RNAs. Shin et al. innovated assist system for localizing nerve injury through the electrodiagnosis.

Along with the two major contributors are data acquired by notes in medical examination and laboratory test reports. There is the distinction of data from visual, gene-related, and electrophysiological (EP) information due to they contain significant portions of texts with unstructured narrations, such as medical notes, that are impossible processes directly. As a result, the respective applications of AI point toward the conversion of the unstructured text to the machine level, easy to understand electronic medical record (EMR) [12–15]. As in the case of Karakülah et al., the worked-on extraction of phonetic attributes using AI technologies from case reports to improve the quality of the diagnosis accuracy of the congenital peculiarity [16].

3.2 Devices Using AI Technologies

The above section summarizes that devices using AI technologies primarily divided into two divisions. The first division comprises techniques of machine learning (ML) that process structured data such as imaging, genetic, and EP data [17]. In the medical applications, the ML procedures try to make groups of individuals' symptoms or come to a decision about the possibility of the disease outbreaks. The second division comprises of methods of natural language processing (NLP) that acquire information from unstructured data such as suggestive journals or medical notes to support and boost structured medical data. The NLP techniques aim at converting texts to machine-understandable structured data, which are to be analyzed by methods of ML [18]. AI techniques well-built as the methods are processed and enriched with clinics' study on the registered issue and to be implemented to help medical practice finally.

3.3 The Focus of Disease

All through tremendous growth in AI literature in health care, the research primarily focuses on some types of diseases such as cancer disease, disease related to the nervous system, and cardiovascular (cardiac arrest) disease. There is a brief discussion related to the literature in AI used in these diseases.

1. Cancer: Somashekhar et al. indicated that the IBM Watson for oncology would be a dependable AI system that supports the cancer diagnosis using a double-blinded validation study [18]. Esteva et al. investigated to identify skin cancer subtypes through provided clinical images as in [19].
2. Neurology: Bouton et al. developed a system using AI technologies to bestow the control of movement in individuals suffering from quadriplegia [20]. Farina et al. devised a method using AI techniques to verify the interface of the offline man/machine medium to control upper-limb prostheses by utilizing the discharge timings of spinal motor neurons [21].
3. Cardiology: Dilsizian and Siegel mentioned the possibility of applicability of the AI system to diagnose disease of the heart through the ECG image [22].

Recently, Arterys application got clearance approval from the US Food and Drug Administration (FDA). To publish its Arterys Cardio DL, application based on AI technique to render ventricle segmentation is both automatic and editable using classical MRI images of the heart. Though these three diseases take precedence over other conditions, primarily if these can be solved, then there will have a solution in the future of other illnesses as well. These diseases are futile because they can lead to death; thus, for diagnoses of these diseases should be early for the prevention of exaggeration ill health of the patient concern for furthermore [23]. The enhanced procedure for imaging, genetic, EP, or EMR can aid in early diagnoses of these diseases, which can add power to the system based on AI. The system, based on AI, can deal with all types of conditions. The recent two examples were Long et al., whose work dealt with cataract disease diagnoses using the visual image data [24] and Gulshan et al., whose work dealt with images of the retinal fundus by detecting respective diabetic retinopathy [25]. The remaining sections of this work arranged as follows. Section 2 deals with the methodology of the AI technique used in the prediction of cardiac arrest. Section 3 relates to the observation, along with a simulation of the proposed system. Section 4 deals with experimental details and results. Section 5 comprises the comparison of the existing method with the proposed system. Final Sect. 5 discusses applications of AI in neurology, from the three aspects of early diagnosis, treatment, result prediction along with prognosis analysis and conclusion.

4 Methodology

Adaptive system is a multilayer network of feedforward style in which each node performs a particular function (node function) on incoming signals as well as a set of parameters about this node [26–30]. The characteristic of the node functions that differ from node to node, and the choice of each node function depends on the overall input–output purpose, which the adaptive network is required to carry out. Note that the links in an adaptive system only indicate the flow direction of signals between nodes; no weights attached to the links. To represent different adaptive capabilities, both circle and square nodes in an adaptive network are used. A square node (adaptive node) has parameters, while a circle node (fixed node) has none. The parameter set of an adaptive system is the union of the parameter sets of each adaptive node. For desired input–output mapping, updating of these parameters according to given training data and a gradient-based learning procedure is done.

Soft computing mainly comprises components like ANN and fuzzy system (FS). These two components have been accepted universally for their contribution to the field of known and unknown possibilities. Ample system-free evaluation data is best-suited for ANNs, but sufficient process data required, and expert knowledge is essential for fuzzy systems to work, as shown in Fig. 1. Therefore, both ANN and FS can provide both numeric, quantitative ability, and symbolic qualitative ability, respectively, can be implemented to meet the issue raised where optimum accuracy of

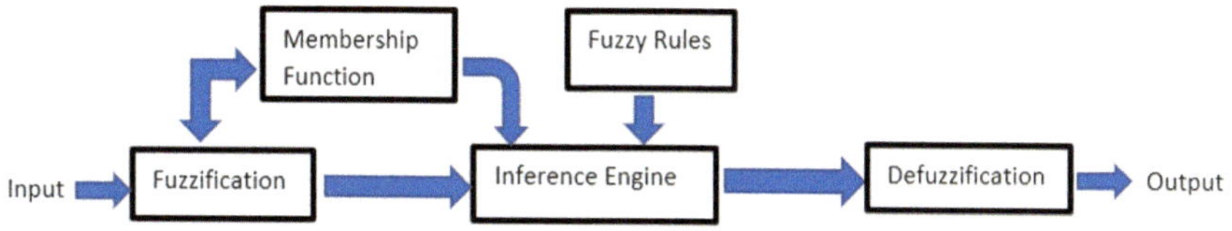

Fig. 1 Block diagram of the fuzzy system

evaluation is not possible [31–35]. The hybrid system containing both ANN and fuzzy designed to render capabilities or properties like adaptability, parallelism, nonlinear processing, robustness, and learning in the massive data arena for the known and unknown systems [36]. The various fields are adopting ANFIS for their numerous applications. Either of two fuzzy models, i.e., Mamdani fuzzy model or TSK fuzzy model, can be used. However, the TSK model widely selected due to its smooth and computationally efficient outcomes than Mamdani's model. In this section, there is a proposal for a class of adaptive networks that function the same as fuzzy inference systems. The system proposed here is equivalently named ANFIS, which stands for the adaptive network-based fuzzy inference system. For the application of the hybrid learning rule, the description of the system deals with the parameter set representation to the node sequentially. Besides this, there is a demonstration of the implementation of the Stone–Weierstrass theorem to ANFIS with simplified fuzzy if-then rules along with the relation of the membership function network to this kind of simplified ANFIS [37–52].

Architecture of ANFIS

Assuming that x, y are two inputs and z as one output for the adaptive network-based fuzzy inference system. Two fuzzy if laws of Takagi and Sugeno rule is the basis for this system rule [53–68].

Law 1: i fx is a1 and y is b1, then f1 = P1x + Q1y + R1

Law 2: i fx is a2 and y is b2, then f2 = P2x + Q2y + R2

where aj and bj = fuzzy variable associated with antecedent inputs as in Fig. 2.

P_j, Q_j, R_j = constant values.

Layer 1: Node variables mathematically calculated for the layer of node j as:

$$M^1_{\mathrm{Pj}} = \mu_{\mathrm{aj}}(x), \quad j = 1, 2$$
$$M^1_{\mathrm{Pj}} = \mu_{\mathrm{bj}}(y), \quad j = 3, 4$$

where $\mu_{\mathrm{aj}}(x)$ and $\mu_{\mathrm{bj}}(y)$ are member functions.

Layer 2 firing function calculated as

$$M_{\mathrm{P2},j} = Wj = \mu_{\mathrm{aj}}(x) * \mu_{\mathrm{bj}}(y) \quad j = 1, 2$$

Layer 3: normalized firing strength can be the ratio of jth firing strength to total firing strength as:

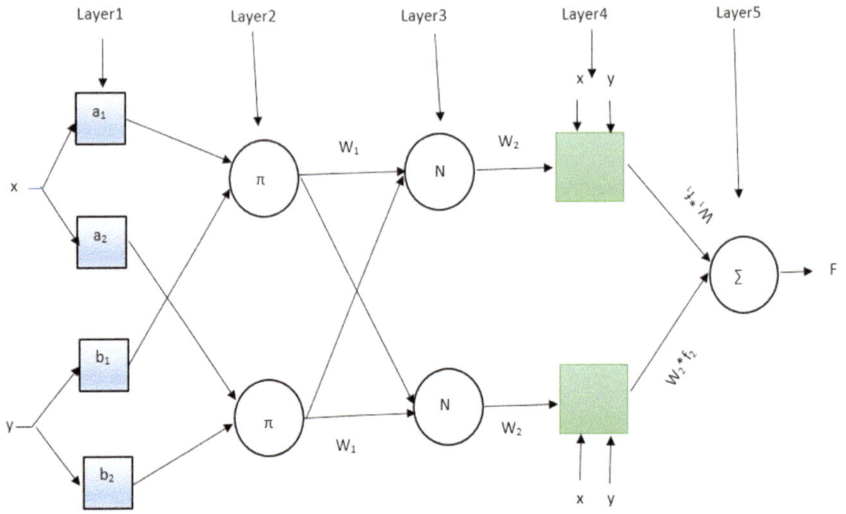

Fig. 2 Structural design of an ANFIS algorithm

$$M_{p3,j} = \overline{W_J} = \frac{W_j}{W_j + W_2} \quad j = 1, 2$$

where $\overline{W_j}$ = normalized firing strength.

Layer 4: For every node *j* in this layer is a square node with a node function

$$M_j^4 = \overline{W_j} f_j = \overline{W_j}\left(P_j x + y + R_j\right),$$

where $\overline{W_j}$ is the output of layer 3, and $\left\{P_j, Q_j, R_j\right\}$ is the parameter set. Parameters in this layer referred to as resultant metrics,

Layer 5: This layer containing a single circular node is labeled as $\sum$ that computes the total output as the addition of all incoming signals.

$$M_1^5 = \text{overall output} = \sum_j \overline{W_j} f_j = \frac{\sum_j w_j f_j}{\sum_j w_j}$$

The ANFIS algorithm explained can be used for the prediction of heart disease in humans for aiding the medical practitioner for the timely treatment of the patient. This process is further elaborated and analyzed with results in the next section of this work, i.e., observation. In this present work, this algorithm is used for the implementation of the heart disease prediction system (HDPS).

5 Observation

In this section of the work, there is a detailed description of how the proposed technique simulated. For simulation, the experiments implemented using MATLAB 2015 b software. The test platform is Intel core 5i 8th generation, 1.6-GHz CPU, 8-GB RAM processor with Windows 10 operating system. A heart disease prediction system (HDPS) designed using ANFIS is shown in Fig. 3.

The flow starts with raw data (sound) acquisition, preprocessing, feature extraction, classification, and prediction decision. The work focuses on the study of the performance of the cardiac arrest prediction system that will provide robust, accurate, and minimized speed in total. The response generated by the system based on the decision delivered by the system. Both for training and testing purposes, the data acquisition task is separately collected. Audio data (i.e., heartbeat recording) from the data repository is preprocessed for subsequent manipulation. The audio data repository is from www.physionet.org.Audio data preprocessing that includes removing silence and the unvoiced part, based on short-time energy followed by noise removing (filtering). In this work, zero first-order finite impulse response (FIR) filter with a 6 dB/octave gain used to remove the noise from the input frames. FIR filter used because of its ability to minimize to zero artifacts along with other associated noise like background noise. The next stage is feature extraction. Heart sound features include cepstral features, temporal features as well as spectral features. In the Mel-frequency cepstral coefficient (MFCC) technique, the coefficients should be between 8 and 20, but the best-suited factor for simulation was 16. The total coefficient, which includes MFCC, delta, and double delta coefficients, has a dimension of 51. The next step of the proposed model is the usage of the technique of principal component analysis (PCA) to get extricated features minimized based on the dimension of these features. PCA consists of a mathematical method that converts a list of

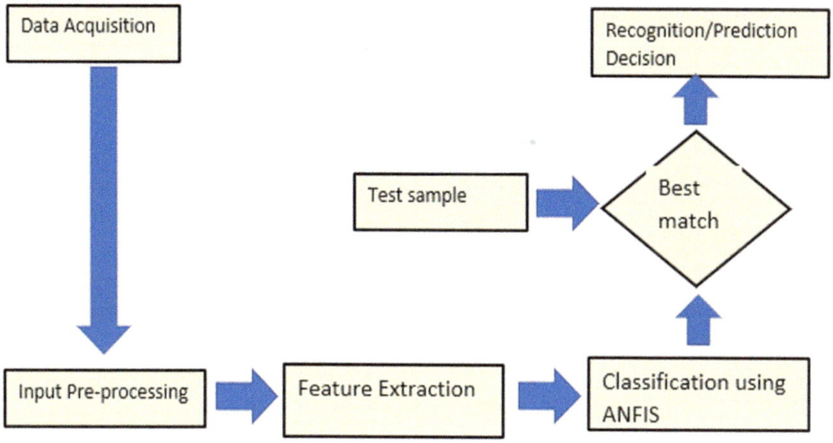

Fig. 3 Design flow of the proposed system

Table 1 Specifications of ANFIS

Input data size	Wav files
SNR	0–3 db
ANFIS type	ANFIS with five layers
ANFIS training method	Backpropagation gradient descent method along with least-squares
Average training epochs	10–200
Total no. of membership functions	10
Type of membership function	Gbellmf, linear
RMSE with 6 Linear at 200 epochs	1.504×10^{-3}

correlated variables into a list of nonrelated variables without losing the information contained in it defined as principal components. The accounts of the first primary component vary as much as the variation can be observed as with forthcoming parts as possible way as in data as well. PCA is a popular procedure for matching patterns and dimension minimizing of high data dimension. The data having feature vectors mainly consist of the exclusive list of information related to the audio signal like wavelength, entropy, and spectral density of the inputs. For this disease prediction system, the feature-length reduced due to the principal component number, which gives the best optimal performance during training. Here, the classifier used for the recognition or prediction is ANFIS. The set of the least-squares (LS) method and the backpropagation gradient descent (BPGD) method, applied for training fuzzy inference system (FIS) membership function (MF) parameters to imitate a training dataset provided. The key specifications of ANFIS, along with five layers, are presented in Table 1. Heart sound feature of reduced length is used for the training of the ANFIS for 10–200 epochs. The results are calculated based on ten trials of epoch on average. The usage of Gbellmf MF is doing fuzzification. Due to which, the Gbellmf MF provides a very realistic real world to fuzzy conversion which is fulfilled from these experiments. The consideration of ten such MFs to give optimum results performed. The root-mean-square error (RMSE) is the measure with which calculation of the cost function parameter showed for training the ANFIS.

6 Experimental Details

The performance evaluation of HDPS is based on computational time and accuracy. The computational time in total minimizes to a large extent by using this proposed system. 100 heart sounds in total have been provided to the system for training, validation, and testing, and from them, a small set of samples with artifacts chosen for testing. The training of these samples in the system is made extensive, and the system is subjected to slight variation in signal to noise ratio (SNR) range between 0 and 3 dB to reach the reliable, robust, and proper prediction. The results obtained are

Table 2 Comparison of average rate accuracy and computational delay in seconds for both ANFIS and ANN-based on the prediction of cardiac arrest

No. of epoch	ANN with MFCC		ANFIS with MFCC	
	Accuracy %	Delay time in Sec	Accuracy %	Delay time in Sec
10	85	0.82	92	0.51
20	85	0.82	93	0.52
50	85	0.85	93	0.55
100	82	0.88	99	0.51
150	81	0.89	96	0.53
200	80	0.91	98	0.54

compared with [19] and inferred that the ANFIS produces relatively RMSE values, but the speed of computation is specifically better. As a result, the RMSE convergence reaches optimum in a few numbers of epochs. The time and minimum RMSE for 200 epochs training with the ANFIS are shown in Table 1. The average accuracy rates found out among the small list of training epochs with ANFIS and ANN for cardiac arrest prediction shown in Table 2. Thus, for better verification and authorization of the design of HDPS, the proposed approach is used. Heart sound sample graph and spectrogram of heart sound are shown in Fig. 4a, b, respectively. During ANFIS training, the average RMSE convergence plot is shown in Fig. 5. The advantage of ANFIS-based approach shows inference on an average 40% fewer epochs than the ANN aided method and yields between 7 and 13% better winning rates. Each sample requires training time in between 2.1 and 3.3 s. The ten trials performed on the sample sets on which the inference and the results on average shown. The equation of root-mean square-error (RMSE) is shown as follows:

$$\mathrm{RMSE} = \sqrt{\frac{1}{D}\sum_{t=1}^{D}(A_t - F_t)^2}$$

where A_t and F_t are actual and fitted values, respectively, and the number of training or testing sample is D. The RMSE is usually the primary measure to make sure that the extent of learning performed by the ANFIS.

7 Results

The simulation results start with an illustration of the test results found while testing the proposed system in this section. The testing begins with ANN along with MFCC.

As observed in Table 2, ANN, along with MFCC, worked at a computation time delay of 0.8–0.9, giving an accuracy of 80–85%. The table depicts ANFIS along with

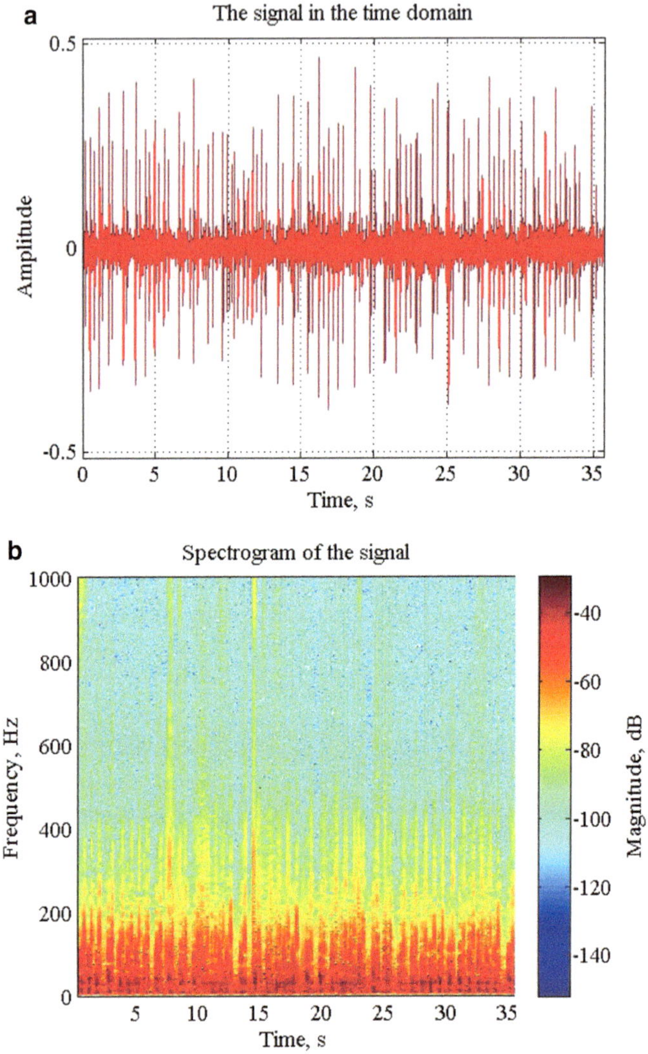

Fig. 4 **a** Heart sound graph, **b** Spectrogram of heart sound

MFCC worked at a computation time delay of 0.5–0.6 s, providing an accuracy of 90–98%.

The classification accuracy increases from 90 to 98% using ANFIS along with MFCC, with the computation time delay between 0.5 and 0.6 s, as reflected in the graphical representation in Figs. 6 and 7, the black line represents the classification accuracy of ANFIS along with MFCC, which lies between 90 and 98% depending upon the number of epoch. As the epoch increases, the black line becomes a straight horizontal line during (epoch 5–epoch 45). It then seems to decrease to some extent

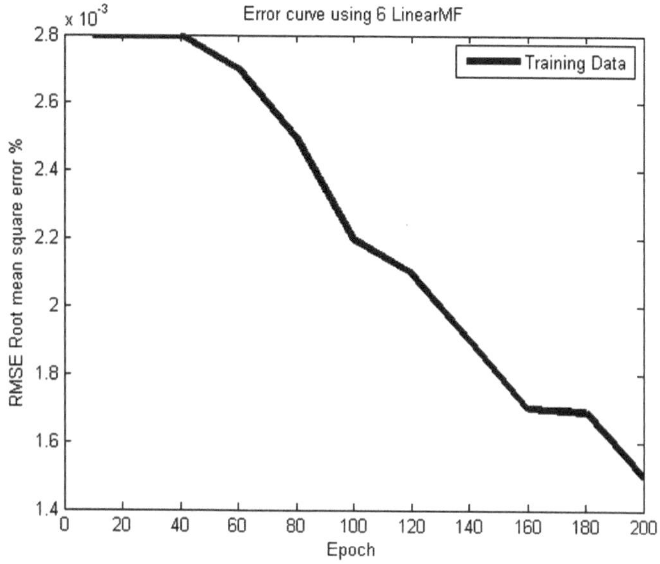

Fig. 5 ANFIS average RMSE convergence

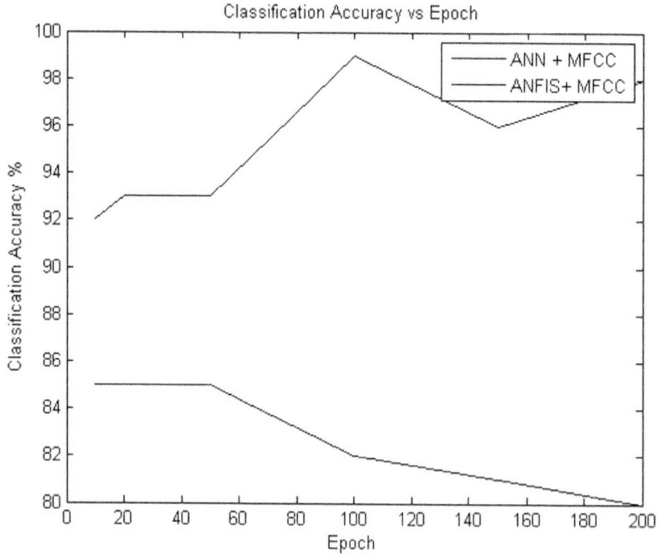

Fig. 6 Graphical representation of ANN and ANFIS based on accuracy percentage versus epoch

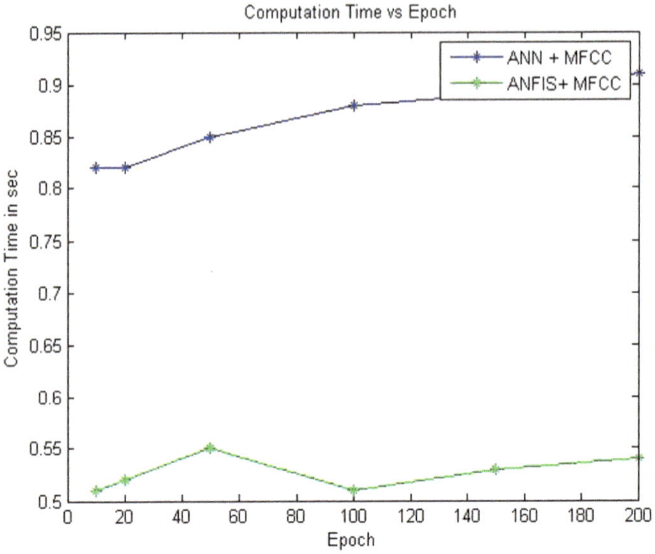

Fig. 7 Graphical representation of ANN and ANFIS based on computation delay versus epoch

after 50–100 epochs and increases after epoch 100 to epoch 200. This black line is better than other technologies used in the context of classification accuracy. The blue line represents ANN along with MFCC is far low than the previous technique.

In the case of Fig. 7, the computation time delay of ANFIS along with MFCC is better than ANN along with MFCC. Here blue with * represents a computational delay of ANFIS along with MFCC and green with * represents ANN along with MFCC in Fig. 7. As the classification accuracy in best in ANFIS with MFCC, the computation time delay is also best and optimum, so it is better than other techniques used. The classification accuracy increases from 90 to 98% using ANFIS along with MFCC, with the computation time delay between 0.5 and 0.6 s, which reflects the graphical representation in Figs. 6 and 7. The black line in Fig. 6 represents the classification accuracy of ANFIS along with MFCC, which lies between 90 and 98% depending upon the numbers of epoch. As the number of epoch increases, the black line becomes a straight horizontal line during (epoch 5–epoch 45). It then seems to decrease to some extent after 50–100 epochs and increases after epoch 100 to epoch 200. This black line is better than other techniques used in the context of classification accuracy. The blue line represents ANN along with MFCC is far low than the previous method.

In the case of Fig. 7, the computation time delay of ANFIS along with MFCC is better than ANN along with MFCC. Here blue with * represents a computational delay of ANFIS along with MFCC and green with * represents ANN along with MFCC in Fig. 7. As the classification accuracy in best in ANFIS with MFCC, the computation time delay is also best and optimum, so it is better than other techniques used. The current work is as shown in Table 2.

8 Comparison with the Existing Method

The comparison of the previous work with present work in this area is as shown in Table 3. As in the current work has the advantage of the classification accuracy along with low computational delay reflected in Table 2. The computational delay is essential for the real-time implementation of this simulation on to the hardware. In this work, ANFIS, along with MFCC, gives the best possible way to get accuracy in classification.

9 Conclusion

The work in this paper described the proposed system of HDPS and in which the critical decision support system is ANFIS. The experimental results inferred based on this proposed approach are more competent in comparison with the ANN-based system, as mentioned in the previous sections. Accuracy rates, speed of computation, reliability, high precision, and robustness are the main powers of the approach that can fulfill any kind of application requirement both in offline and online systems. In the years to come, the work will extend in the inclusion of more features to enhance the architecture of ANFIS to reach the highest recognition accuracy. The overall performance of the system can be improved further by considering a greater number of samples and by using better preprocessing techniques with ANFIS-based blocks. This paper attempts to propose a neuro-fuzzy network for heart disease prediction systems. Many efforts made to solve the disease recognition with the network of fuzzy logic along with the neural network. The system consistently integrated with functionalities of sensory encoding, learning, and decoding. The recognition performance of the system is analyzed with two different techniques used where ANFIS, along with MFCC, turns out to be the best technique with high accuracy and low computation delay. The proposed system structure would also be beneficial for developments in both hardware and software. In the future, this type of adaptive neuro-fuzzy generation neural network would pave the path for further research [69]. It will lead to a great revolution in various areas of science, proving beneficial to humanity as a whole.

Table 3 Tabular comparison of related work to this present work

S. No.	Authors	Feature use	Recognition method	Performance
1	C. N. Gupta et al. (2005)	Daubechies-2 wavelet along with principal component analysis (PCA)	Multilayer perceptron–backpropagation neural network (MLP-BP)	Accuracy (94%)
2	Ashwin R. Jadhav et al. (2017)	Peak detection using Shannon energy envelope calculation algorithm	Backpropagation neural networks with adaptive learning technique	Precision (0.9107)
3	Mohammad Nassralla et al. (2017)	MFCC	Random forest hidden Markov model (HMM)	12% better than state-of-the-art method HMM
4	Ignacio J. Diaz Bobillo (2016)	Energy envelope calculation	Tensor flow	Accuracy 84.54%
5	Zaiton Sharg et al. (2000)	Instantaneous energy and frequency estimations	HMM	High recognition accuracy (78%)
6	Sukryool Kang et al. (2015)	MFCC and descriptive parameters	Support vector machine	Accuracy (98%)
7	Jonathan Rubin et al. (2017)	Combination of MFCC and descriptive parameters	Deep convolutional neural network (CNN)	Precision: 0.84
8	Lijuan Jia et al. (2012)	Wavelet decomposition and normalized average Shannon energy	Fuzzy neural network method	Classification accuracy better than the state of the art
9	Michael Tschannen et al. (2016)	Wavelet transform	Deep convolutional neural network (CNN)	Obtained a score, sensitivity, and specificity of 0.812, 0.848, and 0.776
10	In this work	MFCC	Adaptive neuro-fuzzy inference system (ANFIS)	98% accuracy classification along mean delay of 0.5 s

References

1. Murdoch, T.B., Detsky, A.S.: The inevitable application of big data to health care. JAMA **309**, 1351–1352 (2013)
2. Kolker, E., Özdemir, V., Kolker, E.: How healthcare can refocus on its super-customers (Patients, n = 1) and customers (Doctors and Nurses) by leveraging lessons from Amazon, Uber, and Watson. OMICS **20**, 329–333 (2016)
3. Dilsizian, S.E., Siegel, E.L.: Artificial intelligence in medicine and cardiac imaging: harnessing big data and advanced computing to provide personalized medical diagnosis and treatment. Curr. Cardiol. Rep. **16**(1), 441 (2014)
4. Patel, V.L., Shortliffe, E.H., Stefanelli, M., et al.: The coming of age of artificial intelligence in medicine. Artif. Intell. Med. **46**, 5–17 (2009)
5. Jha, S., Topol, E.J.: Adapting to artificial intelligence: radiologists and pathologists as information specialists. JAMA **316**, 2353–2354 (2016)
6. Pearson, T.: How to Replicate Watson Hardware and Systems Design for Your Own Use in Your Basement (2017)
7. Weingart, S.N., Wilson, R.M., Gibberd, R.W., et al.: Epidemiology of medical error. BMJ **320**, 774–777 (2000)
8. Graber, M.L., Franklin, N., Gordon, R.: Diagnostic error in internal medicine. Arch. Intern. Med. **165**, 1493–1499 (2005)
9. Winters, B., Custer, J., Galvagno, S.M., et al.: Diagnostic errors in the intensive care unit: a systematic review of autopsy studies. BMJ Qual. Saf. **21**, 894–902 (2012)
10. Lee, C.S., Nagy, P.G., Weaver, S.J., et al.: Cognitive and system factors contributing to diagnostic errors in radiology. AJR Am. J. Roentgenol. **201**, 611–617 (2013)
11. Neill, D.B.: Using artificial intelligence to improve hospital inpatient care. IEEE Intell. Syst. **28**, 92–95 (2013)
12. Administration UFaD.: Guidance for Industry: Electronic Source Data in Clinical Investigations (2013)
13. Gillies, R.J., Kinahan, P.E., Hricak, H.: Radiomics: images are more than pictures, they are data. Radiology **278**, 563–577 (2016)
14. Li, C.Y., Liang, G.Y., Yao, W.Z., et al.: Integrated analysis of long noncoding RNA competing interactions reveals the potential role in progression of human gastric cancer. Int. J. Oncol. **48**, 1965–1976 (2016)
15. Shin, H., Kim, K.H., Song, C., et al.: Electrodiagnosis support system for localizing neural injury in an upper limb. J. Am. Med. Inf. Assoc. **17**, 345–347 (2010)
16. Karakülah, G., Dicle, O., Koşaner, O., et al.: Computer based extraction of phenoptypic features of human congenital anomalies from the digital literature with natural language processing techniques. Stud. Health Technol. Inf. **205**, 570–574 (2014)
17. Darcy, A.M., Louie, A.K., Roberts, L.W.: Machine learning and the profession of medicine. JAMA **315**, 551–552 (2016)
18. Murff, H.J., FitzHenry, F., Matheny, M.E., et al.: Automated identification of postoperative complications within an electronic medical record using natural language processing. JAMA **306**, 848–855 (2011)
19. Somashekhar, S.P., Kumarc, R., Rauthan, A., et al.: Abstract S6-07: double blinded validation study to assess performance of IBM artificial intelligence platform, Watson for oncology in comparison with Manipal multidisciplinary tumour board? First study of 638 breast cancer cases. Cancer Res. **77**(4 Suppl.), S06–S07 (2017)
20. Esteva, A., Kuprel, B., Novoa, R.A., et al.: Dermatologist-level classification of skin cancer with deep neural networks. Nature **542**, 115–118 (2017)
21. Bouton, C.E., Shaikhouni, A., Annetta, N.V., et al.: Restoring cortical control of functional movement in a human with quadriplegia. Nature **533**, 247–250 (2016)
22. Farina, D., Vujaklija, I., Sartori, M., et al.: Man/machine interface based on the discharge timings of spinal motor neurons after targeted muscle reinnervation. Nat. Biomed. Eng. **1**, Article no. 0025 (2017)

23. Marr, B.: First FDA approval for clinical cloud-based deep learning in healthcare (2017)
24. Long, E., Lin, H., Liu, Z., et al.: An artificial intelligence platform for the multihospital collaborative management of congenital cataracts (2017)
25. Gulshan, V., Peng, L., Coram, M., et al.: Development and validation of a deep learning algorithm for detection of diabetic retinopathy in retinal fundus photographs. JAMA **316**, 2402–2410 (2016)
26. James, G., Witten, D., Hastie, T., et al.: An Introduction to Statistical Learning with Applications in R, 1st edn. Springer (2013)
27. Goodfellow, I., Bengio, Y., Courville, A.: Deep Learning, 1st edn. The MIT Press (2016)
28. Kantor, P.: Foundations of Statistical Natural Language Processing, pp. 91–92. MIT Press (1999)
29. Bishop, C.M. (ed.): Pattern recognition and machine learning. Inf. Sci. Stat. (2007)
30. Orrù, G., Pettersson-Yeo, W., Marquand, A.F., et al.: Using support vector machine to identify imaging biomarkers of neurological and psychiatric disease: a critical review. Neurosci. Biobehav. Rev. **36**, 1140–1152 (2012)
31. Sweilam, N.H., Tharwat, A.A., Abdel Moniem, N.K., Moniem, N.K.A.: Support vector machine for diagnosis cancer disease: a comparative study. Egypt. Inf. J. **11**, 81–92 (2010)
32. Khedher, L., Ramrez, J., Girriz, J.M., et al.: Early diagnosis of Alzheimer's disease based on partial least squares, principal component analysis and support vector machine using segmented MRI images. Neurocomputing **151**, 139–150 (2015)
33. Mirtskhulava, L., Wong, J., Al-Majeed, S., Pearce, G., et al.: Artificial neural network model in stroke diagnosis. In: 17th UKSim-AMSS International Conference on Modelling and Simulation (UKSim). IEEE (2015)
34. Khan, J., Wei, J.S., Ringnér, M., et al.: Classification and diagnostic prediction of cancers using gene expression profiling and artificial neural networks. Nat. Med. **7**, 673–679 (2001)
35. Dheeba, J., Albert, Singh N., Tamil Selvi, S.: Computer-aided detection of breast cancer on mammograms: a swarm intelligence optimized wavelet neural network approach. J. Biomed. Inf. **49**, 45–52 (2014)
36. Hirschauer, T.J., Adeli, H., Buford, J.A.: Computer-aided diagnosis of Parkinson's disease using enhanced probabilistic neural network. J. Med. Syst. **39**, 179 (2015)
37. Ravi, D., Wong, C., Deligianni, F., et al.: Deep learning for health informatics. IEEE J. Biomed. Health Inf. **21**, 04–21 (2017)
38. Lecun, Y., Bottou, L., Bengio, Y., et al.: Gradient-based learning applied to document recognition. Proc. IEEE Inst. Electr. Electron. Eng. **86**, 2278–2324 (1998)
39. Research BA. Caffe.: http://caffe.berkeleyvision.org/ (2017)
40. Seide, F., Agarwal, A. (eds.): CNTK: Microsoft's open-source DeepLearning toolkit. In: ACM SIGKDD International Conference on Knowledge Discovery and Data Mining (2016)
41. Abadi, M., Agarwal, A., Barham, P., et al.: TensorFlow: Large-Scale Machine Learning on Heterogeneous Distributed Systems (2016)
42. Afzal, N., Sohn, S., Abram, S., et al.: Mining peripheral arterial disease cases from narrative clinical notes using natural language processing. J. Vasc. Surg. **65**, 1753–1761 (2017)
43. Fiszman, M., Chapman, W.W., Aronsky, D., et al.: Automatic detection of acute bacterial pneumonia from chest X-ray reports. J. Am. Med. Inf. Assoc. **7**, 593–604 (2000)
44. Miller, T.P., Li, Y., Getz, K.D., et al.: Using electronic medical record data to report laboratory adverse events. Br. J. Haematol. **177**, 283–286 (2017)
45. Castro, V.M., Dligach, D., Finan, S., et al.: Large-scale identification of patients with cerebral aneurysms using natural language processing. Neurology **88**, 164–168 (2017)
46. Saenger, A.K., Christenson, R.H.: Stroke biomarkers: progress and challenges for diagnosis, prognosis, differentiation, and treatment. Clin. Chem. **56**, 21–33 (2010)
47. Heeley, E., Anderson, C.S., Huang, Y., et al.: Role of health insurance in averting economic hardship in families after acute stroke in China. Stroke **40**, 2149–2156 (2009)
48. Villar, J.R., González, S., Sedano, J., et al.: Improving human activity recognition and its application in early stroke diagnosis. Int. J. Neural Syst. **25**, 1450036–1450055 (2015)

49. Mannini, A., Trojaniello, D., Cereatti, A., et al.: A machine learning framework for Gait classification using inertial sensors: application to Elderly, Post-Stroke and Huntington's disease patients. Sensors **16**, 134–148 (2016)
50. Rehme, A.K., Volz, L.J., Feis, D.L., et al.: Identifying neuroimaging markers of motor disability in acute stroke by machine learning techniques. Cereb. Cortex **25**, 3046–3056 (2015)
51. Griffis, J.C., Allendorfer, J.B., Szaflarski, J.P.: Voxel-based Gaussian Naïve Bayes classification of ischemic stroke lesions in individual T1 weighted MRI scans. J. Neurosci. Methods **257**, 97–108 (2016)
52. Kamnitsas, K., Ledig, C., Newcombe, V.F., et al.: Efficient multi-scale 3D CNN with fully connected CRF for accurate brain lesion segmentation. Med. Image Anal. **36**, 61–78 (2017)
53. Rondina, J.M., Filippone, M., Girolami, M., et al.: Decoding post-stroke motor function from structural brain imaging. Neuroimage Clin. **12**, 372–380 (2016)
54. Thornhill, R.E., Lum, C., Jaberi, A., et al.: Can shape analysis differentiate free-floating internal carotid artery Thrombus from atherosclerotic plaque in patients evaluated with CTA? for stroke or transient ischemic attack? Acad. Radiol. **21**, 345–354 (2014)
55. Bentley, P., Ganesalingam, J., Carlton Jones, A.L., et al.: Prediction of stroke thrombolysis outcome using CT brain machine learning. Neuroimage Clin. **4**, 635–640 (2014)
56. Love, A., Arnold, C.W., El-Saden, S., et al.: Unifying acute stroke treatment guidelines for a Bayesian belief network. Stud. Health Technol. Inf. **192**, 1012 (2013)
57. Ye, H., Shen, H., Dong, Y., et al.: Using Evidence-Based Medicine Through Advanced Data Analytics to Work Toward a National Standard for Hospital-Based Acute Ischemic Stroke treatment. Mainland China (2017)
58. Zhang, Q., Xie, Y., Ye, P., et al.: Acute ischaemic stroke prediction from physiological time series patterns. Australas. Med. J. **6**, 280–286 (2013)
59. Asadi, H., Dowling, R., Yan, B., et al.: Machine learning for outcome prediction of acute ischemic stroke post intra-arterial therapy. PLoS One **9**(1), 1–11, e88225 (2014)
60. Asadi, H., Kok, H.K., Looby, S., et al.: Outcomes and complications after endovascular treatment of Brain Arteriovenous Malformations: a prognostication attempt using artificial intelligence. World Neurosurg. **96**, 562–569 (2016)
61. Birkner, M.D., Kalantri, S., Solao, V., et al.: Creating diagnostic scores using data-adaptive regression: an application to prediction of 30 day mortality among stroke victims in a rural hospital in India. Ther. Clin. Risk Manag. **3**, 475–484 (2007)
62. Ho, K.C., Speier, W., El-Saden, S., et al.: Predicting discharge mortality after acute ischemic stroke using balanced data. In: AMIA Annual Symposium Proceedings 2014, pp. 1787–796 (2014)
63. Chen, Y., Dhar, R., Heitsch, L., et al.: Automated quantification of cerebral edema following hemispheric infarction: application of a machine learning algorithm to evaluate CSF shifts on serial head CTs. Neuroimage Clin. **12**, 673–680 (2016)
64. Siegel, J.S., Ramsey, L.E., Snyder, A.Z., et al.: Disruptions of network connectivity predict impairment in multiple behavioral domains after stroke. Proc. Natl. Acad. Sci. U.S.A. **113**, E4367–E4376 (2016)
65. Hope, T.M., Seghier, M.L., Leff, A.P., et al.: Predicting outcome and recovery after stroke with lesions extracted from MRI images. Neuroimage Clin. **2**, 424–433 (2013)
66. Lohr, S.: IBM is Counting on Its Bet on Watson, and Paying Big Money for It (2016)
67. Otake, T.: IBM Big Data Used for Rapid Diagnosis of Rare Leukemia Case in Japan (2016)
68. Graham, J.: Artificial Intelligence, Machine Learning, and the FDA (2016)
69. Kayyali, B., Knott, D., Kuiken, S.V.: The big-data revolution in US health care: accelerating value and innovation (2013)

Stress and Depression in the Korean College Students: Mediated Effects of Self-differentiation and Self-efficacy

Weon-Hee Moon and Jeong-Yeon Kim

Abstract *Background/Objectives*: This study aims to identify the relationship between stress, depression, and self-differentiation among Korean college students and to verify the effect of self-differentiation in the relationship between stress and depression. *Methods/Statistical analysis*: The data were collected from 1,192 college students attending college and analyzed using SPSS 23. *Findings*: Self-differentiation and self-efficacy showed negative correlations with stress and depression. Explanatory power of self-efficacy with stress was 22.9% on depression, demonstrating partial mediated effect of self-efficacy in the relationship between stress and depression. Explanatory power of self-differentiation with stress was 23.9% on depression, demonstrating partial mediated effect of self-differentiation in the relationship between stress and depression. *Improvements/Applications*: Based on the study results, it is necessary to develop and operate the educational programs on self-efficacy and self-differentiation to lower the depression under the stress condition by proper coping with the individual situations.

Keywords Korean college students · Stress · Depression · Self-efficacy · Self-differentiation

1 Introduction

The Korean college students are considered as adults to be expected to solve the problems independently as matured human beings without limitation of alcohols, smoking, and so on. However, in a survey on the college adaptation and psychological states among 2600 students, it was found that they suffered from the difficulties in

W.-H. Moon
Department of Nursing Science, Pai Chai University, 155-40 Baejae-ro, Seo-gu, Daejeon 35345, South Korea
e-mail: whmoon@pcu.ac.kr

J.-Y. Kim (✉)
Department of Medical and Information, Health Institute of Technology, 21 Chungjeong St., Dong-gu, Daejeon 34504, South Korea
e-mail: jykim@hit.ac.kr

A. K. Bhoi et al. (eds.), *Bio-inspired Neurocomputing*, Studies in Computational Intelligence 903, https://doi.org/10.1007/978-981-15-5495-7_8

adapting the college life (24.7% on study adaptation and 73.3% on relationship adaptation) and 74.5% of the total respondents were confirmed to be potential risk group on the anxiety. They showed to be suffered from psychological depression and greater sense of deprivation due to severe stress on the study, and pressure and anxiety on the uncertain future, complaining the difficulty due to depression in 43.2% [1], which requires the supports urgently for psychological stabilization and mental health of the college students.

Although depression level may result in the drastic options that could affect suicidal thinking and trial significantly in the adolescents [2, 3], proper stress managements can prevent the depression level of the college students [4], with the evidences that direct effects of stress were significant on depression [5]; living stress of the college students showed the high explanatory power on suicidal thinking [6]; and so on. It is almost impossible to avoid stress during the daily life. Stress is the individual recognition and emotional response on the stressors which are causes of stress [7]. With the same stress, the level of individual depression varies by the ways that the individual assesses and responds under the stress condition rather than the stress itself [8, 9].

In this perspective, multiple studies have been conducted on the mediated effects of the stress on the depression, social supports [10, 11], self-esteem [10, 12], and so on were reported as the mediated variables.

Self-efficacy was reported that it was a protective element to control the stress not to cause the severe mental health problems [13]. As self-efficacy was higher, they perceived the stress less [14]; showed higher satisfaction in their lives and stress coping capability to lower the frequency of stress symptoms and deviant behaviors [15]; and lower level of mal-adjustment [16] and depression [17].

Self-differentiation is the process to individualize himself or herself from undifferentiated family ego mass so as to establish the identity and acquire self-impulsive thinking and freedom in the behaviors [18].

Self-differentiated person can think independently free from emotional fusion so as to respond the stress efficiently, hence, he or she is considered to live goal-oriented life [19]. Upon reviewing the previous studies in the subjects with Korean adolescents, they reported to adapt the college life more as their level of self-differentiation was higher. The level of self-differentiation was a critical factor to affect depression/anxiety and level of interpersonal relationship problem [20]. Self-differentiation mediated parents' psychological control on depression [21], and self-esteem was confirmed to be a mediated variable between self-differentiation and mental health in the college students [22]. However, few studies had been conducted on the mediated effects of self-differentiation in the relationship between stress and depression. In this regard, this study was aimed to prepare the effective programs to respond the students' depression and to provide the fundamental data for systemic counseling supports upon revealing the effects of stress, self-efficacy, and self-differentiation on depression, and mediated elements of self-differentiation in the relationship between stress and depression.

2 Materials and Methods

2.1 *Study Subjects and Data Collection Methods*

The subjects of this study are Korean college students who understand the purpose of the study and agree to participate in it. The survey conducted in the form of self-indulgence and 92.6% of the first questionnaire distributed utilized in the final analysis, except for the questionnaires with insufficient answers and 88 questionnaires showing an ideal value of ±3 or more. Statistical analyses were performed using SPSS23 Program including frequency analysis with means and standard deviations, t-test, ANOVA, Pearson correlation analysis, regression analysis, and so on for the collected data. For verifying how self-efficacy and self-differentiation affect between stress and depression in Korean college students, three stages of regression performed in accordance with the analysis procedure proposed by Baron and Kennedy [23].

2.2 *Variables and Measurement Methods*

2.2.1 Stress

For the stress measurement, Chon's [24] college student life stress scale was used, which consists of eight sub-factors: Relations with the opposite sex (six questions), relations with professors (six questions), relationships with family (seven questions), relationships with friends (five questions), future concerns (seven questions), life values (five questions), and academic issues (seven questions). The higher the score, and used a total of 50 questions about the three-point scale means stress experienced college life. Cronbach α on sub-variables of stress was 0.82 ~ 0.89 in this study.

2.2.2 Self-differentiation

For measuring self-differentiation, the measure developed by Je [25] was used based on the family system theory of Bowen. The scale consists of five sub-factors: the emotional function of the region (seven questions), self-integration (six questions), family questionnaire (six questions), emotional disconnect (six questions), and family retreat (11 questions). It measured on a four-point scale for a total of 36 questions, and the higher the score, the higher the self-differentiation level. Cronbach α on sub-variables of self-differentiation index was 0.56 ~ 0.82 in this study.

2.2.3 Self-efficacy

General self-efficacy index developed by Kim [26] was used. It consisted of sub-factors including 12 questionnaires on self-efficacy, seven questionnaires on confidence, five questionnaires on preferred level of task difficulty, being a total of twenty-four 7-point scale questionnaires. As the score was higher, self-efficacy meant higher. In this study, Cronbach α of self-efficacy was 0.86.

2.2.4 Depression

To measure the level of depression, ten questionnaires which had higher factor loadings among negative affect self-statement questionnaire (NASSQ) on depression developed by Ronan [27] and translated in Korean by Lee and Kim [28] were used. In this study, Cronbach α of depression was 0.91.

3 Results and Discussion

3.1 Relationship Between Stress, Self-differentiation, and Depression Among College Students

The results of correlative analysis to identify the level of stress, self-efficacy, self-differentiation, and depression in Korean college students and to check the relevance between each variable are shown in Table 1. Depression had a statistically significant positive correlation with stress ($r = 0.437$, $p < 0.001$), there was a statistically significant negative correlation with self-efficacy ($r = -0.334, p < 0.001$), and with self-differentiation ($r = -0.41, p < 0.001$) (Fig. 1).

Upon the correlation analysis results among stress, self-efficacy, self-differentiation, and depression; self-efficacy and self-differentiation showed the negative correlations with stress and depression but the positive correlations between self-efficacy and self-differentiation. Higher self-efficacy and self-differentiation meant less stress and depression. This is consistent with the results of previous studies that college life stress of the students showed negative correlation with self-efficacy and positive correlation with depression [29]; the groups with higher level of self-differentiation experienced less stress [30]; and the level of depression was lower as the level o f self-differentiation was higher [20, 31].

Table 1 Correlations among variables

Variables	Mean	SD	Pearson correlation: $r(p)$			
			Stress	Self-efficacy	Self-differentiation	Depression
Stress	1.947	0.489	1			
Self-efficacy	3.109	0.448	−0.337*	1		
Self-differentiation	3.499	0.456	−0.489*	0.374*	1	
Depression	1.860	0.626	0.437*	−0.334*	−0.408*	1

*: $p < .001$

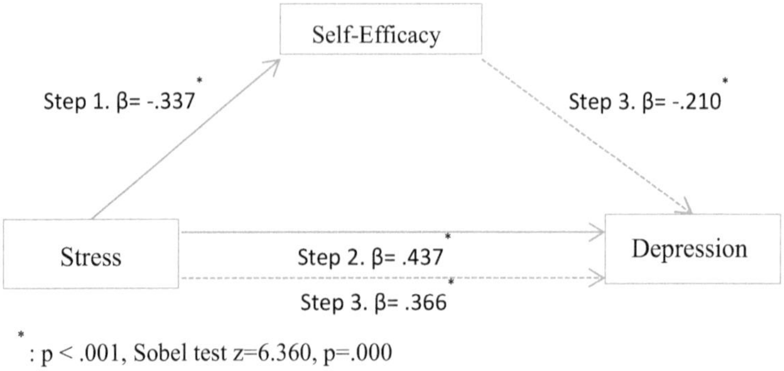

Fig. 1 Mediated effect of self-efficacy in the relationship between stress and depression

3.2 Mediated Effects of Self-efficacy in the Relationship Between Stress and Depression

Table 2 shows the verification results of mediated effect of self-efficacy in the relationship between stress and depression.

Self-efficacy has set as a parameter. For stress, the regression coefficient was −0.34 in Step 1, and the negative effects could be seen. The regression coefficient in Step 2 was 0.44, and in Step 3, the independent variable was 0.37, and the parameter was −0.21, indicating a significant statistic. Also, the effects of independent variables in the second phase were significantly higher than the effects of independent variables in the third phase. In order to verify the significance of the parametric effect, a Sobel test conducted, and the result value of 6.36 was shown to confirm that the intermediate effect was significant.

In conclusion, self-efficacy in the relationship between stress and depression has a partial mechanism effect.

Upon the results of mediated effect of self-efficacy in the relationship between stress and depression, stress and self-efficacy of the college students explained 23% of depression, demonstrating partial mediated effect of self-efficacy in the relationship

Table 2 Mediated effect of self-efficacy

Step	Independent variable	Dependent variable	*B*	SE	β	*t*	Adj. R^2	*F*(*p*)
1	Stress	Self-efficacy	−0.308	0.026	−0.337	−11.876	0.113	141.046*
2	Stress	Depression	0.559	0.035	0.437	16.141	0.190	260.521*
3	Stress	Depression	0.469	0.036	0.366	13.050	0.229	164.813*
	Self-efficacy		−0.294	0.039	−0.210	−7.489		

*: $p < .001$

between stress and depression. This is consistent with the previous study results that coping capability on the stress situation was higher as self-efficacy was higher [32]; depression symptoms were low upon well management of stress [33]; and those with low level of self-efficacy perceived the problems or situations worse due to feeling of helplessness that he or she did not respond properly to the situations, causing severe depression [34]. Since the college period is the time that the students should achieve multiple tasks such as trials of psychological and economic independency from parents, establishment of social network, preparation of jobs, and so on, they face lots of stress, which requires the programs and supporting system to prevent from depression by overcoming stress upon enhancement of self-efficacy and multiple trainings to relieve stress.

3.3 Mediated Effects of Self-differentiation in the Relationship Between Stress and Depression

Table 3 shows the verification results of mediated effect of self-differentiation in the relationship between stress and depression.

Self-differentiation has set as a parameter. For stress, the regression coefficient was −0.490 in Step 1, and the negative effects could be seen. The regression coefficient in Step 2 was −0.437, and in Step 3, the independent variable was 0.312, and the parameter was −0.255, indicating a significant statistic. Also, the effects of independent variables in the second phase were significantly higher than the effects of independent variables in the third phase. In order to verify the significance of the parametric effect, a Sobel test conducted, and the result value of 7.805 was shown to confirm that the intermediate effect was significant.

In conclusion, self-differentiation in the relationship between stress and adaptation has a partial mechanism effect (Fig. 2).

Self-differentiation, along with stress, explains depression 23.9%, and has been confirmed to be related to the effects of stress on depression. This suggests the need to pay attention to self-differentiation as an important factor in stress and depression. People with low self-differentiation cannot deal with severe stress objectively and consistently, and treat it with subjective and instinctive judgment, thus, revealing many problems in interpersonal relationships [35] and choosing degenerative methods such as violence, neglect of responsibility, assertions of unilateral rights, homogeneity among family members, and reverse-functional communication types [36], so we need to develop a program that can promote self-differentiation.

Table 3 Mediated effect of self-differentiation

Step	Independent variable	Dependent variable	B	SE	β	t	Adj. R^2	$F(p)$
1	Stress	Self-differentiation	−0.456	0.024	−0.490	−18.642*	0.239	347.507*
2	Stress	Depression	0.559	0.035	0.437	16.130*	0.190	260.179*
3	Stress	Depression	0.399	0.039	0.312	10.360*	0.239	174.362*
	Self-differentiation		−0.351	0.041	−0.255	−8.475*		

*: $p < 0.001$

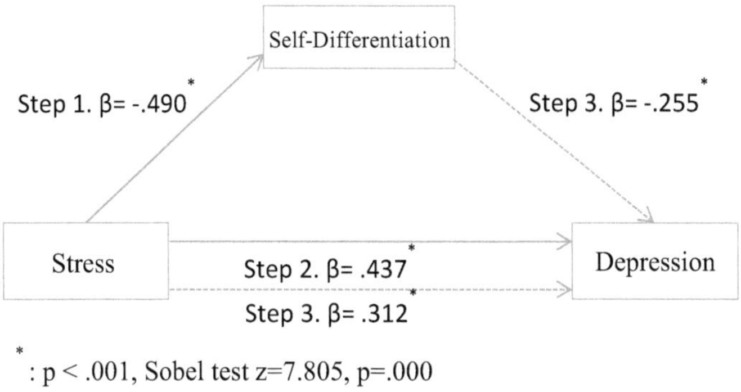

Fig. 2 Mediated effect of self-differentiation in the relationship between stress and depression

4 Conclusion

This study was conducted to identify the mediated effects of self-differentiation and self-efficacy in the relationship between stress and depression among the college students. Upon the analysis results, first, self-differentiation and self-efficacy showed the negative correlations with stress and depression. Second, explanatory power of self-efficacy with stress was 22.9% on depression, demonstrating partial mediated effect of self-efficacy in the relationship between stress and depression. Third, explanatory power of self-differentiation with stress was 23.9% on depression, demonstrating partial mediated effect of self-differentiation in the relationship between stress and depression.

According to the study, depression in Korean college students has a statistically significant positive correlation with stress and a statistically significant negative correlation with self-efficacy and self-differentiation. Moreover, in the relationship between stress and depression of the Korean college students, each of self-efficacy and self-differentiation has been confirmed to have a partial mechanism effect. It also confirmed that groups with high levels of self-differentiation were less stressed, and low levels of depression, which directly affected college students' depression. Therefore, it will be necessary to develop and operate self-differentiation promotion programs for Korean college students for healthy college life.

Since self-efficacy and self-differentiation showed to control depression by partial mediated effect on stress, it seems to be necessary to provide with the educational programs to enhance self-efficacy and self-differentiation so as to lower the level of depression upon responding individual stress situations properly. This study has the limitation to generalize the results due to random sampling; however, it has the meaning to find out the mediated effects of self-efficacy and self-differentiation in the relationship between stress and depression. Further studies are required to verify causal relationship among stress, self-efficacy, self-differentiation, and depression and to investigate whether the depression symptoms will be lowered after the interventional programs applied with enhanced self-efficacy and self-differentiation.

Acknowledgements This work was supported in part by the research grant of Pai Chai University in 2020.

References

1. Korean Council for University Education. (2018, April). College students is in crisis they've gotten warning signals with mental health. Seoul: Author. Available from http://www.kcue.or.kr/bbs/view.php?gb=news&page=6&idx=615&kind=&culm=bbs_title&word=(website)
2. Hong, Y.S., Jeon, S.Y.: The effects of life stress and depression for adolescent suicidal ideation. Ment. Health Soc. Work (19), 125–149 (2005)
3. Park, E.O.: The influencing factors on suicide attempt among adolescents in South Korea. J. Korean Acad. Nursing **38**(3), 465–473 (2008). Available from http://www.ndsl.kr/ndsl/commons/util/ndslOriginalView.do?dbt=JAKO&cn=JAKO200828056666474&oCn=JAKO200828056666474&pageCode=PG11&journal=NJOU00290378
4. Baumeister, R.F.: Suicide as escape from self. Psychol. Rev. **97**, 90–113 (1990)
5. Park, K.H.: Development of a structural equation model to estimate university student's depression. J. Korean Acad. Nursing **38**(6), 779–788 (2008). https://doi.org/10.4040/jkan.2008.38.6.779
6. Kang, S.H.: A study on the factors influencing the suicide ideation of university students. Doctor's thesis, University of Cheongju, Chungbuk (2008)
7. Park, S.C., Baek, K.I.: Effects of self-differentiation and self-efficacy on interpersonal relationship and stress coping strategy. J. Buddhist Couns. **3**(1), 79–96 (2011)
8. Cho, S.K., Choi, Y.S.: Effects of stress and coping style on depression and anxiety of male and female university students. Korean J. Fam. Therapy **22**(1), 53–74 (2014). https://doi.org/10.21479/kaft.2014.22.1.53
9. Lazarus, R.S., Folkman, S.: Stress Appraisal and Coping. Springer Publication Co, New York (1984)
10. Kim, B.G.: The effect of stress on depression of the elderly: focusing on the moderating effect of social support and mediated effect of sense of self-respect. J. Korea Contests Assoc. **19**(6), 281–291 (2019). https://doi.org/10.5392/JKCA.2019.19.06.281
11. Kang, M.G., Lee, S.I., Lee, H.S.: The social support of teachers as a moderator of the relationship between depression and economic stress of specialized vocational high school students. Korean J. Technol. Educ. **18**(2), 120–147 (2018). Available from http://www.dbpia.co.kr/pdf/pdfView.do?nodeld=NODE07524001
12. Oh, M.S., Shim, W.C.: The effects of academic stress on children's depression: focusing on moderating effects of self-esteem. J. Sch. Soc. Work **29**(29), 237–261 (2014)
13. Eom, T.W.: The effects of self-efficacy and social support in the relationship between economic stress and depression of the indigent population. Ment. Health Soc. Work. **28**, 36–66 (2008). Available from http://www.dbpia.co.kr/journal/articleDetail?nodeId=NODE06082092
14. Jun, Y.J.: A study on the relationships between self-efficacy, stress and stress coping behavior. Master's dissertation, University of Chungnam National University, Daejeon (2004)
15. Lee, K.S.: A study on the stress between junior high school students and incarcerated adolescent: focused on the relationship among social support, self-efficacy, life-satisfaction, delinquent-behavior. Master's dissertation, In ha University, Incheon (2000)
16. Kim, M.J.: The relationship between junior high students' maladjustment and life stress and coping. Master's dissertation, Chungnam National University, Daejeon (1994)
17. Min, K.S.: The relationships of academic achievement pressure and academic self-efficacy on depression among foreign language high school students. Master's dissertation, Kookmin University, Seoul (2012)

18. Vanderkooi, I.K., Handelsman, N.M.: Toward an integration of Jungian and Bowen perspectives on psychotherapy. Fam. Therapy **6**, 615–623 (1984)
19. Bowen, M.: Theory in the practice of psychotherapy. In: Guerin, P.J. (ed.) Family Therapy: Theory and Practice. Gardner Press, New York (1976)
20. Bae, M.Y., Lee, E.H.: The relationships between self-differentiation, interpersonal problems and depression/anxiety levels in adolescence: the mediated effects on self-control ability. Stud. Korean Youth **20**(1), 85–112 (2009). Available from http://www.dbpia.co.kr/pdf/pdfView.do?nodeId=NODE01152394
21. Shim, S.G., Hong, J.S.: The mediated effect of self-differentiation between parental psychological control and depression of college students. J. Parent Educ. **10**(1), 163–182 (2018)
22. Kim, S.O., Jeon, Y.J.: A study on relationships among university students' self-differentiation, self-esteem and mental health: focused on depression and anxiety. Korean J. Hum. Ecol. **22**(4), 539–558 (2013)
23. Baron, R.M., Kenny, D.A.: The moderator-mediator variable distinction in social psychological research: conceptual, strategic and statistical considerations. J. Pers. Soc. Psychol. **51**, 1173–1182 (1986)
24. Chon, K.K., Kim, K.H.: Development of the life stress scale for college students: a control theory approach. Korean J. Clin. Psychol. **10**(1), 137–158 (1991)
25. Je, S.B.: The relationship between differentiation of self and dysfunctional behavior. Doctor's dissertation, Pusan National University, Pusan (1989)
26. Kim, A.Y.: A study on the academic failure-tolerance and its correlates. Korean Soc. Educ. Psychol. **11**(2), 1–19 (1997)
27. Ronan, K.R., Kendall, P.C., Rowe, M.: Negative affectivity in children: development and validation of a self-statement questionnaire. Cogn. Therapy Res. **18**, 509–528 (1994)
28. Lee, J.Y., Kim, J.H.: Negative self-statements of depression and anxiety in youth: a validation study of Korean negative affect self-statement Questionnaire (NASSQ). Korean J. Clin. Psychol. **21**(4), 871–89 (2002). Available from http://www.dbpia.co.kr/pdf/pdfView.do?nodeId=NODE06371020
29. Yoo, E.J., Kim, S.K., Juong, S.K.: A study on stress of college life and self-efficacy on depression among students from beauty-related colleges. J. Digit. Policy Manag. **11**(9), 279–287 (2013). https://doi.org/10.14400/JDPM.2013.11.9.279
30. Bae, O.H., Hong, S.O.: The study of the effect of self-differentiation degree on stress level and stress coping strategies in college students. Korean J. Hum. Ecol. **17**(1), 27–34 (2008). https://doi.org/10.5934/KJHE.2008.17.1.027
31. Lee, W.O.: The relationship between family system and single adults' self-differentiation and depression. Master's dissertation, Seoul Women's University, Seoul (2003)
32. Lee, S.Y.: A study on the stress and the psychological traits between high school students and college students: focused on the relationship among social support, self-efficacy, life-satisfaction and delinquent-behavior. Master's dissertation, In ha University, Incheon (2000)
33. Bae, Y.S.: The influence of the level of self-efficacy on stress and depression among male high school students. Master's dissertation, In ha University, Incheon (2014)
34. Hong, H.Y.: The relationship of perfectionism, self-efficacy and depression. Master's dissertation, Ewha Women's University, Seoul (1995)
35. Bowen, M.: Family Therapy in Clinical Practice. Jason Aronson, New York (1982)
36. Yun, S.M.: The effect of the self-differentiation on interpersonal behavior and dysfunctional career thoughts. Master's dissertation, Ajou University, Suwon (2010)

An Automated Segmentation of Brain MR Image Through Fuzzy Recurrent Neural Network

Jalluri Gnana SivaSai, P. Naga Srinivasu, Munjila Naga Sindhuri, Kola Rohitha, and Sreesailam Deepika

Abstract The human brain is the major controller of the humanoid system. The abnormal growth and division of cells in the brain lead to a brain tumor, and the further growth of brain tumors leads to brain cancer. In the area of human health, Computer Vision plays a significant role, which reduces the human judgment that gives accurate results. CT scans, X-Ray, and MRI scans are the common imaging methods among magnetic resonance imaging (MRI) that are the most reliable and secure. MRI detects every minute objects. Our paper aims to focus on the use of different techniques for the discovery of brain cancer using brain MRI. In this study, we performed pre-processing using the adaptive bilateral filter (ABF) for removal of the noises that are present in an MR image. This was followed by the binary thresholding and Fuzzy Recurrent Neural Network (FR-Net) segmentation techniques for reliable detection of the tumor region. Training, testing, and validation datasets are used. Based on our machine, we will predict whether the subject has a brain tumor or not. The resultant outcomes will be examined through various performance examined metrics that include accuracy, sensitivity, and specificity. It is desired that the proposed work would exhibit a more exceptional performance over its counterparts.

Keywords Brain tumor · Magnetic resonance imaging · Adaptive bilateral filter · Fuzzy recurrent neural network

---

J. G. SivaSai (✉) · P. N. Srinivasu · M. N. Sindhuri · K. Rohitha · S. Deepika
Department of CSE, Anil Neerukonda Institute of Technologies and Sciences, Visakhapatnam, India
e-mail: siva.jalluri18@gmail.com

P. N. Srinivasu
e-mail: parvathanenins@gmail.com

M. N. Sindhuri
e-mail: sindhu.sg1224@gmail.com

K. Rohitha
e-mail: rohithakola425@gmail.com

S. Deepika
e-mail: deepikasreesailam@gmail.com

A. K. Bhoi et al. (eds.), *Bio-inspired Neurocomputing*, Studies in Computational Intelligence 903, https://doi.org/10.1007/978-981-15-5495-7_9

1 Introduction

The brain tumor is one all the foremost common and, therefore, the deadliest brain diseases that have affected and ruined several lives in the world. Cancer is a disease in the brain in which cancer cells ascends in brain tissues. Conferring to a new study on cancer, more than one lakh people are diagnosed with brain tumors every year around the globe. Regardless of stable efforts to overcome the complications of brain tumors, figures show unpleasing results for tumor patients. To contest this, scholars are working on computer vision for a better understanding of the early stages of tumors and how to overcome using advanced treatment options.

Magnetic resonance (MR) imaging and computed tomography (CT) scans of the brain are the two most general tests to check the existence of a tumor and recognize its position for progressive treatment decisions. These two scans are still used extensively for their handiness, and the capability to yield high-definition images of pathological tissues is more. At present, there are several other conducts offered for tumors, which include surgery, therapies such as radiation therapy, and chemotherapy. The decision for which treatment relies on the many factors such as size, kind, and grade of the tumor present in the MR image. It's conjointly chargeable for whether or not cancer has reached the other portions of the body.

Precise sighting of the kind of brain abnormality is enormously needed for treatment operations with a resolution to diminish diagnostic errors. The precision is often makeshift utilizing computer-aided diagnosis (CAD) systems. The essential plan of computer vision is to produce a reliable output, which is an associate estimation to assist medical doctors in image understanding and to lessen image reading time. These advancements increase the steadiness and correctness of medical diagnosis—however, segmenting an MR image of the tumor and its area itself a very problematic job. The occurrence of tumors in specific positions within the brain image without distinguishing picture intensities is an additional issue that makes a computerized detection of brain tumor and segmentation a problematic job.

The MRI is the most regularly utilized strategy for imaging brain tumors and the identification of its vicinity. The conventional technique for CT and MR image classification and detection of tumor cells remains largely supported for the human reviewing apart from different other methods. MR images are mainly used because there are non-destructive and non-ionizing. MR imaging offers high-definition pictures that are extensively utilized in discovering brain tumors. MRI has diverse schemes such as flair, T1-weighted, T2-weighted images. There are many image processing techniques such as pre-processing, segmentation of images, image improvements, feature extraction, and classifiers.

Sathya et al. [3], provided a different clustering algorithm such as K-means, Improvised K-means, C-means, and improvised C-means algorithms. Their paper presented an experimental analysis for massive datasets consisting of unique photographs. They analyzed the discovered consequences using numerous parametric tests.

Roy e tal. [18] calculated the tumor affected area for proportioned analysis. They confirmed its software with numerous statistics groups with distinctive tumor sizes, intensities, and location. They showed that their algorithm could robotically hit upon and phase the brain tumor from the given photo. Image pre-processing consists of fleeting that pictures to the filtering technique to remove distractors found in given pictures.

Kaur et al. [6] defined a few clustering procedures for the segmentation process and executed an assessment on distinctive styles for those techniques. Kaur represented a scheme to measure selected clustering techniques based on their steadiness in exceptional tenders. They also defined the diverse performance metric tests, such as sensitivity, specificity, and accuracy.

Sivaramakrishnan et al. [1] projected an efficient and innovative discovery of the brain tumor vicinity from an image that turned into finished using the Fuzzy C-approach grouping algorithm and histogram equalization. The disintegration of images is achieved by the usage of principal factor evaluation is done to reduce the extent of the wavelet coefficient. The outcomes of the anticipated FCM clustering algorithm accurately withdrawn tumor area from the MR images.

Sudharani et al. [5] presented a K-nearest neighbour algorithm to the MR images to identify and confine the hysterically full-fledged part within the abnormal tissues. The proposed work is a sluggish methodology but produces exquisite effects. The accuracy relies upon the sample training phase.

Marroquin et al. [10] presented the automated 3D segmentation for brain MRI scans. Using a separate parametric model in preference to a single multiplicative magnificence will lessen the impact on the intensities of a grandeur. Brain atlas is hired to find non-rigid conversion to map the usual brain. This conversion is later cast-off segment the tumor region.

Kwok et al. [7] delivered wavelet-based photograph fusion to easily cognizance at the object with all focal lengths as several vision-related processing tasks can be carried out more effortlessly when wholly substances within the images are bright. In their work Kwok et al. investigated with different datasets, and results show that presented work is extra correct as it does not get suffering from evenness at different activity stages computations.

Mukambika et al. [15] proposed methodology for the subsequent stage's classification of the tumor, whether it is present or not. Their proposed work represents the comparative study of strategies used for tumor identification from MR images, namely the Level set approach and discrete wavelength transforms (DWT) and K-method segmentation algorithms. After that phase, feature extraction is done followed SVM classification.

Astina et al. [11] implemented an operative automatic classification approach for brain image that projected the usage of the AdaBoost gadget mastering algorithm. The proposed system includes three main segments. Pre-processing has eradicated noises in the datasets and converted images into grayscale. Median filtering and thresholding segmentation are implemented in the pre-processed image.

Kumar and Mehta [8] proposed the texture-based technique in this paper. They highlighted the effects of segmentation if the tumor tissue edges aren't shrill. The performance of the proposed technology may get unwilling results due to those edges. The texture evaluation and seeded region approach turned into executed inside the MATLAB environment.

Pereira et al. [17] presented that magnetic resonance prevents physical segmentation time in the medical areas. So, an automatic and reliable segmentation technique for identifying abnormal tissues by using Convolutional Neural Network (CNN) had been proposed in the research work. The massive three-dimensional and underlying roughness amongst brain images makes the process of segmenting the image a severe issue, so a robust methodology such as CNN is used.

Devkota et al. [4] have proposed that a computer-aided detection (CAD) approach is used to spot abnormal tissues via Morphological operations. Amongst all different segmentation approaches existing, the morphological opening and closing operations are preferred since it takes less processing time with the utmost efficiency in withdrawing tumor areas with the least faults.

Vinotha et al. [23] proposed brain tumor detection and the usage of the Histogram Equalization (HE) and the Fuzzy Support Vector Machine (FSVM) classification techniques. The brain MR image is pre-processed with histogram equalization and segmented the apprehensive components from the photo primarily based on the MRF algorithm for segmentation technique. MRF approach expanded the tumor segmentation accuracy through which the overall performance of the proposed approach changed into advanced.

Mahmoud et al. [9] presented a model using Artificial Neural Networks for tumor detection in brain images. They implemented a computerized recognition system for MR imaging the use of Artificial Neural Networks. That was observed that after the Elman community was used during the recognition system, the period time and the accuracy level were high, in comparison with other ANNs systems. This neural community has a sigmoid characteristic which elevated the extent of accuracy of the tumor segmentation.

Sufyan et al. [2] has presented a detection using enhanced edge technique for brain-tumor segmentation that mainly relied on Sobel feature detection. Their presented work associates the binary thresholding operation with the Sobel approach and excavates diverse extents using a secure contour process. After the completion of that process, cancer cells are extracted from the obtained picture using intensity values.

Kumar and Varuna Shree [22] proposed work for the detection tumor region using discrete wavelength transforms (DWT). This work consists of three phases, namely an image enhancement using filtering technique, gray-level co-incidence matrix (GLCM) feature extraction of tumor in addition to DWT based tumor location developing segmentation. It is used to improve overall performance and reduce complexity. The denoised accompanied by the aid of morphological filtering operations which put off the noises that can be even shaped subsequent segmentation technique. The PNN classifier is to use for classifying the abnormality, which is trained by different datasets, and the accuracy is measured within the detection of tumor region of mind MR images.

Paul and Bandyopadhyay [20] has presented the brain segmentation that has automated the use of the Dual Localization technique. In the initial phase, the skull masks are generated for the brain MR images. The tumor areas are improvised using the K-manner procedure. In the final step of their proposed work, they evaluated by its dimensions such as length and breadth.

Vaishali [21] proposed a method that includes step by step procedure starting with image pre-processing followed by extraction of useful objects and finally classification of tumor region. Pre-processing is completed to enhance the image using eliminating the noise via making use of Gaussian filters from the authentic ones. The next step is feature extraction, in which a magnified image is used to extract the feature using a symlet wavelet technique. The very last step is the classification of tumors by the use of a Support vector machine (SVM).

Pan et al. [16] has used brain MRI pix for getting useful statistics for classifying brain tumor. In their proposed method, they used Convolutional Neural Networks (CNN) algorithms for developing a brain tumor detection system. The performance of their CNN report is measured primarily based on sensitivity and specificity parameters, which have stepped forward when in comparison to the Artificial Neural Networks (ANN).

Subashini and Sahoo [12] has suggested a technique for detecting the tumor commencing the brain MR images. They also worked on different techniques, which include pulse-coupled Neural Network and noise removal strategies for reinforcing the mind MRI images and backpropagation network for classifying the brain MRI images from tumor cells. They observed image enhancement and segmentation of the usage of their proposed technique, and the backpropagation network helps in the identification of a tumor in a brain MR image.

Sankari and Vigneshwari [19] has proposed a Convolutional Neural Network (CNN) segmentation, which principally based on the brain tumor classification method. The proposed work used the non-linearity activation feature that's a leaky rectified linear unit (LReLU). They primarily focused on necessary capabilities, which include mean and entropy of the image and analyzed that the CNN algorithm is working higher for representing the complicated and minute capabilities of brain tumor tissues present in the MR Images.

2 Objective of Paper

A tumor is a mass of tissues that grow on body parts out of control of regular forces, which regulates the growth. Tumors can, directly and indirectly, damage all the healthy brain cells. Tumors can be of dissimilar sizes, positions, and locations. They can also overlap their intensities over healthy tissues. Tumors may be primary and may not be primary too. The most common primary tumors related to the brain are Gliomas and Malignant. Identification of brain tumors in MRI scans during earlier stages are essential in medical diagnosis. So, the core objective of our project is to develop an automated segmentation of brain MR image for a tumor detection system that can detect a tumor from magnetic resonance brain images. In our paper, we proposed a tumor detection system that presented good accuracy and less computational time when compared to its contour parts.

The proposed system begins with reading the MRI image from the dataset, and then image pre-processing is done by using filtering techniques such as adaptive bilateral for removal of noise pixels present in the original brain tumor image. After applying filters, the image gets smoothened, and edges cannot be detected. To overcome this, we use an edge detection technique. So, the brain MRI image enhancement is done using the Sobel filter. The brain MRI segmentation is done with the support of Fuzzy Recurrent Neural Networks based on the concept of the feed-forward framework. After the segmentation stage, various features are extracted using shape and texture primarily based on the brain MR image. After feature extraction, brain tumor MRI image classification is done using Convolutional Neural Network, and classification accuracy is measured.

3 Preprocessing Using Adaptive Bilateral Filter and Image Enhancement

The Brain MRI image dataset has been downloaded from the Kaggle. The MRI dataset consists of around 1900 MRI images, including normal, benign, and malignant. These MRI images are taken as input to the primary step. The pre-processing is an essential and initial step in improving the quality of the brain MRI Image. The critical steps in pre-processing are the reduction of impulsive noises and image resizing. In the initial phase, we convert the brain MRI image into its corresponding gray-scale image. The removal of unwanted noise is done using the adaptive bilateral filtering technique to remove the distorted noises that are present in the brain picture. This improves the diagnosis and also increase the classification accuracy rate.

Bilateral filtering smooth images while conserving edges utilizing a nonlinear grouping of neighbouring image pixels. This filtering technique is simple, local, and concise. It syndicates a grey level grounded on their likeness and the symmetrical nearness and chooses near vales to farther values in both range and domain. So, we proposed a new smoothing and sharpening algorithm, Adaptive Bilateral Filter.

MR image segmentation with the aid of preserving the threshold information, which is convenient to identify the broken regions extra precisely. It was a trendy surmise that the objects that are placed in close propinquity might be sharing similar houses and characteristics. Here for noise removal, the intensity is taken into consideration as the top characteristic. So, when the pixel is captivated via the noise, then such pixel will showcase an indigent correlation with the neighboring pixels in such a case that precise pixel is smoothened by using thinking about all eight neighboring pixels.

The inceptive threshold value is evaluated and is designated by a variable Th between two pixels I_m and I_n. The pixel I_m is assigned to the respective segments most effective if the pixel I_m is keeping a darkish insensitive than the evaluated threshold. When the intensity value of I_n is brighter than the pixel is allocated to the adjacent segment, which is calculated by means of the objective function.

The initial threshold value is approximated by considering the intensities P_x and P_y of pixels I_m and I_n by using the following equation.

$$\mathrm{Th} = W \times \sum_{i=0}^{255} \frac{P_x + P_y}{2}$$

where W is the Laplacian second-order differential equation that spread pixel over the segments of the MR image by isotropic derivative. The value of W is derived as

$$W = \frac{\partial^2 w}{\partial \mathrm{I}_m} + \frac{\partial^2 w}{\partial \mathrm{I}_n}$$

After the completion of the pre-processing, the image will be free from the noises, but we still need to enhance the image since the obtained image is smoothened, edges may not be preserved, and the image will be dull. To overcome all these, we use edge detection called Sobel filtering technique. The whole thing is done by calculating the gradient of image intensities at each pixel within the image. It is widely used in image analysis to help locate edges in images. It combines Gaussian smoothing and differentiation. It will also enhance the darker areas of the image, slightly increase contrast and as sharp as possible.

4 Fuzzy Recurrent Neural Network (FR-Net) Based Image Segmentation

In this paper, we have proposed a different and robust automated segmentation mechanism for the segmentation of the brain MR image through Recurrent Neural Network based on the concept of the feed-forward framework that works based on the numerous image pyramids, that works almost like a Level set based deformable Model. In the proposed approach at each phase, the pace of the curves is being assessed with q quite lesser network features that would consume minimal training set when compared to that of the conventional Convolutional neural network-based approach which needs a rigorous training of the data for better optimal results, and the execution time of the traditional approach is more when compared to that of the proposed FR-Net approach.

The essences of the Recurrent Neural Network have further improved through the fuzzifying the resultant outcome that would yield an optimal segmentation of the MR image when compared to that of the traditional MR segmentation approaches. The proposed approach could be better understood by the formulations stated below. Let is assumed that the input image sequences that are denoted by $I_s = \{I_s | t_f = 1, 2, 3 \ldots, \max_t\}$ that are concerning the predetermined time frames t_f and the resultant prognosticated label sequence is represented through the variable $L = \{L_s | t_f = 1, 2, 3 \ldots \max_t\}$. The proposed algorithm would work concerning the Fuzzy Membership evaluation and the RNN through Fully Convolutional Neural networks.

Recurrent Neural Networks is an approach that is being customized to accept the outcomes of the previous inputs and outputs by storing the patterns in the network. The fully convolutional sublayer of the over the input image I and the key features that are assumed as k_f. The U-net structure is being used in the process of identification of the elements for the automated segmentation of the MR image. The pixel-wise feature map f_m at every given point is being assumed through various units, Convolutional Long Short-Term Memory(C-LSTM) that is being capable of analyzing the composed as following

$$x_s = \sigma(i_s * p_{mi} + z_{s-1} * p_{zi} + b_x)$$

$$y_s = \sigma\left(i_s * p_{my} + z_{s-1} * p_{zy} + b_y\right)$$

$$u_s = u_{s-1} \odot y_s + x_s \odot \tanh(i_s * p_{mc} + z_{s-1} * p_{zc} + b_u)$$

$$v_s = \sigma(i_s * p_{mo} + z_{s-1} * p_{z0} + b_0)$$

$$p_s = v_s \odot \tan z(u_s)$$

In the above statements, the symbol * represents convolution and represents the variable wise multiplication, and the symbol represents the sigmoid function, and the variables are the list of input where the variable (i) denotes the input gate, (f) represents the forget gate, (c) signifies memory cell and (o) represents the output gate, and the variable p designates the convolutional kernel and b designates the bias at each gate. And the variable represents the input vector map and designate the output vector space. The output variable concerning the time stamp that demonstrates both the input at the current moment and the output at the previous state denoted by and that are being used for the process of prediction. However, from Fig. 1, we can better understand that the proposed approach would be using both the foreground and background information for the operation of prediction.

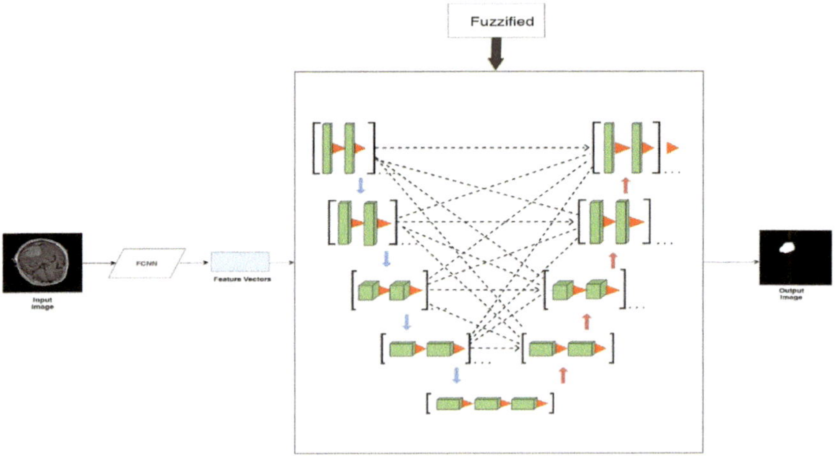

Fig. 1 Represents the architecture diagram of FR-Net

5 Experimental Results and Discussion

The proposed approach has been practically implemented through the real-time dataset that is available in the Brain Web, the images of sizes 256 × 256 and 512 × 512 has been used for practical implementation of the algorithm, And the proposed algorithm has been assessed through various performance evaluation metrics that include True Positive, True Negative the former one that designates how many times does the proposed algorithm is able to correctly recognize the damaged region as damaged region and the later one designates how many times does the proposed algorithm correctly identified non-damaged region as non-damaged region. And the False Positive (FN) and False Negative (FN) the former one designates how many times does the proposed algorithm fails to recognize the damaged region correctly, and the later represents how many times does the proposed algorithm fails to identify the non-tumors region as non-tumors regions. Basing on values of TP, TN, FP, and FN, the values of Sensitivity and Specificity of the proposed algorithm.

On experimentation, it was observed that the proposed methodology seems to be outperformed when compared to that of Convolutional Neural Network (CNN), Heuristic approach for real-time segmentation (HARIS) proposed by Naga Srinivasu et al. [14] and Genetic Algorithm with Teacher Learner Based Optimization (TLBO) as stated by Naga Srinivasu et al. [13]. Among all the aforementioned approaches, the proposed Fuzzy Recurrent Neural Network (FR-Net) based approach seems too much better in terms of quality of the output and optimized computational time when compared to its counterparts presented in Table 3.

It is observed from the above Tables 1 and 2 upon performing the pre-processing for the noise removal using the proposed filtering technique so the quality of image has been improved that would be having a substantial influence on the eminence of the outcome, we have used Adaptive Bilateral filter for restoring the image by

Table 1 Represents the performance evaluation of an adaptive bilateral filter for a 256*256 size image

Noise variance	PSNR	MSE	RMSE	IQI
9	33.3	29.0	5.3	0.562
8	35.1	26.2	5.1	0.599
7	36.4	23.1	4.8	0.659
5	38.9	20.4	4.5	0.701
4	40.6	17.0	4.2	0.742
3	41.9	14.8	3.9	0.785
1	44.2	11.9	3.5	0.863

Table 2 Represents the performance evaluation of an adaptive bilateral filter for 512*512 size image

Noise variance	PSNR	MSE	RMSE	IQI
9	35.4	27.1	5.2	0.411
8	37.7	23.4	4.8	0.472
7	40.1	21.2	4.5	0.501
5	45.7	16.9	4.1	0.572
4	48.2	14.1	3.7	0.606
3	49.3	12.2	3.4	0.676
1	50.6	10.5	3.2	0.704

preserving the edge related information, and the outcome of the proposed approach seems to be more pleasing with better IQI and PSNR value. Moreover, the proposed noise removal technique also displayed minimal values for both MSE and RMSE. It is observed on the practical implementation that the proposed algorithm has exhibited better performance over the smaller image when compared to the smaller size image. Moreover, the computational time is considerably high for larger size MR images (Figs. 2 and 3).

It is observed in the practical implementation of the proposed approach the tumor region is being correctly identified, and the proposed method can recognize the growth of the region of the tumor with almost a negligible difference with the actual ground facts. It has been observed for both the 256 × 256 and 512 × 512 size MR images, and it seems that the proposed approach has outperformed for a smaller size image over the larger size one, In Table 4, the first column represents the actual core that is at the center of the tumor and the whole tumor that also considers the partially

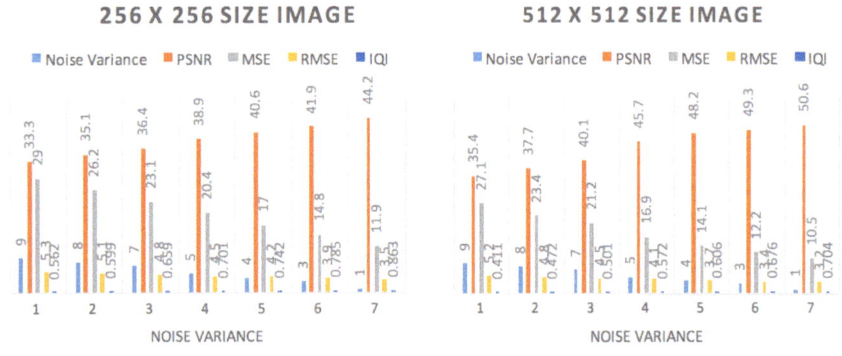

Fig. 2 Represents the performance of Adaptive Bilateral Filter in image de-noising

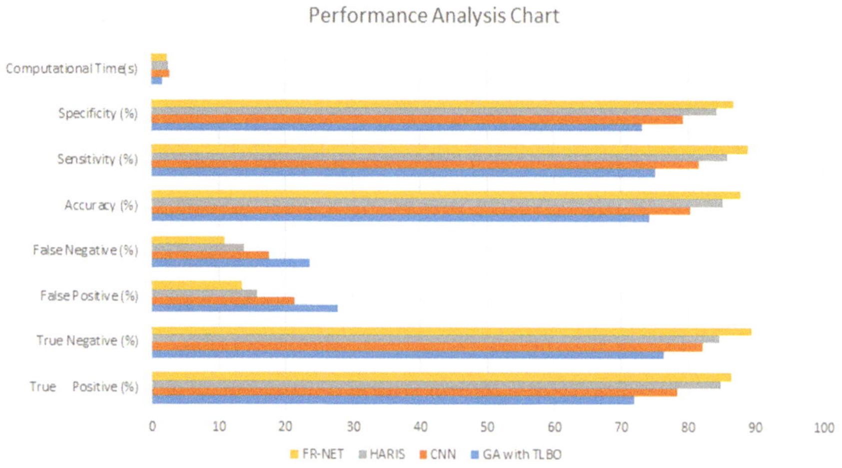

Fig. 3 Represents the performance of proposed Fuzzy RNN

effected tissues based in the texture value and the last column represents the enhances tumor from the previous examination.

The resultant outcomes presented in Figs. 4 and 5 the image from left to right represents in a step by step process of the proposed algorithm. The first image represents the original image, which is read from the dataset. The second image is the denoised image, which is carried out by applying the adaptive bilateral filter. The third image is the enhanced image obtained after binary thresholding. The fourth image is an image that is by means of after the Fuzzy Recurrent Neural Network (FR-Net). The final image represents the tumor location from the given segmented image.

It is observed from Table 3 upon performing various segmentation techniques that have the ability to recognize the isolated region from the MR images that are used to analyze the shape and size of the denoised image. We have used Fuzzy Recurrent Neural Networks for segmentation, and the output of our proposed work is pleased with better accuracy, sensitivity, and computational time.

6 Conclusion

We proposed a computerized method for the segmentation and identification of a brain tumor using the Fuzzy Recurrent Neural Network. The input MR images are read from the local device using the file path and converted into grayscale

Table 3 Represents the accuracy, sensitivity, and specificity of the proposed approach in comparison with its counterparts

Segmentation techniques	True positive (%)	True negative (%)	False positive (%)	False negative (%)	Accuracy (%)	Sensitivity (%)	Specificity (%)	Computational time(s)
GA with TLBO	71.9	76.4	28.1	23.6	74.2	75.2	73.1	1.7
CNN	78.3	82.1	21.7	17.9	80.2	81.4	79.3	2.7
HARIS	83.7	84.5	16.3	15.5	84.1	84.4	83.8	2.6
FR-NET	86.4	89.2	13.6	10.8	87.8	88.8	86.7	2.3

Table 4 Represents the precise location and the approximated evaluation of tumor

	Tumor core (TC)	Whole tumor (WT)	Enhanced tumor (ET)
T1-Weighted MR Image of size 256 × 256 (Without Adaptive Bilateral Filter)	9.236 ± 0.981	3.468 ± 0.561	8.312 ± 0.769
T1-Weighted MR Image of size 256 × 256 (With Adaptive Bilateral Filter)	8.436 ± 0.689	4.642 ± 0.881	7.924 ± 0.489
T1-Weighted MR Image of size 512 × 512 (Without Adaptive Bilateral Filter)	9.102 ± 0.819	4.432 ± 0.598	7.282 ± 0.533
T1-Weighted MR Image of size 512 × 512 (With Adaptive Bilateral Filter)	8.272 ± 0.622	5.121 ± 0.932	6.963 ± 0.416

images. These images are pre-processed using an adaptive bilateral filtering technique for the elimination of noises that are present inside the original image. The binary thresholding is applied to the denoised image, and Fuzzy Recurrent Neural Network segmentation is applied, which helps in figuring out the tumor region in the MR images. The proposed model had obtained an accuracy of 87.8% and yields promising results without any errors and much less computational time.

7 Future Scope

It is observed on extermination that the proposed approach needs a vast training set for better accurate results; in the field of medical image processing, the gathering of medical data is a tedious job, and, in few cases, the datasets might not be available. In all such cases, the proposed algorithm must be robust enough for accurate recognition of tumor regions from MR Images. The proposed approach can be further improvised through in cooperating weakly trained algorithms that can identify the abnormalities with a minimum training data and also self-learning algorithms would aid in enhancing the accuracy of the algorithm and reduce the computational time.

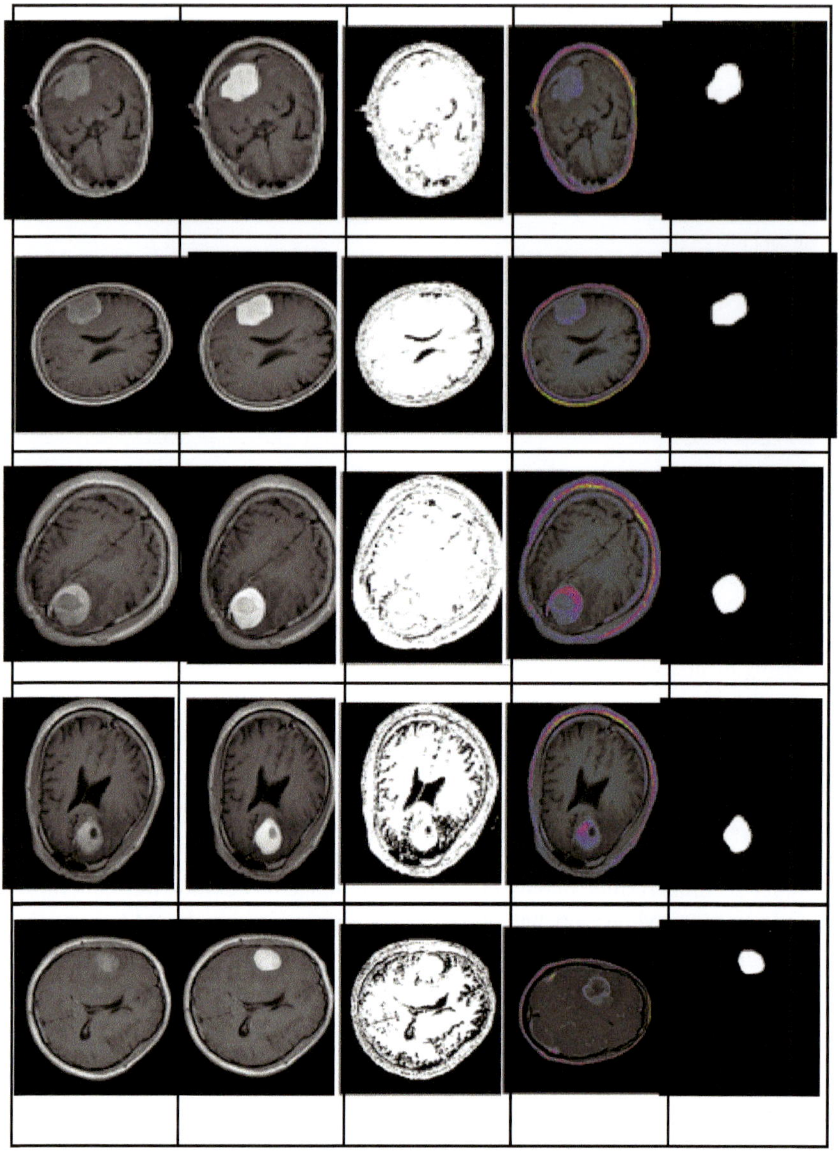

Fig. 4 Experimental results of Fuzzy RNN for MR image of size 256 × 256

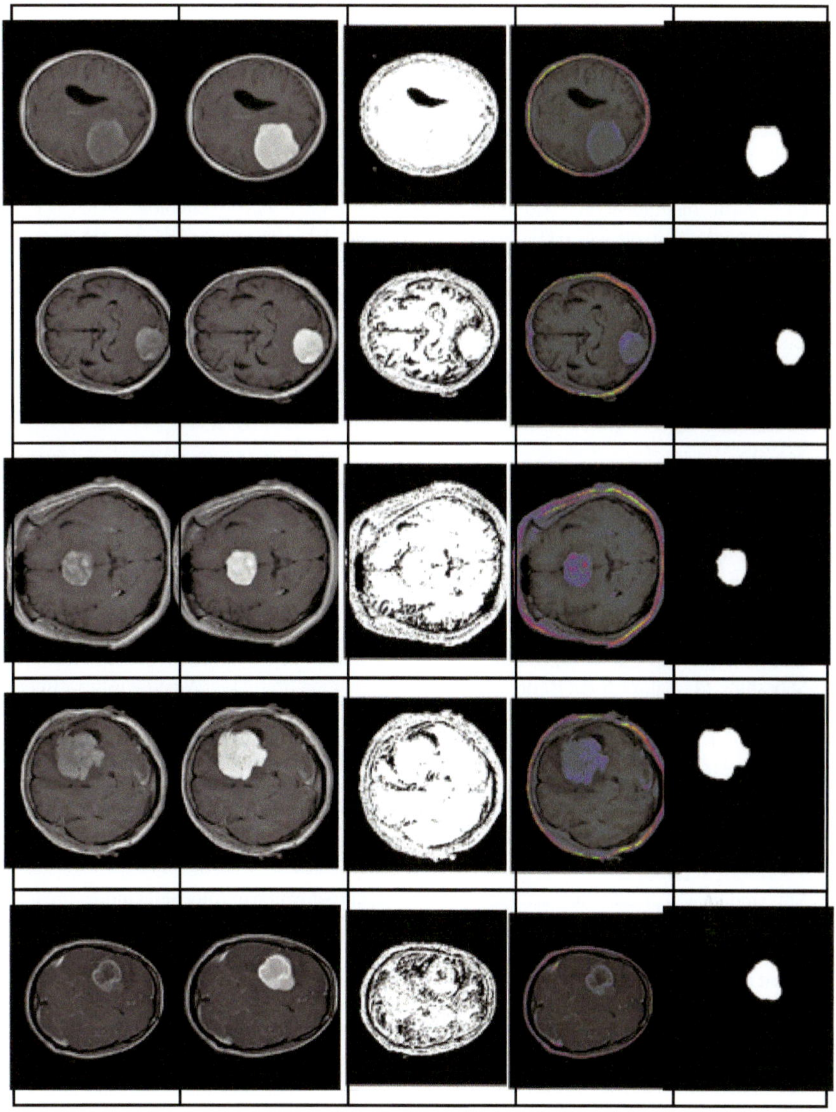

Fig. 5 Experimental results of Fuzzy RNN for MR image of size 512 × 512

References

1. Sivaramakrishnan, A., Karnan, M.: A novel based approach for extraction of brain tumor In MRI images using soft computing techniques. Int. J. Adv. Res. Comput. Commun. Eng. **2**(4) (2013, April)
2. Aslam, A., Khan, E., Beg, M.M.: Improved edge detection algorithm for brain tumor segmentation. Procedia Comput. Sci. **58**, 430–437 (2015). ISSN 1877-0509

3. Sathya, B., Manavalan, R.: Image segmentation by clustering methods: performance analysis. Int. J. Comput. Appl. **29**(11), 0975–8887 (2011, September)
4. Devkota, B., Alsadoon, Abeer, Prasad, P.W.C., Singh, A.K., Elchouemi, A.: Image segmentation for early stage brain tumor detection using mathematical morphological reconstruction. Procedia Comput. Sci. **125**, 115–123 (2018). https://doi.org/10.1016/j.procs.2017.12.017
5. Sudharani, K., Sarma, T.C., Satya Rasad, K.: Intelligent Brain Tumor lesion classification and identification from MRI images using k-NN technique. In: 2015 International Conference on Control, Instrumentation, Communication and Computational Technologies (ICCICCT), Kumaracoil, 2015, pp. 777–780 (2015). https://doi.org/10.1109/iccicct.2015.7475384
6. Kaur, J., Agrawal, S., Vig, R.: A comparative analysis of thresholding and edge detection segmentation techniques. Int. J. Comput. Appl. **39**(15), 29–34 (2012). https://doi.org/10.5120/4898-7432
7. Li, S., Kwok, J.Y., Tsang, I.H., Wang, Y.: Fusing images with different focuses using support vector machines. IEEE Trans. Neural Netw. **15**(6), 1555–1561 (2004)
8. Kumar, M., Mehta, K.K.: A Texture based tumor detection and automatic segmentation using seeded region growing method. Int. J. Comput. Technol. Appl, **2**(4), 855–859 (2011). ISSN: 2229-6093
9. Mahmoud, D., Mohamed, E.: Brain tumor detection using artificial neural networks. J. Sci. Technol. **13**, 31–39 (2012)
10. Marroquin J.L., Vemuri B.C., Botello S., Calderon F.: An accurate and efficient Bayesian method for automatic segmentation of Brain MRI. In: Heyden A., Sparr G., Nielsen M., Johansen P. (eds) Computer Vision—ECCV 2002. ECCV 2002. Lecture Notes in Computer Science, vol. 2353. Springer, Berlin, Heidelberg (2002)
11. Minz, A., Mahobiya, C.: MR image classification using adaboost for brain tumor type. In: 2017 IEEE 7th International Advance Computing Conference (IACC), pp. 701–705. IEEE (2017)
12. Subashini, M., Sahoo, S.K.: Brain MR image segmentation for tumor detection using artificial neural networks. Int. J. Eng. Technol. **5**(2), 925–933 (2013)
13. Srinivasu, P.N., Srinivas, G., Rao, T.S.: An automated Brain MRI image segmentation using Generic Algorithm and TLBO. IJCTA, **9**(32), 233–241 (2016)
14. Srinivasu, P.N., Rao, T.S., Balas, V.E.: A systematic approach for identification of tumor regions in the human brain through HARIS algorithm. In: Deep Learning Techniques for Biomedical and Health Informatics, pp. 97–118. Academic Press (2020). https://doi.org/10.1016/B978-0-12-819061-6.00004-5
15. Mukambika, P.S., Uma Rani, K.: Segmentation and classification of MRI brain tumor. Int. Res. J. Eng. Technol. (IRJET), **4**(07), 683–688 (2017). ISSN: 2395–0056
16. Pan, Y., Huang, W., Lin, Z., Zhu, W., Zhou, J., Wong, J., Ding, Z.: Brain tumor grading based on neural networks and convolutional neural networks. Conference proceedings: annual international conference of the IEEE engineering in medicine and biology society. IEEE Eng. Med. Biol. Soc. Conf. **2015**, 699–702 (2015). https://doi.org/10.1109/EMBC.2015.7318458
17. Pereira, S., Pinto, A., Alves, V., Silva, C.A.: Brain tumor segmentation using convolutional neural networks in MRI images. IEEE Trans. Med. Imaging **35**(5), 1240–1251 (2016)
18. Roy, S., Bandyopadhyay, S.K.: Detection and qualification of brain tumor from MRI of brain and symmetric analysis. Int. J. Inf. Commun. Technol. Res. **2**(6), 584–588 (2012, June)
19. Sankari, A., Vigneshwari, S.: Automatic tumor segmentation using convolutional neural networks. In: 2017 Third International Conference on Science Technology Engineering & Management (ICONSTEM), pp. 268–272. IEEE (2017)
20. Paul, T.U., Bandhyopadhyay, S.K.: Segmentation of brain tumor from brain MRI images reintroducing K-Means with advanced dual localization method. Int. J. Eng. Res. Appl. **2**(3), 226-231 (2012). ISSN 2278-0882
21. Vaishali et al.: Wavelet-based feature extraction for brain tumor diagnosis—a survey. Int. J. Res. Appl. Sci. Eng. Technol. (IJRASET) **3**(V) (2015) ISSN: 2321-9653

22. Varuna Shree, N., Kumar, T.N.R.: Identification and classification of brain tumor MRI images with feature extraction using DWT and probabilistic neural network. Brain Inf. **5**, 23–30 (2018). https://doi.org/10.1007/s40708-017-0075-5
23. Vinotha, K.: Brain tumor detection and classification using histogram equalization and fuzzy support vector machine approach. Int. J. Eng. Comput. Sci. **3**(5), 5823–5827 (2014). ISSN 2319-7242

Use of Deep Learning for Disease Detection and Diagnosis

Sushruta Mishra, Anuttam Dash, and Lambodar Jena

Abstract A dynamic research area which is progressively growing its importance in the field of medical diagnosis and detection is computer-aided diagnosis and detection. Suppose we are in a locality which happens to be very far away from healthcare centre or we are not financially strong enough to pay our hospital bill or we lack time to take sick leaves from our workplaces. In such scenario, the diagnosis of diseases by the help of high-end sophisticated tools can be very fruitful. A lot of AI algorithms have been proposed and developed by computer science scientists for the detection and diagnosis of diseases such as cancer, diseases of lung, rheumatoid arthritis, diabetic retinopathy, diseases of heart, Alzheimer's disease, hepatitis, dengue, liver disease and Parkinson's disease. Improved perception accuracy and disease diagnosis are some of the points put up by the recent machine learning researchers. Compared to the traditional computation algorithms, deep learning algorithms are way more effective in disease detection and diagnosis. Deep learning involves the usage of large neural networks that have neurons connected to each other that have the ability to modify their hyper-parameters whenever updated new data comes in. It is that technology which makes the computer systems able to learn things themselves without explicit programming from human side. In this study, we have mentioned the recent developments and trends in the deep learning field which can make a great impact for efficient detection and diagnosis of several types of diseases. Our chapter deals with exploring deep learning usage in efficient diagnosis of certain disease risk factors thereby assisting medical experts in precise decision-making. Also, two different case studies are discussed in detail to highlight the contribution of deep learning techniques in disease diagnosis.

S. Mishra · A. Dash (✉)
KIIT University, Bhubaneswar, Odisha, India
e-mail: anuttam.dash@gmail.com

S. Mishra
e-mail: mishra.sushruta@gmail.com

L. Jena (✉)
Siksha 'O' Anusandhan University, Bhubaneswar, Odisha, India
e-mail: jlambodar@gmail.com

A. K. Bhoi et al. (eds.), *Bio-inspired Neurocomputing*, Studies in Computational Intelligence 903, https://doi.org/10.1007/978-981-15-5495-7_10

Keywords Deep learning · Disease diagnosis · Machine learning · Medical field · Artificial neural networks

1 Evolution of Machine Learning in Disease Detection

Suppose we are in a locality which happens to be very far away from healthcare centre or we are not financially strong enough to pay our hospital bill or even we lack time to take sick leaves from our workplaces. In such scenario, the diagnosis of diseases by the help of high-end sophisticated machines can be very fruitful. There are a lot of AI algorithms that have been proposed and developed by computer science scientists for the detection and diagnosis of diseases such as cancer, diseases of lung, rheumatoid arthritis, diabetic retinopathy, diseases of heart, Alzheimer's disease, hepatitis, dengue, disease of liver and Parkinson's disease [1]. Deep learning involves the usage of large neural networks that have neurons connected to each other that have the ability to modify their hyper-parameters whenever updated new data comes in [2].

A dynamic research area which is progressively growing its importance in the field of medical diagnosis and detection is computer-aided diagnosis and detection. Recent research studies and works in the field of machine learning show that we can achieve better perception and disease diagnosis accuracy. This technology allows computer models to think intelligently like humans. In order to solve classification problems, several machine learning algorithms are used such as supervised, unsupervised, semi-supervised, evolution-based learning, reinforcement learning and deep learning algorithms [2] as shown in Fig. 1.

- **Supervised learning**: This learning method contains datasets with data labels predefined on training example upon which the model is trained upon which is in correlation to this training the estimator that responds precisely to all possible inputs. It can also be thought as a technique of learning through examples. The regression and classification problems fall under this category.
- **Unsupervised learning**: In unsupervised learning, the data points within the dataset belonging to similar groups are clustered distinctively by learning the similarities, patterns and structures among them. Density estimation is its other name. Clustering problems fall under this category.
- **Semi-supervised learning**: This signifies a subset of supervised learning technique. It consists of both the labelled and unlabelled data. Generally, the amount of unlabelled data is more than that of the labelled data. It lies somewhere between the unsupervised (unlabelled data) and supervised learning (labelled data).
- **Reinforcement learning**: Behaviourist psychology is the backbone of reinforcement learning. The model is clued-up about the error in the answer set, but it is not clued-up about how to remove the error. It uses a combination of trial and error with exploration capabilities to reach the correct answer. In other words, the model suggests the actions based on the previous experiences which do not necessarily indicate improvement.

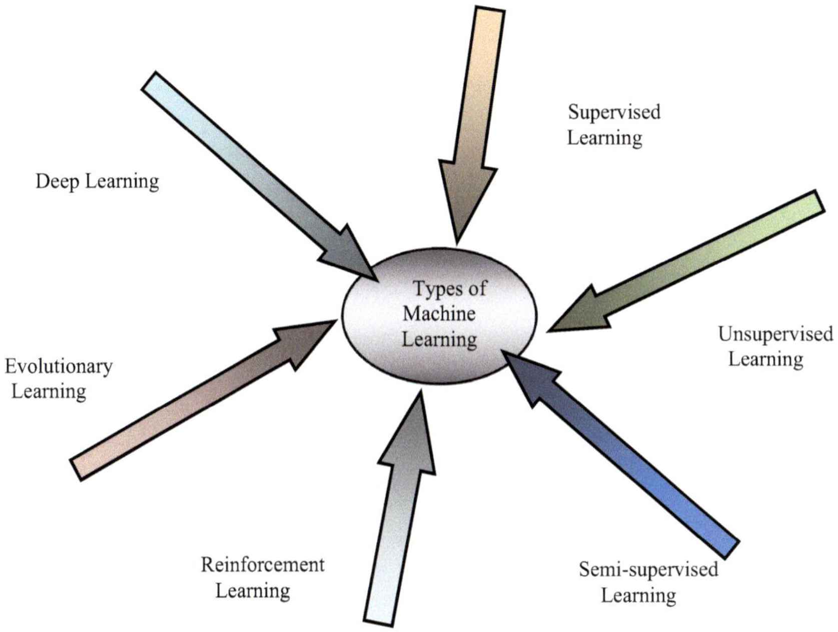

Fig. 1 Types of machine learning techniques

- **Evolutionary learning**: It is based on the natural phenomenon of natural selection, reproduction and survival of the fittest. It works in correlation with the biological process of evolution of organisms. The feasibility of a solution is checked on the basis of the fitness value associated with it which can determine the accuracy of the model.
- **Deep learning**: The hyper-parameters associated with the model is continuously updated until its optimality through the method of back-propagation. Deep learning is just the implementation of a neural network with many hidden layers, causing it to be deep.

Classification of data and recognition of pattern becomes important from a way back. Human beings have a good and well-defined ability to sense the environment. They are capable of mapping their perceptions to actions. Big-data is combined multidisciplinary technology of statistics, probability, databases and machine learning. The diagnostics and detection of diseases is a very serious topic in the medical field. The patient's disease should be diagnosed exactly according to the results of his medical assessments and tests. The usage of decision support systems in computer systems can be very useful for efficient diagnosis of diseases. Medical centres produce huge data about patient's health reports, medical assessments, follow-ups and other medication data. A suitable way to arrange all these data is complex in nature. The improper management of all these data has affected the data organization system. In order to process the data effectively and efficiently, it can only be done in the enhancement in the quality of the data. Techniques of machine learning are therefore needed to arrange data according to the similarity in their attributes. In the case of multi-class distribution of data, the classifier models are used to analyze the medical data in order to efficiently perform disease detection and diagnosis.

Initially, machine learning (ML) algorithms were designed and used for observing medical datasets. Nowadays, for efficiently analyzing data, ML recommends a variety of tools. Eminently in the previous years, digital evolution has endeavoured comparatively less cost and easily obtained mode for collecting and storing data. Data collecting and examining devices were being kept at different hospitals to have capability for collecting and dividing data in huge info systems. The technology of machine learning is the most efficient one to analyze medical-related datasets, and the task is carried out in diagnostics problems. Accurate diagnostics data are represented in a medicinal record or reports in hospitals and specific data sections. For executing the algorithms, appropriate diagnosis records of the patient are input in the system. Recent cases which are solved are helpful in obtaining results. The derived classifiers might assist physicians during performing of diagnosis of patients at good speed and precision. The above classifiers can become helpful for training inexperienced doctor or students to diagnose the disease.

Previous research of ML had provided the facility of self-driven cars, speech detection facilities, good Web searching, and enriched visualization of the current generation. Nowadays, there is widespread use of machine learning in numerous fields in such a way that sometimes we may need repeatedly in our life. Various studies are considering it as a superior and wonderful method for social progress. The technique explores electronics health records usually containing high dimensional patterns and number of datasets. Pattern recognizing methods are the basics of MLT that are offering help in anticipating and helping in making decisions for diagnosing and to choose treatment. ML algorithms make it possible to handle huge amount of data, as well as to make combination data from divergent source and to assimilate the backdrop information during study.

Different researchers have performed their research on different ML algorithms for diagnosing diseases. It has been approved by the researchers that the machine learning algorithm performs well in diagnosing various diseases. A diagrammatic approach of diseases diagnosed by machine learning techniques is shown in Fig. 2 [3].

1.1 Significance of Deep Learning in Disease Diagnosis

There are a lot of advantages of deep learning in the healthcare industry [3]. Some of them are:

- Deep learning helps in finding out a vital relationship in the data as well as it also records the information regarding existing clients that might help patients having similarities in symptom or diseases.

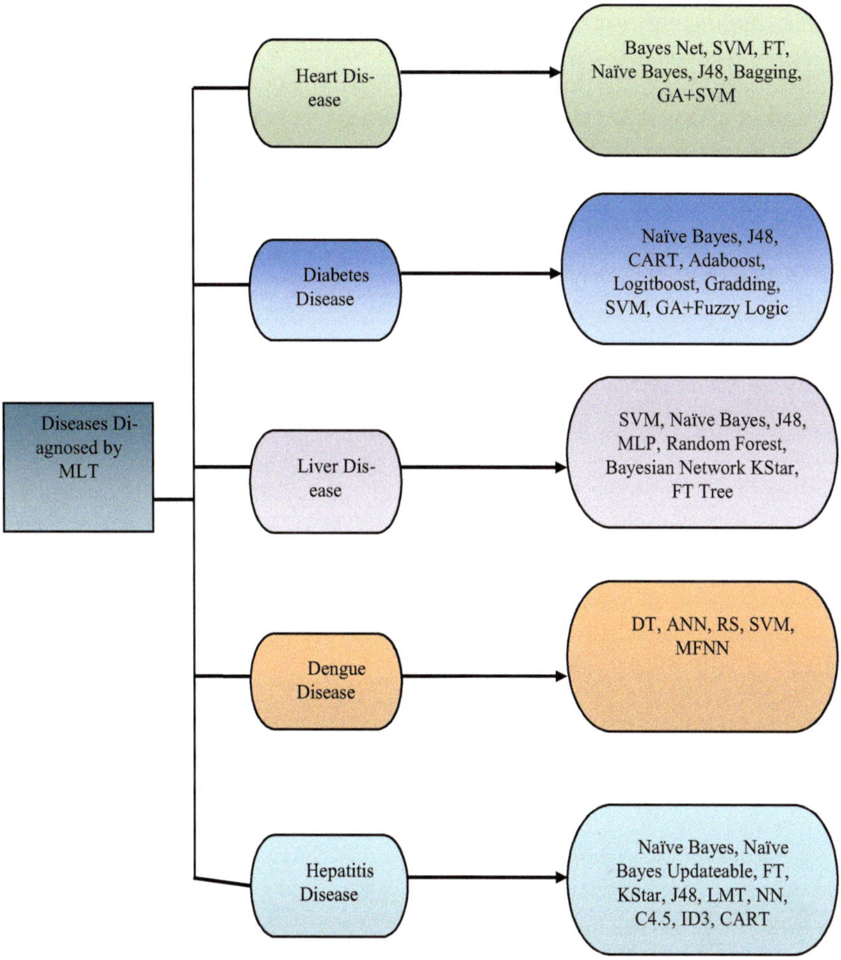

Fig. 2 Some diseases diagnosed by ML techniques

- Deep learning helps us to make some model-establishing on information available from various sources on the requirement of risks scored on the administrations rather than discharge.
- Deep learning provides accuracy along with well-timed risks score on providing the ability of the confidence and approximating resource allocating methods.
- The approach for deep learning leads to less price and helps in providing improvised outcomes. While deep learning algorithms interact with the data used for training, they become clear enough and decisive which allows person to get unprecedented insight into care process, variable analysis and diagnostic.
- Graphics processing units or GPU are performing more efficiently and are becoming fast.

- The blowing-up of innovating models of deep learning algorithms is at a chunk of the previous cost and so becomes the techniques that got more sophistication.
- Electronic health records or EHR and other digitization endeavours help in providing facility for accessing healthcare data to use qualified algorithm which was successfully done previously.
- Diagnostic got better accuracy and becomes fast with the help of deep learning that identifies patterns by connecting the tools.
- Deep learning helps in determining if the skin lesions got cancerous or not as recommended by any other board certificated dermatologist.

1.2 Scope of Deep Learning Techniques for Disease Diagnosis

Various ML diagnostic applications basically come under the following categories:

- **Chatbots**: Company generally prefers AI-based chatbots with speech recognition capacity for identifying patterns in the symptom for forming a potential diagnosis, preventing disease and/or recommending as a proper course of action [4].
- **Oncology**: Scientists are taking the help of deep learning methodology to perform the training of algorithms for identifying cancerous tissue similar to trained physicians [5].
- **Pathology**: Pathology signifies specific medical stream which is connected along with diagnosing of disease from the analyzed laboratory reports of fluids present in the body such as blood and urine, as well as tissues. Machine vision and other machine learning technology can make advancements in the efforts basically dedicated only to a pathologist with microscope. Now, the present scenario applied voice pathology, and it plays a vital role in advanced pathological system [6].
- **Rare diseases**: Face recognition software can be connected with machine learning for helping clinical diagnosis of rarely found diseases. Specific patient's photographs are investigated using face analysing and deep learning mechanism for detecting phenotype which are correlated with rarely appeared genetically diagnosed diseases [6].
- **Imaging**: Latest advancements have been suggesting about the requirement of artificial intelligence to elucidate and figure out the result of the maxillofacial operation or surgery of cleft patients' in relation to face attractively or age appearing characteristics [7].
- **Telehealth**: Advancement of telemedicine has set out the acceleration of attainable AI application. It makes able for performing monitoring of patients using artificial intelligence which may let on communicating of data to doctors on likely disease-related activity which have been taken place. The wearing devices or computer might help for performing regular observation of a patient and also allowing to figure out the change that may be least distinguished by human beings [8].

- **Electronic health records**: Electronic health record becomes imperative to the digitization and spreading info of the healthcare industries. Nonetheless, log on the operation of various data comes with its self-issues, especially subjective overload and burnout for user. EHR developer is recently doing automation of many of the processes along with started using natural language processing (NLP) tools for improvising the processes. A study conducted by the Centerstone Research Institute was telling about the predictive modelling of EHR data achievement which is 70–72% accurate in foretelling of individual therapeutics response at ground level. Use of AI tool performs scanning EHR data and can perform accurate prediction about the sign of disease in human beings. Many disease prediction and diagnosis are efficiently solved by use of deep learning technique [9].
- **Radiology**: It is a special branch that has taken the maximum consideration which is the stream of radiology. It makes able to perform an interpretation of imaging results that might aid clinical people in finding a very minute change in an image that could be missed by a clinician in accidental manner. Research done at Stanford results in an algorithm which is able to detect pneumonia in a more accurate way in comparison with radiologist [10]. Conference of Radiological Society held at North America has dedicated maximum percentage of time to the use of AI with the field of imaging. Development of AI technology in radiology has been understood as a threat by few doctors, as the techniques can show better performance in particular jobs in comparison with doctors, by doing change in the recent role of radiologist.
- **Personalized treatment**: Response of various patients to different drugs and treatments varies individually. Therefore, specialized treatment has powerful potential for ascending lifetime of the patients. It seems extremely difficult to find out the aspects that affect the choice of therapeutics. Deep learning performs automation on the complex statistical task as well as reveals the features that tell that the patient may give a specific acknowledgement to a specific treatment [11]. Therefore, the algorithm can forecast the specific response of patient to a specific treatment.
- **Improved editing of gene**: Clustered regularly interspaced short palindromic repeats (CRISPR), specifically the CRISPR-Cas9 technique, for editing of gene signifies a good jump in the efficiency for editing DNA in a cost-effective manner—absolutely as being a surgeon. The method trusts on short guide RNAs (sgRNA) to be destined as well as editing a particular point located on the DNA [12]. But, the guide RNA may be suitable for different DNA location which might have an unforeseen side effect. Therefore, one must take proper care before choosing the guide RNA to avoid side effect which could be hazardous, and this seems to be a considerable flaw in the application of the CRISPR system.

Deep learning has been extensively used in various domain of social significance like telecommunication, healthcare and other socially relevant fields. In [12], authors developed a machine learning model to effectively predict academic performance of students with the application of decision tree technique. Similarly, an automated and efficient deep learning framework model for smart phone purchase pattern is presented in [13] which provided optimum result using multi-classifier models. Many such works have been undertaken in several significant domains of knowledge.

2 Demonstration of Deep Learning Application in Detecting Diabetic Retinopathy

Enduring diabetes leads to a very dangerous eye disease called diabetic retinopathy (DR). Often in developed nations, diabetic retinopathy is found to be the main cause which leads to visual imparity among the working age group. Approximately one out of every two diabetic persons has been diagnosed with some stage of diabetic retinopathy. In majority number of cases, early detection and diagnosis of symptoms of DR can prevent the person from visual impairment. However, with the use of present methods and tools, such revelation can be difficult. Therefore, there is a need for automated methods and tools for detection of diabetic retinopathy (DR). Promising results have been achieved from the previous methods which used machine learning, image classification and pattern recognition. In this study, colour fundus pictures have been utilized. Finally, the outcomes of the experiments show that the model has reasonable medical potential. Comparison of normal retina with diabetic retinopathy is shown in Fig. 3, while Fig. 4 demonstrates the variations in the features like illumination and contrast that can be seen from a random image dataset used for training phase.

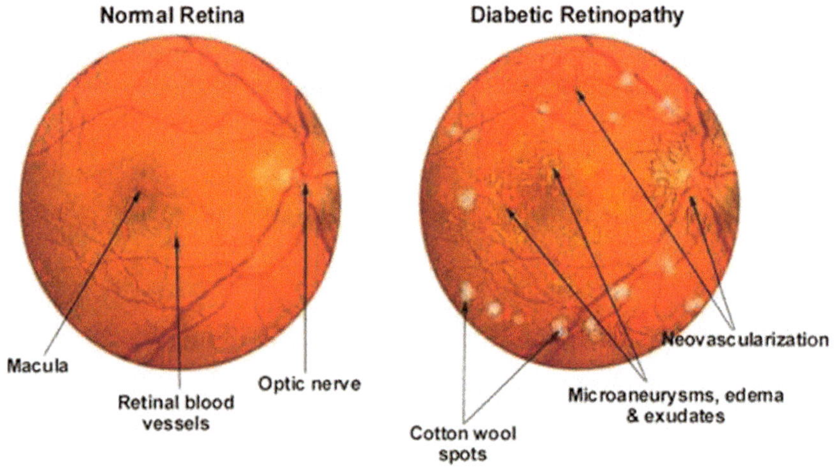

Fig. 3 Comparing normal retina with diabetic retinopathy

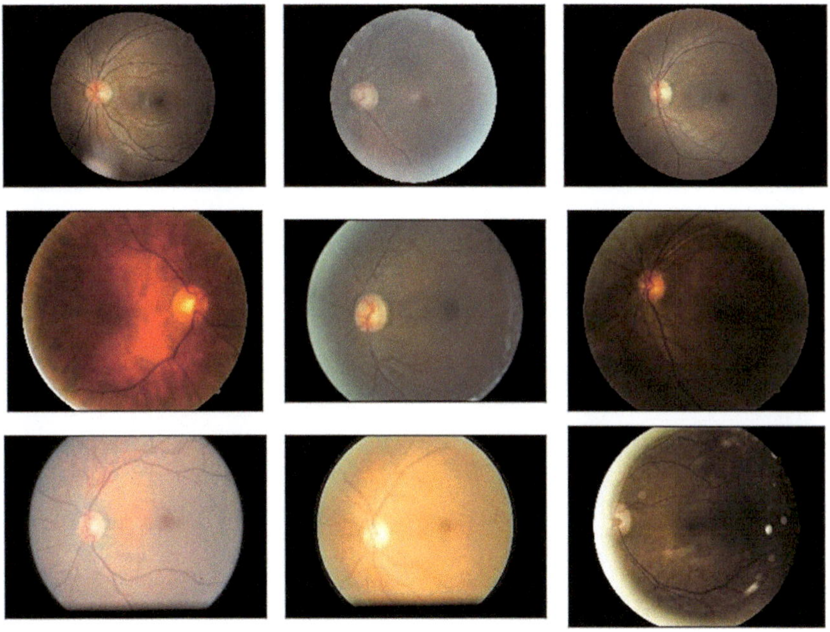

Fig. 4 Variations in the illumination, contrast etc. that can be observed from a random image dataset used for training

2.1 Discussion of Deep Learning Model

ConvNet Model Architecture: The most significant foundation of a neural network model is its architecture. It is basically the blueprint of how the model has been designed in correlation to its specifications. The architecture of the model has been described in the subsequent sections.

Description: The proposed model has been adapted from the Inception-V3 architecture model which uses inputs generated from multiple convolution filters on the same set of input. Pooling is also done at the same time. All the results of each layer are then cumulated. This architecture thus aims at multi-level extraction of features from every single input. Thus, the architecture is capable of extracting a local (1×1) and general (3×3) feature in one moment. The Inception-V3 data is given in Table 1.

Inception V-3 architecture design principles: The architecture of the Inception-V3 model is illustrated in Fig. 5. The proposed architecture aims at not only improvising the model's statistical performance and accuracy but also on a proficient mode to achieve so bearing in mind the memory track and time required for training. Some of the principles used in developing the architecture have been mentioned below:

1. *No extreme compression on information*: There should be a gradual reduction in the size of illustration from input end towards the output end in such a way that

Table 1 Inception-V3 data

Type	Patch-1	Patch-2	Patch-3	Size	Depth
Conv	3	3	2	[299,299]	3
Conv	3	3	1	[149,149]	32
Conv padded	3	3	1	[147,147]	32
Pool	3	3	2	[147,147]	64
Conv	3	3	1	[73,73]	64
Conv	3	3	2	[71,71]	80
Conv	3	3	1	[35,35]	192
3 Inception				[35,35]	288
5 Inception				[17,17]	768
2 Inception				[8,8]	1280
Pool	8	8		[8,8]	2048
Linear	Logits			[1,1]	2048
Softmax	Classifier			[1,1]	1000

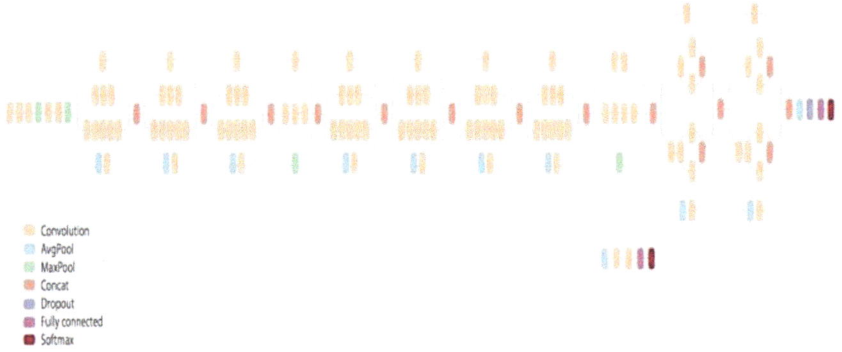

Fig. 5 Schematic diagram of Inception-v3

whenever a cross-section between output and input end is examined there should not be a delay with excessive compression.

2. *Representations in higher dimensions*: In case of representation in a higher dimension, outcomes of local processing have been found more effective. When we increase the number of activation layers in a tile, the number of unscrambled features is developed in the convolutional neural network, thus reducing the time for training.
3. *Spatial combination*: Spatial combinations are gained over vectors in lower dimensions without losing the representational power. If there are enough evidence and limitations suggesting that these outcomes can be compressed easily, the reduction in dimension even enables lesser learning time.

4. *Achieving network balance by optimizing depth and width*: For high-performance networks, there is a need for optimizing the width (total number of nodes in a layer) and depth (total number of layers) of the neural network. High-class networks do not always refer to mounting both depth and width only. Best possible enhancement can be achieved only by parallel increment of the depth and width. Hence, the computational budget should be distributed between both depth and width. These points when considered as a whole are used to increase the network quality. Therefore, efficient and judicious usage of the resources can be fruitful in increasing the complete functionality of the network.

Experiments and results: Different datasets are used to perform training of a convolutional neural network. However, for testing purpose, the kaggle images test dataset was used. Following is the illustration of the results obtained by changing the hyper-parameter in isolation and individually.

1. **Batch size**

The network is trained by minimizing the gradient matrix by dividing the training set into batches, so that comparatively less memory will be required for the purpose of training the network. The training set is continuously divided into batches for that many numbers of iterations for which a satisfactory validation loss and validation accuracy are achieved. Finally, the effect on the final weights caused due to different batch sizes was observed. Keeping all other parameters the same, two batch sizes were used, i.e., 64 and 128. The results have been illustrated in the below table (Table 2).

2. **Epoch**

The total number of iterations required for covering the training set is referred to as an epoch. Suppose your dataset consists of 1000 training elements with a batch size of 500, then in order to complete an epoch, two iterations will be required. Usually, with more number of epochs, better results are met. Results have been illustrated in Table 3.

Table 2 Changes in batch size

Batch size	Sensitivity	Specificity	Accuracy	Unclassified
64	94	81	82	20
128	97	87	88	25

Table 3 Epoch

Epoch	Sensitivity	Specificity	Accuracy	Unclassified
100	72	69	70	30
200	97	87	88	25

3. **Pre-processing**

Pre-processing happens usually after the training phase. It aims at defining a standard that needs to be followed for the image set by performing image augmentation. Through this, some of the features are even highlighted before the training phase. principal component analysis (PCA), independent component analysis (ICA), Normalization of image and mean subtraction are commonly used methods of data pre-processing. Results are illustrated in Table 4.

4. **Train dataset size**

The training is done using the image dataset that contains all the labelled images. Results are illustrated in Table 5. The following inferences can be made from the experiments.

- It was found that the larger batch sizes performed with more efficiency than the corresponding smaller ones, but they need more computational resources.
- With more number of epoch, better accuracy can be achieved because the model trains on a similar set of images for number of rounds and thus likelihood of doing error reduction. However, there will be point beyond which any change in the number of epochs will not increase the performance and accuracy.
- Pre-processing helps in the improved understanding of the input data by reducing the network bias.
- Bigger training datasets enables the estimator to learn various kinds of edge cases and variations concerning the problem making it versatile and thereby increases the performance (Figs. 6 and 7).

It can be derived by observing the results obtained from the experiments that the hyper-parameters played an important role in deciding the model's accuracy and performance. Thus, finding the optimal values of the hyper-parameters is a really important task. The model achieved a promising score bearing in mind the restricted amount of training data available. The factors such as the quantity and quality of training dataset as well as the values of hyper-parameters proved to be important

Table 4 Pre-processing

	Sensitivity	Specificity	Accuracy	Unclassified
No pre-processing	93	85	82	17
Pre-processing	97	87	88	25

Table 5 Train data size

Dataset size	Sensitivity	Specificity	Accuracy	Unclassified
10,000	75	65	67	28
40,000	97	87	88	25

Fig. 6 ROC testing curve

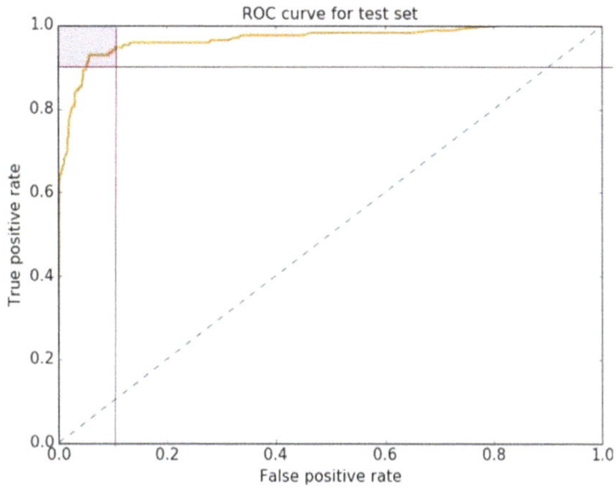

Fig. 7 ROC validation curve

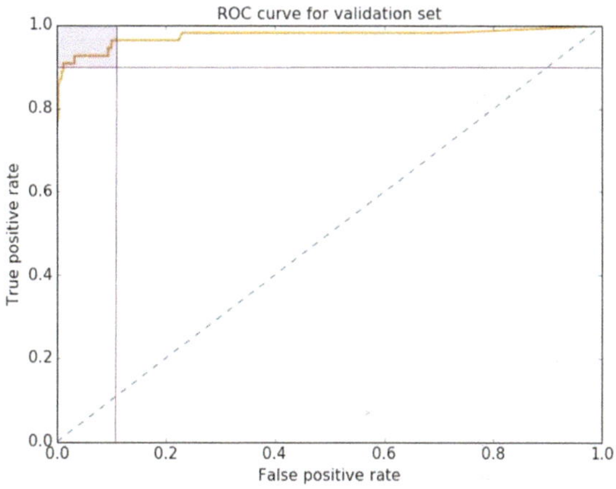

for the Inception-v3 model for achieving better scores on the ImageNet challenge. Discrepancies that occur during the human evaluation of corner medical situations can be yet another issue that changes the facts and figures of the data and thus leads to sub-optimal results.

3 Discussion of 3D Deep Learning in Multi-modal Imaging-Guided Survival Time Prediction of Brain Tumour Patients

3.1 Analysis of the Demonstrated Model

One of the most life-threatening and dangerous forms of brain tumour is high-grade glioma which has been found to be the cause of deaths of almost 50% patients in a period of 1–2 years. Thus, there is a need for efficient guidelines for treatment planning of patients suffering from glioma through accurate prognosis. Conventional methods for predicting the survival period of glioma patients utilize limited features and medical information from magnetic resonance imaging (MRI) which is found to be subjective, time-consuming and laborious. This work proposes techniques to perform extraction of features by multi-modal pre-operative image collected from patients of high-grade glioma. In this work 3D CNN networks have been adopted, and also an architecture for learning supervised features and handling multichannel data has been proposed. An support vector machine (SVM) model has been trained with the pivotal medical data for doing prediction if the patient has a large survival or small survival time. Experiments make it clear that the proposed methods are capable of achieving an accuracy of around 89.9%. It was found that in order to accurately predict the OS time, learnt features from DTI and fMRI proved to be very useful. Therefore, precious information about functional neuro-oncological applications was provided.

The utilization of 3D deep learning approach along with the handcrafted features has been done to find out the features to better typify the asset of various brain tumours. Firstly, high-level tumour features from DTI, fMRI, T1 MRI images for differentiating the short and long survival time patients were extracted using a supervised technique, i.e., convolutional neural network (CNN). Then, an SVM with selected and extracted features is trained. These steps are shown in Fig. 8. Since the MRI images are 3D images, a 3D CNN with 3D patches has been adopted. Both the DTI and fMRI images contain multiple channels. Hence, a multichannel CNN (CNN) has been mentioned in the subsequent sections to concatenate the information obtained from all the channels of the DTI and fMRI images.

Acquisition of data and pre-processing: Data collected from various patients of brain tumour consist of the DTI and fMRI images in resting state along with the T1 MRI images. The sample images of a patient are shown in Fig. 9 in which we can observe the multichannel images for DTI and fMRI along with single-channel images for T1 MRI. A conventional pipeline system has been adopted for pre-processing these three modalities of images. They are briefly aligned first. Normalization of intensity is then done for T1 MRI. Diffusion tensor modelling is done for DTI, which is followed by the calculation of diffusing metrics (i.e., MD, lambda 1/2/3, FA, RD) along with the B0 image. Fluctuation powers of blood-oxygen-level-dependent

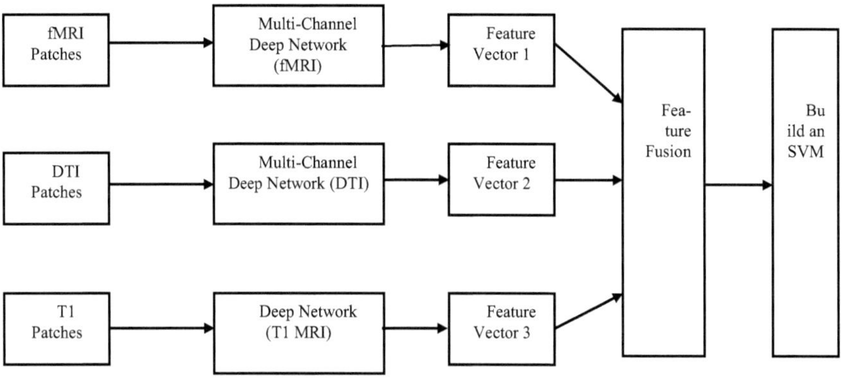

Fig. 8 Survival time prediction system

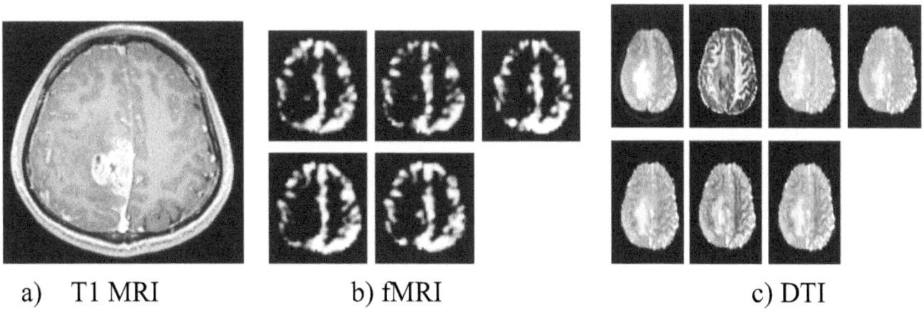

Fig. 9 Glioblastoma sample subject

(BOLD) in five (non-overlapped) bands of frequency (0–0.24 Hz) are then found out for fMRI images.

Data were extracted from approximately 69 patients who were diagnosed with WHO-III and IV. Barring criterion is (1) any form of operation of brain tumour chemotherapy, radiation therapy before being used in the research, of (2) incomplete and missing data and (3) presence of artefacts or too much motion head. Based on the proposed medical guidelines, all the patients were treated equally. A local ethical committee approved the entire study.

Single-channel feature extraction: As already discussed, a 3D CNN model having a set of 3D trained filters has been proposed to handle the 3D T1 MRI images. The architecture of the CNN model is shown in Fig. 10. It consists of four convolutional groups of layers (conv1 to conv4. Two convolutional layers have been used to develop the conv1 and conv2) and three layers of full connection (fc1, fc2, fc3). The outputs of the convolutional layers will be the inputs to the local three-dimensional regions in which every layer performs a 3D filter (with size $3 \times 3 \times 3$) convolutional operation (i.e., between the 3D regions and the weights in which they are functioning). The results are in 3D format with a size similar to their inputs. These results then undergo a

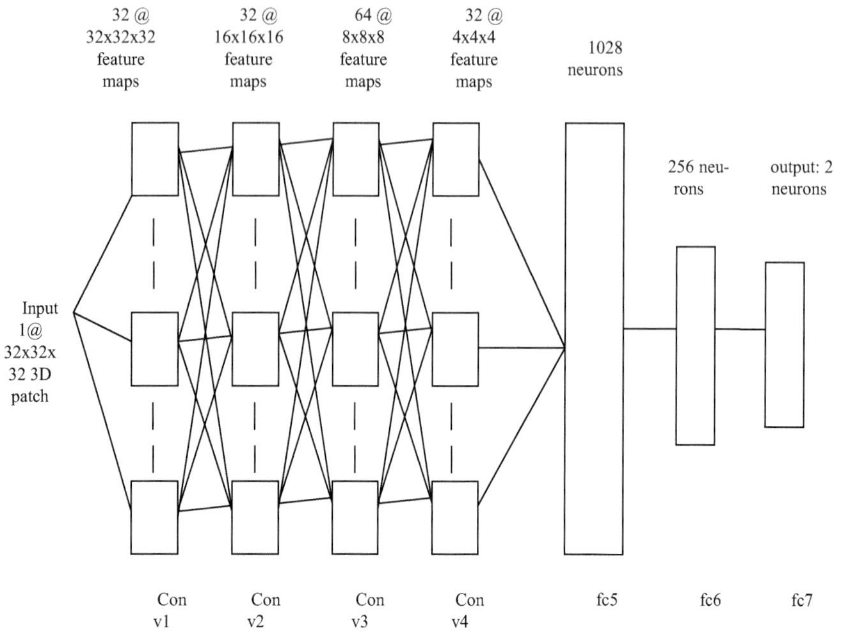

Fig. 10 Feature extracting feature by using single-channel from 3D patches using a CNN model

max-pooling phase in order to do a downsampling action along the three dimensions. In the fc layers, the neurons are all attached to their respective activation functions as is the case in the convolutional layers. The class scores are determined by using the outputs of the last layer, i.e., fc7. In order to learn the significant features in the fully connected layer, the back-propagation procedure is required after the supervision on the class scores. The effectiveness and efficiency of the model depend on the outputs of the final two layers: fc6 and fc7 which provide the characteristics which are totally supervised.

Multichannel feature extraction: As shown in Fig. 9, both the DTI and fMRI images are made up of multiple channels. A new kind of multichannel CNN model (mCNNs) was proposed for dealing with the multichannel data with the purpose of training an mCNN for a modality with consideration of the multichannelling images that provides insight about brain tumour. Different acquisition methods can lead to direct complementary information in the original image spaces. The proposed 3D model was improved by taking useful insights from the multi-modal deep Boltzmann machine. Specifically, each channel of data is applied in the same convolution layers in the proposed model, followed by an implementation of the fusion layer to add the outputs of the conv layers in all the channels. Finally, the features are extracted by using three fc layers. The new CNN model is shown in Fig. 11. The lower layer networks (i.e., the layers for individual pathways) varies as mentioned in the architecture which is the main reason for difference in the input distribution among the channels. The outputs of these streams are then combined in the fusion layer. The

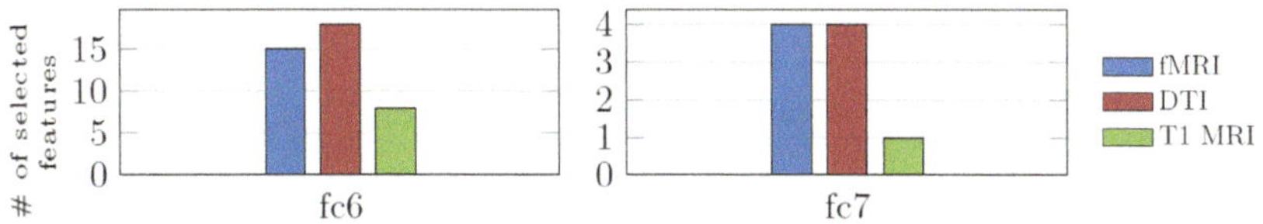

Fig. 11 Model demonstrating the average number of feature selection from each modality

layer groups described are similar to the single-channel data layers. It is hard for a model of single channel to find out the correlations in the channels because of the variance in the statistical properties of different channels. The differences of the multichannel data are met by fusing the features of their respective higher-layers as proposed in the mentioned mCNN model.

Classification: The raw 3D images from three modalities to the high-level representation can be mapped once to mCNNs for fMRI and DTI images and CNN (Fig. 10) for T1 images which are trained. This is achieved when every image-patches are fed to the corresponding mCNN or CNN model, in correlation with the currently extracted patch's modality. The learned features of the input images are fc layers' outputs. Finally, feature vectors having the same length will be used to represent each subject. An SVM classifier is then trained for survival classification using the survival labels.

3.2 *Result Analysis of the Discussed Framework*

Attributes are learned through data samples of multi-modal brain tumour by supervised learning. Later, these attributes are extracted and used in training of SVM algorithm in prediction of survival analysis. Back-propagation model is adopted to train neural network and in updation of network metrics. The network weights are initialized by Xavier algorithm. The initial learning rate and weight decay parameter are determined by conducting a coarse line search, followed by decreasing the learning rate during training. Various distinct attributes are incorporated in the training process for evaluation of the presented framework. Specifically, it may be noted here that two generally applied attribute set concerning CNN model become the output of fc6 and fc7 layers which are discussed below.

Training of every modality of network is done on corresponding modality samples. Subsequently, the respective attributes are obtained by feeding every testing pool into the trained model network. It is to be noted that fc6 which denotes the layer preceding output layer constitutes 256 neurons and fc7 which represents the output layer consists of two neurons. Besides these attributes, the handcrafted features (HF) also is a part of our analysis. Attributes corresponding to brain tumour, gender, location of tumour, age during diagnosis, tumour size and other generic attributes of brain tumour are included in handcrafted features (HF).

Table 6 Evaluating the performance of various features along with selection/reduction techniques

	ACC (%)	SEN (%)	SPE (%)	PPR (%)	NPR (%)
HF	62.96	66.39	58.53	63.18	65.28
HF + SIFT	78.35	80.00	77.28	67.59	87.09
HF + 2D-CNN	81.25	81.82	80.95	74.23	88.35
fc7	80.12	85.60	77.64	71.71	87.50
fc6-PCA	80.55	84.85	76.92	75.68	85.71
fc6-SR	76.39	86.67	69.05	66.67	87.88
HF + fc7	89.58	92.19	88.22	84.44	95.57
HF + fc6-PCA	89.85	96.87	83.90	84.94	93.93
HF + fc6-SR	85.42	92.60	80.39	75.36	96.83

In this study, tenfold cross-validation technique was applied where in every fold of testing, nine different folds were implemented in training of both CNN as well as SVM which is used with the trained attributes. Then, for all the tenfold of validation, an average for all these folds are computed and generated as given in Table 6.

As seen in the Table 6, combining both 3D CNN and 3D mCNN framework along with handcrafted (HF) attributes produces an optimal prediction rate of accuracy with 89.85%. But when HF is applied alone or even when integrated with attributes of unsupervised learning set, the accuracy rate significantly dropped to 62.96% or 78.35%, respectively. Also, it can be clearly observed that the traditional 2D model of CNN is outperformed by 3D model which proves that feature training can be better performed in 3D filters. As far as specificity and sensitivity are concerned, it is well known that more the value of sensitivity, less is the probability to misclassify the patients with short survival rate and vice versa. The presented attribute extraction model resulted in almost 30% more specificity and sensitivity as compared to the conventional handcrafted attributes. Also as seen in the analysis, the presented framework helps in the prediction of patients with less survival rate with increased confidence than patients with long survival rate. Also almost identical power in OS time prediction is exhibited by the attributes in various layers of CNN and mCNN frameworks. The quantity of attributes chosen from each modality are visualized and used in the analysis of significance of attributes in prediction of OS.

As far as attributes of fc6 are concerned, features are counted and are selected through sparse denotation, and as far as fc7 layer is concerned, ℓ_1-regularized SVM is used to classify the attributes from fc7 layer in order to select the most dominant attributes. Then, all chosen attributes are averaged in cross-validation folds which are shown in Fig. 11.

The above analysis is an effort to propose a unique deep learning CNN model for predicting the minimum and maximum survival time (OS) of the glioma patients. In this work, an mCNN model was deployed to handle the multichannel DTI and fMRI images while on the other hand a 3D CNN was used to handle the single-channel T1 MRI images. Finally, a binary support vector machine classifier was fed up with the features extracted. By observing the results of the experiments, it was derived that the supervised learnt features of the proposed work lead to better prediction accuracy of the OS time of the glioma patients. Thus, it was found that the computational analysis models can be provoked by such deep learning models to extract helpful features that can be used in various neuro-oncological problems. However, it was found that the DTI images were proved to be more useful to build predictive models than fMRI images individually, but the combination of DTI and fMRI images can be better than the T1MRI images.

4 Conclusion

Machine or computer vision is proving itself more and more useful in the field of medical applications, and thus, the improvement in this ground can be more fruitful to the medical field. However, this is a probabilistic study, and the process of trial and error is the foundation on which the final value of these applications depend on the practical environment and are levelled high so that they can be used in medical diagnostic area.

Though the artificial intelligence applications lie in the field of medical diagnostics which is in its initial development phase with very fewer data of patients available, it still holds a great potential to improve the way disease diagnosis which is currently being handled by medical professionals and healthcare systems. It also provides the facility of imparting the individuals with the capability to understand their health changes in real time.

Though artificial intelligence holds great power to deal with real-time medical diagnostics problem, there is still a dilemma about its sensitivity, reliability and how can it be implemented into clinical diagnosis field without any clinical expertise. Continual scrupulous application testing of AI models will be essential to authenticate their usefulness united with the instructions to medical professionals as well as surrounding medical system that how to efficiently apply these techniques in medical exercise.

We went through many deep learning literatures and works to find out the impact and usage of these method frameworks in the area of diagnosing and detecting diseases. The following questions enlighten us on the conclusion of the survey.

- *Diversification of this application in clinical diagonalizing*:
  The medical field is one of the areas, whereas this method has a lot of applications. In this study, the medical diagnosis was performed by the implementation and deployment of deep neural networks. As we have already discussed, the deployment phase comprises of the segmentation, detection, prediction, classification, etc. The experiments and results of the survey show that the usage of deep learning methods and frameworks provides better results than other integer-based computation algorithms. Therefore, it can be said that deep learning has and will keep on to have diverse applications in the area of clinical diagnosing method.

- *Regarding the substitution of this technique in the course of time*:
  The future works in the area of deep learning guarantee a lot of development in the field of disease detection and diagnosis. However, we cannot clearly comment on the feasibility of the deep learning models replacing the doctors or healthcare professionals as of now. It can just be concluded that deep learning is and will keep on supporting and helping healthcare professionals for diagnosis of the diseases.
- *Future of the above technique*:
  All the developments and works happening in the area of deep learning point out that it will keep widening its utilities in a lot of domains. Deep learning is currently being used in control on the emission of greenhouse gases and transportation, classification of text, traffic control, detection of objects, translation, detection of speech and in various other fields. The usages of these applications were not popular in the past. Usual high-end computation algorithms, when applied to the same domains, proved way less efficient as when compared with deep learning algorithms. Our survey study thus points out that deep neural networks and deep learning algorithms will continue to expand their usage in many other domains in the future through continuous development and betterment.

5 Future Scope

Medical science is a field which has a lot of scope for AI. However, to guarantee the certainty and feasibility of the AI models in the medical domain, clinical data and trials are needed. But, it can be said that ML techniques will unavoidably turn out to be one of the most vital proponents of prediction and diagnosis of diseases (both population-wise as well as individually). AI can be a game-changer in the medical industry by the introduction of personalized healthcare. Custom health treatment, accurate disease diagnosis, plans of clinical care, tailor of efficient and effective drugs are some of the capabilities that will be served by deep learning very soon. The achievements currently made by AI models in the healthcare domain suggest its healthy future, though it is still in its early stage of development.

References

1. Bhardwaj, R., Nambiar, A. R., Dutta, D.: A study of machine learning in healthcare. In: 2017 IEEE 41st Annual Computer Software and Applications Conference (COMPSAC). vol. 2. IEEE (2017)
2. LeCun, Y., Bengio, Y., Hinton, G.: Deep learning. Nature **521**(7553), 436–444 (2015)
3. Fatima, M., Pasha, M.: Survey of machine learning algorithms for disease diagnostics. J. Intell. Learn. Syst. Appl. **9**(01), 1 (2017)
4. Fadhil, A., Gabrielli, S.: Addressing challenges in promoting healthy lifestyles: the al-chatbot approach. In: Proceedings of the 11th EAI International Conference on Pervasive Computing Technologies for Healthcare, pp. 261–265. ACM (2017)

5. Beam, A.L., Kohane, I.S.: Big data and machine learning in health care. JAMA **319**(13), 1317–1318 (2018)
6. Hossain, M.S., Muhammad, G.: Healthcare big data voice pathology assessment framework. IEEE Access **4,** 7806–7815 (2016)
7. Lee, J.G., Jun, S., Cho, Y.W., Lee, H., Kim, G.B., Seo, J.B., Kim, N.: Deep learning in medical imaging: general overview. Korean J. Radiol. **18**(4), 570–584 (2017)
8. Strode, S.W., Gustke, S., Allen, A.: Technical and clinical progress in telemedicine. JAMA **281**(12), 1066–1068 (1999)
9. Shickel, B., et al.: Deep EHR: a survey of recent advances in deep learning techniques for electronic health record (EHR) analysis. IEEE J. Biomed. Health Inf. **22**(5), 1589–1604 (2017)
10. Wang, S., Summers, R.M.: Machine learning and radiology. Med. Image Anal. **16**(5), 933–951 (2012)
11. Papadakis, G.Z., et al.: Deep learning opens new horizons in personalized medicine. Biomed. Rep, **10**(4), 215–217 (2019)
12. Mishra, S., Panda, A.R.: Predictive evaluation of student's performance using decision tree approach. J. Adv. Res. Dyn. Control Syst. 511–516 (2018)
13. Panda, A.R., Mishra, M.: Smart Phone Purchase Prediction with 3-NN Classifier. J. Adv. Res. Dyn. Control Syst. 674–680 (2018)

Review and Comparison of Commonly Used Activation Functions for Deep Neural Networks

Tomasz Szandała

Abstract The primary neural networks' decision-making units are activation functions. Moreover, they evaluate the output of networks neural node; thus, they are essential for the performance of the whole network. Hence, it is critical to choose the most appropriate activation function in neural networks calculation. Acharya et al. (2018) suggest that numerous recipes have been formulated over the years, though some of them are considered deprecated these days since they are unable to operate properly under some conditions. These functions have a variety of characteristics, which are deemed essential to successfully learning. Their monotonicity, individual derivatives, and finite of their range are some of these characteristics. This research paper will evaluate the commonly used additive functions, such as swish, ReLU, Sigmoid, and so forth. This will be followed by their properties, own cons and pros, and particular formula application recommendations.

Keywords Deep learning · Neural networks · Activation function · Classification · Regression

1 Introduction

There exist many applications of deep learning neural network, such as voice analysis, speech or pattern recognition, and object classification. This has been demonstrated in its excellent performance in many fields. Moreover, it comprises a structure with hidden layers; this implies that the layers are more than one [1, 2]. However, from the study of deep learning structure, it can adapt in terms of real-world application with natural scenes due to its salient features. There exist only a few layers in the first deep learning model used for classification tasks; for example, there were only five layers in LeNet5 model [3].

T. Szandała (✉)
Wroclaw University of Science and Technology,Wrocław, Poland
e-mail: Tomasz.Szandala@pwr.edu.pl

A. K. Bhoi et al. (eds.), *Bio-inspired Neurocomputing*, Studies in Computational Intelligence 903, https://doi.org/10.1007/978-981-15-5495-7_11

Furthermore, there has been an increased depth due to the need for more advanced network models, applications, and available computing power rise [4]. For example, twelve layers in AlexNet [5], nineteen, or sixteen layers in VGGNet depending on variants [6], GoogleNet has twenty-two layers [7], and the largest ResNet architecture has one hundred and fifty-two layers [8]. Finally, Stochastic Depth networks have more than one thousand two hundred layers, which have been trainable proved. Thus, getting deep into neural networks will provide a better understanding of the hidden layers, which will help in boosting their performance and training. A neural cell output in the neural network is calculated by the activation unit. The derivative of the activation function is later used by the backpropagation algorithm. Thus, a differentiable activation function must be picked for analysis. This will enable smooth submission of the function to backpropagation weight updates hence avoiding zigzag formation as in the sigmoid function. Furthermore, Alom et al. [9] suggest that it should be easy to calculate an activation function spare completing power, an essential property in huge neural networks with millions of nodes. Thus, it is evident that the artificial neural network requires activation function as a critical tool in mapping response variable and inputs for nonlinear complex and complicated functions. It, therefore, shows that there is an introduction of systems with nonlinear properties in different fields [10]. However, the primary role of the activation function is the conversion of an input signal of a node in an A-NN to signal output.

The training on the neural network process becomes difficult and challenging when it has multiple hidden layers. Zigzagging weight, vanishing gradient problem, too complicated formula, or saturation problem in the neural network of the activation function are some of these challenges. This leads to a continuous learning process which might take a long time [11, 12]. Byrd et al. [13] discuss a comparison of different activation functions is made in this research paper both practically and theoretically. These activation functions include softplus, tanh, swish, linear, Maxout, sigmoid, Leaky ReLU, and ReLU. The analysis of each function will contain a definition, a brief description, and its cons and pros. This will enable us to formulate guidelines for choosing the best activation function for every situation.

Thus, this paper is unique since it entails real-world applications of activation functions. Hence, it has a summary of the current trends in the usage and use of these functions against the state of the art research findings in practical deep learning deployments. The complication in this paper will enable suitable decisions about how to choose the appropriate and best activation function and implementing it in any given real-world application. As indicated earlier, it is challenging to manage large test datasets of different activation functions. Thus, Banerjee et al. [14] discuss that tracking experiment process, running different experiments across several machines, and managing training data are some real-world applications of these functions. The analysis of different activation functions with individual real-world applications followed by a summary is shown below.

2 Base Activation Functions

In the real sense, activation functions can be expressed mathematically in a neural network's node. However, only a few of these functions are commonly used and well known for neural network analysis. Backpropagation algorithm [11, 12] multiplies the derivatives of the activation function. Hence, the picked-up activation function has to be differentiable [15]. Furthermore, the function should smoothly submit to backpropagation weight updates to avoid zigzagging, for example, in a sigmoid function [16]. Last but not least, an activation function should easy calculate the spare computing power, which is an important feature in extremely big neural networks consisting of millions of nodes. Below is the analysis of some functions with individual pros and cons, including real-world applications, which will lead to a critical comparison between them.

2.1 *Step Function*

The first question that the classification of the activation function should answer is whether or not the activation of the neuron should take place. An individual can only activate the neuron in a situation where the input value is higher than a given threshold value or leave it deactivated when the condition is not met. The figure below demonstrates how a step function can either be activated or deactivated.

$$f(x) = \begin{cases} 0, & \text{for } x \leqslant T \\ 1, & \text{for } x > T \end{cases}$$

Figure 1 follows a function $F(x) = 1$, with conditions: for x and for $x > T$. Thus, neural functions require logical functions to be implemented in them for effective functioning. Moreover, primitive neural networks commonly implement step function, and in this case, they must be with hidden layers, preferably a single layer. He et al. [17] suggest that this is done because the learning value of the derivative is not represented and also it will not have any effect in the future. Hence, the classification of this type of network is under linear separable problems, such as XOR gate, AND gate, and so forth. Therefore, a single linear line can be used to separate all classes 0 and 1 of neural networks. Thus, this function is not flexible hence has numerous challenges during training for network model analysis. Hence, it can only be used as an activation function of a single output network for binary classifications. These functions can only be represented by their derivatives, as shown in the graph below.

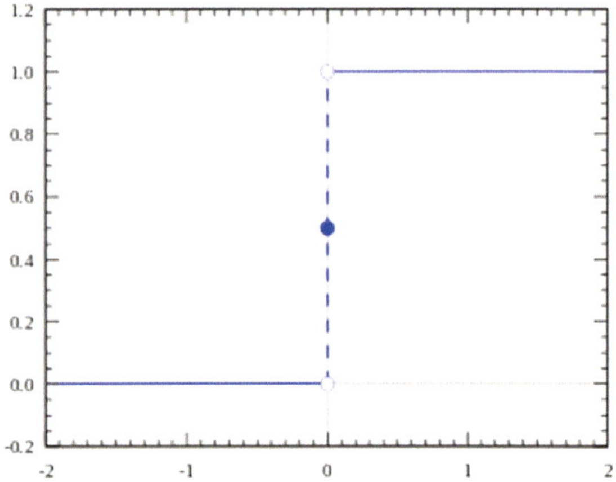

Fig. 1 Step function

2.2 Linear Activation Function

The zero gradients are the major problem in step function; thus, it does not update gradient during the backpropagation as a result of gradient/slope descent being unable to be progressive. Therefore, we can try applying a linear function instead of the step function. In consideration to the simplest example of a linear equation, the equation output can be equivalent to the input, but in more cases, the value of "a" vary with 1, where "a" is the proportional input activation. Equation 1 demonstrates a simple linear equation with the required variables.

$$f(x) = a * x, \text{ where } a \in \mathbf{R} \tag{1}$$

Many and various neurons can be activated at the same time using linear activation. In this case, if we decide to choose multiple classes, our interest should be on the one with the maximum value. However, there still exists an activation problem in this situation too. For instance, consider the derivative of the function as shown in Eq. 2.

$$f'(x) = a \tag{2}$$

Also, it appears that issue with gradient descent for training occurs in this function too. Thus, it has a constant derivative function in all situations. This implies that there is no relationship between its x and the gradient. Hence, the constant gradient makes the descent a constant too. Hara, Kataoka, and Satoh [18] suggest that in the case where we have a prediction error, changes are made by the backpropagation, which is constant and is independent of the respective weights used for analysis. Furthermore, this function is of no use in a multilayer network for deep learning. Thus, a linear function activates each layer. The activation interns are taken to the next level as an input. The weighted sum is calculated by the second layer in the input. This is

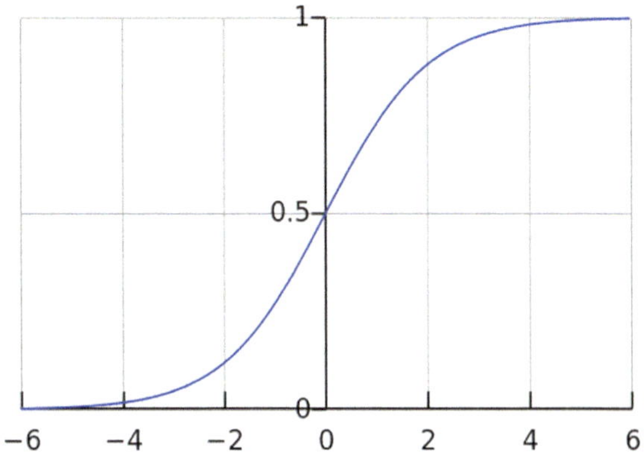

Fig. 2 The demonstration of a sigmoid activation function [2]

followed by a firing based on another linear function. In the case where we have a linear function, the last activation function of the final layer remains linear of the input of the first layer, no matter how many layers are added. Therefore, the multiple layers used in activation can be replaced by a single layer, which is a combination of a linear function and will remain linear. However, linear activation even make sense in the regression prediction of the output layer of networks. Thus, the linear function and its derivatives are expressed on the graph, as shown in Fig. 2.

However, there is still one place where linear activation makes sense: the output layer of networks used for predicting regression [19].

3 "S"-shaped Activation Functions

These types of activation functions used to be very common among artificial neural networks. The output of these activation functions is not linear. Thus, the range of the function is 0–1 forming an S shape. These activations functions are discussed below.

3.1 Sigmoid Function

A mathematical function with the features of a sigmoid curve is referred to as a sigmoid function. The combinations of the sigmoid function are not linear since this type of activation function is nonlinear. Figure 2 represents a demonstration of this function.

Thus, it makes sense to stack layers. This applies to non-binary activation as well. It also has a smooth gradient value. Hence, this makes it suitable for shallow networks like functions simulation. Equation 3 shows how sigmoid functions can be used in deep learning.

$$\sigma(x) = \frac{1}{1 + e^{-x}} \tag{3}$$

From the above equation, it can be observed that the Y values are very steep when the X values are between −2 and +2. Thus, there will be a significant change in Y for any slight change in X within this range [2]. This makes the function to have a tendency of bringing the Y values to either end of the curve; thus, enhancing the classifier by making clear prediction distinctions. This makes it even more suitable for shallow networks like logic functions simulation [20].

$$\frac{\mathrm{d}}{\mathrm{d}x}\sigma(x) = \sigma(x)(1 - \sigma(x)) \tag{4}$$

Moreover, another advantage of this function is that its range is (0, 1), unlike the linear function whose range is $(-\infty, \infty)$. Therefore, the range of activation is bounded, and it is easy to prevent blowing up during the activations. Hence, this is the most widely used activation functions, and it has been used for a long time. But research shows that sigmoid function is not perfect since the values of Y tend to respond slightly to the changes in X values toward either end of the function. This implies that there will be a small gradient at this point. Thus, this leads to a vanishing gradient problem, which is realized toward the near-horizontal part of the activation functions of the curve on either side. Hence, the gradient vanishes or reduces in size and cannot lead to any significant change since its value is minimal. In this case, the network can either learn drastically slowly or refuse to learn further; this will depend on how it is used until the gradient comes closer to the value limits of the floating-point.

Another problem with the sigmoid function is its non-zero centrality [21]; in other words, it always gives a positive result. Consider a sigmoid unit y with inputs x_1 and x_2, both from the sigmoid neuron, weighted w_1 and w_2 respectively. From the definition of a sigmoid function, it is clear that x_1 and x_2 are always positive, same as y. This generates the error signal which comes into y is during backpropagation, the error gradient will be either positive for w_1 and positive w_2 or both negative for w_1 and w_2. Liu et al. [22] demonstrate that when the optimal weight vector calls for an increase in w_1 and a decrease in w_2, the backpropagation procedure cannot improve both weights simultaneously in a single step, because it must either increase both or reduce both. Thus, the entire training procedure takes more steps to converge than it might need with a better activation function. The sigmoid activation function and its derivatives can also be represented in a graph, as shown in Fig. 3.

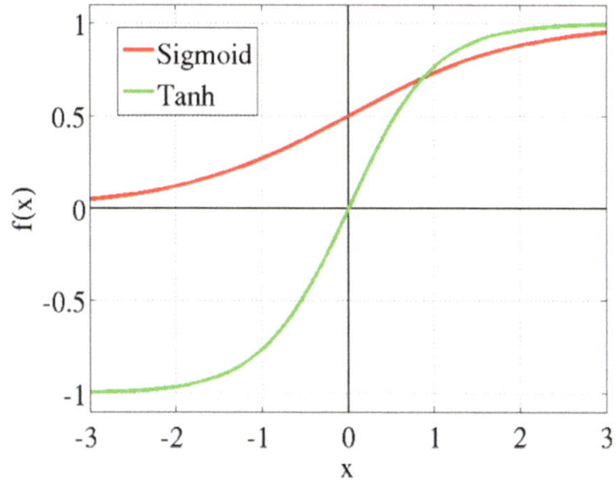

Fig. 3 Sigmoid and tanh functions comparison diagram

3.2 Hyperbolic Tangent Activation Function

The stuck of the neural network on the edge value may occur if we use only the sigmoid activation function. Thus, we need to apply the hyperbolic function as an alternative, which is also known as the tanh function. Figure 3 shows a comparison between the sigmoid and tanh function.

Mollahosseini et al. [23] suggest that the hyperbolic function is sigmoidal, just like in the case of sigmoid functions. However, for tanh the output values range between -1 and 1, which is just the sigmoid function curve extended. Hence, negative inputs of the hyperbolic functions will be mapped to a negative output as well as the input values that are nearing zero will also be mapped to output values nearing zero. Therefore, the network doesn't stuck due to the above features during training.

$$f(x) = \tanh(x) = \frac{e^x - e^{-x}}{e^x + e^{-x}} \tag{5}$$

Another reason why tanh is preferred to sigmoid is that the derivatives of the tanh are noticeably larger than the derivatives of the sigmoid near zero. However, when it comes to big data, an individual usually struggles to quickly find the local or global minimum, which is a beneficial feature of tanh derivative. In other words, an individual is supposed to minimize the cost function faster if tanh is used as an activation function. Last but not least, tanh diminishes the zigzagging problem of the only positive sigmoid [16].

$$\frac{\mathrm{d}}{\mathrm{d}x}\tanh(x) = 1 - \tanh^2(x) \tag{6}$$

However, a commonly used case is binary classification. Thus, using either sigmoid or tanh activation in the final layer produces a quantity that can be scaled from

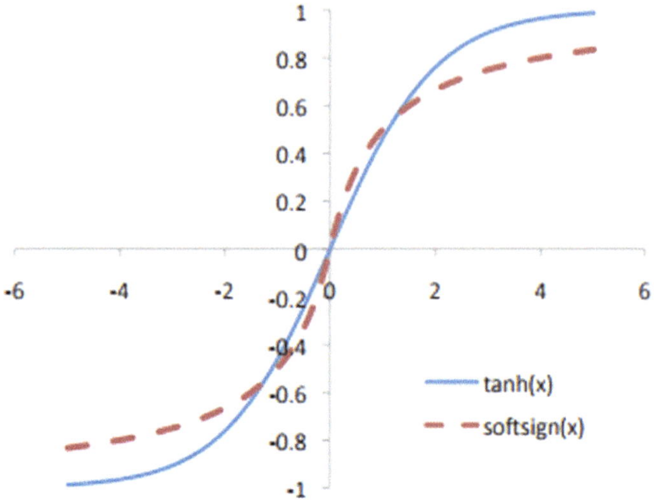

Fig. 4 Softsign versus tanh function

0 to 1. Hence, it is suitable to use a function that requires probabilities as inputs, such as cross-entropy. Therefore, tanh, just like sigmoid struggles with the vanishing gradient problem.

3.3 *Softsign Activation Function*

It is worth to mention a few attempts to upgrade "s"-shaped functions [24, 25]. Softsign is one of the most important functions, which is smoother than tanh activation function.

$$f(x) = \frac{x}{1 + |x|} \tag{7}$$

Moreover, this function grows poly-nominally rather than exponentially Eq. 7. This gentler nonlinearity results in better and faster learning due to lack of struggling with vanishing gradient. Thus, researchers have found that Softsign has prevented neurons from being saturated, resulting in more effective learning. However, it is more expensive to compute than tanh; in other words, it has more complex derivatives. Additionally, its gradient sometimes yields extremely low/high values, such that we can consider it as a sigmoid on steroids [26–28] (Fig. 4).

$$f(x) = \frac{1}{(1 + |x|)^2} \tag{8}$$

Table 1 Accuracy per amount of hidden layer quantity

Number of hidden layers	Accuracy for sigmoid (%)
1 hidden layer	96.81
2 hidden layers	97.18
3 hidden layers	97.29
4 hidden layers	97.10

3.4 Vanishing Gradient Problem

Just like all "s"-shaped functions, vanishing gradient function has a burden of vanishing gradient [29]. For instance, a sample is taken from MNIST dataset and is classified using 1, 2, 3, and 4 hidden layers each with 30 nodes in it. The sigmoid function is used as the vanishing gradient problem since it is the best visible with this function. Thus, after each training phase, the network is validated against a validating set. Table 1 shows the accuracy per amount of the hidden layer quantity.

Firstly, the accuracy growths a little for 2 and 3 layers, but at four hidden layers it drops down, close to an extended network. Job classification is performed better due to an extra hidden layer, which enables the network to more complex classification functions [8]. The network is analyzed to determine the insight of what might be wrong with the function. The error or the final layer is propagated back to the previous layer. Each time propagation occurs, a gradient of the gradient is passed, and so a smaller value is also being passed. For this to be proven, there must be defined a common technique of comparing the speed of learning in the hidden layers.

At first, accuracy growths a little for 2 and 3 layers, but at 4 hidden layers it drops down, close to a flat network. This behavior seems wrong. Intuitively, extra hidden layers are supposed to make the network able to learn more complex classification functions, therefore performing a better job assigning the classes. To get some insight into what goes wrong, we have to analyze how the network learns. The error or final layer is propagated back to previous layers. Each time propagation occurs, a gradient of gradient is passed, and so a smaller value is also being passed.

To prove this, we need to have a global way of comparing the speed of learning in each subsequent hidden layer. To do this, we denote the gradient as a derivative of the cost function $\delta_j^l = \partial C/\partial b_j^l$, i.e., the gradient value for the j-th neuron in the l-th layer. We consider the gradient δ^1 as a vector whose inputs determine how quickly the first hidden layer trains and δ^2 as a vector whose entries determine how quickly the second hidden layer learns. Now, the lengths of these vectors can be utilized as global measures of the speed at which the layers learn. For instance, the length $||\delta^1||$ measures the time at which the first hidden layer learns, while the length $||\delta^2||$ measures the velocity at which the second hidden layer learns. With this definition, we can compute $||\delta^1|| = 0.07$ and $||\delta^1|| = 0.31$. This confirms our previous assumption: The neurons in the subsequent hidden layer really learn much faster than the neurons in the prime hidden layer. Finally, for all 4 hidden layers the respective speeds of

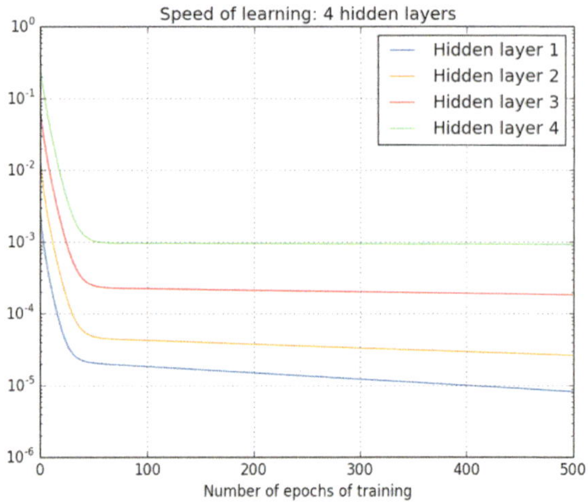

Fig. 5 Learning speed ($||\delta^l||$) for each layer during the learning process

training are 0.005, 0.029, 0.091, and 0.345. The pattern is evident: Earlier layers learn much slower than later layers. This is even better visible if we analyze learning rate throughout the entire learning process (Fig. 5).

To sum up, deep neural networks suffer from a problem known as vanishing gradient, which can be avoided by using functions with static or wider-range derivatives.

4 Rectified Linear Unit Activation Functions

This type of activation function is responsible for transforming the weighted input that is summed up from the node to the strict output or proportional sum. These functions are piecewise linear functions that usually output the positive input directly; otherwise it the output is zero. Types of rectified linear unit activation functions are discussed below.

4.1 Basic Rectified Linear Unit (ReLU)

The rectifier is an activation that assigns zero to values and value itself is above zero (Eq. 9). This is also known as a ramp function and is analogous to half-wave rectification in electrical engineering (Fig. 6). However, the activation function was first introduced in a dynamical network with strong biological motivations and mathematical justifications [30, 31].

$$f(x) = x^+ = \max(0, x) \tag{9}$$

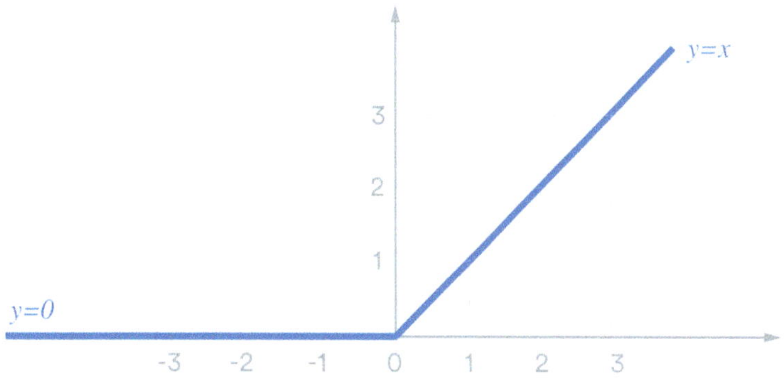

Fig. 6 Rectifier linear unit function

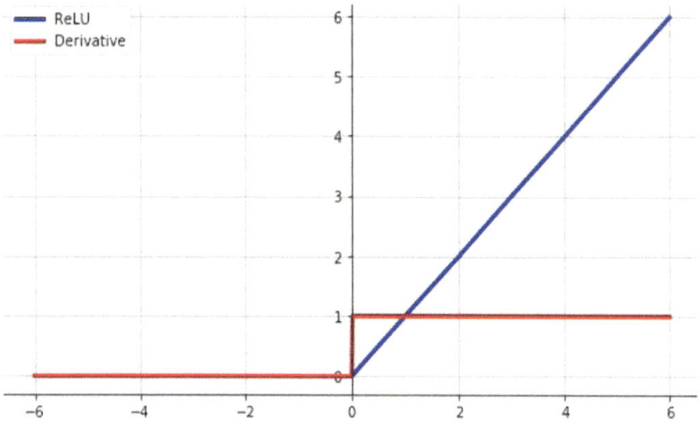

Fig. 7 ReLU function and its derivative

Moreover, it was proven for the first time in 2011 as enabling the more efficient training of deeper networks, compared to the commonly used activation functions before 2011, for instance, the sigmoid or the hyperbolic tangent. Ping et al. [32] suggest that the rectifier was the most popular activation function for the deep neural network as per research in 2018. Finally, calculating the function result and gradient is an easy task because the forward and backpropagation steps quickly. The results from the research show that ReLU is six times faster than other well-known activation functions [5]. It is noted that for simple regression problems with neural networks, tanh can be superior to ReLU [33]. However, any function approximated with ReLU activation function is always piecewise linear.

Moreover, it takes a lot of piecewise linear functions to fit a smooth function like sine. Meanwhile, tanh is very smooth and does not take as many tanh primitives to build something that closely resembles a sine wave. Thus, for classification problems, especially using multiple convolution layers, ReLU and its minor variants are hard to beat. The graph below shows, that the ReLU function and its derivative (Fig. 7).

4.2 Dying ReLU Problem

Unfortunately, ReLU has a significant problem known as "dying ReLU." However, this occurs under certain conditions [34]. Suppose there is a neutral network with weights distribution in the form of a low-variance Gaussian centered at +0.1. Thus, by this condition, most inputs are positive and cause ReLU node to fire. Then, suppose that during a particular backpropagation, there is a large magnitude of gradient passing back to the node. Since the gate is open, it will move this large gradient backward to its inputs. This causes a relatively significant change in the function that computes input [35]. Implying that the distribution has moved representing a low-variance Gaussian centered at −0.1. All the inputs are negative, making the neuron inactive, omitting the weight updates during backpropagation (Eq. 10). A relatively small change in "R"s input distribution leads to a qualitative difference in the nodes' behavior. However, the problem is that a close ReLU cannot update its input parameters. One could also consider trying to "recover" dead ReLUs by detecting them and randomly re-initializing their input parameters. However, this would slow down learning but might encourage more efficient use of metrics. Therefore, to bypass this problem, without an additional mechanism, a modification has been proposed, called Leaky ReLU.

$$f'(x) = \begin{cases} 0, & \text{for } x \leqslant 0 \\ 1, & \text{for } x > 0 \end{cases} \tag{10}$$

4.3 Dying ReLU—Example

Dying ReLU does not have to appear always, but let's take an example where it may occur. One of the simplest networks is XOR gate network. However, it is evident that the dying ReLu problem has indeed occurred in the simulations shown in Fig. 8.

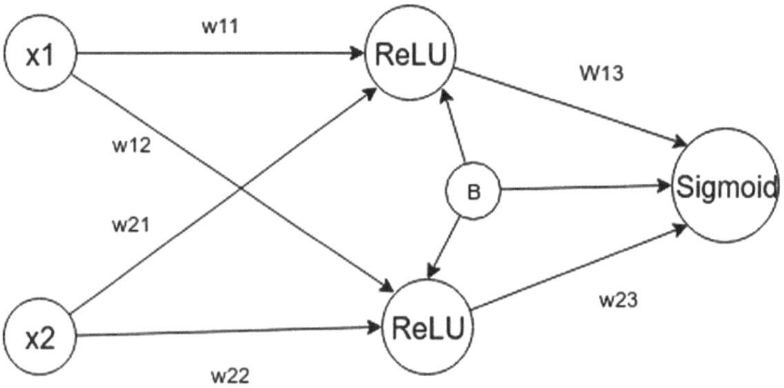

Fig. 8 Sample network fulfilling XOR gate behavior

Table 2 Potential random assigned weights values in the sample XOR network

w11	w12	b1
0.55	0.55	−1
w21	w22	b2
1	1	−1
w31	w32	b3
1	1	0.5

Table 3 The output of the pre-trained XOR network

Input 1	Input 2	Expected (exp)	Real output ($real$)
1	1	0	0.83
1	0	1	0.62
0	1	1	0.62
0	0	0	0.62

Because we wanted the output to be either 0 or 1, for output function, we chose sigmoid and ReLUs that were put in the hidden layer nodes. Let's set initial weights as Table 2.

Weights are chosen randomly for each network, but in this particular example, we adjusted them to invoke "dying ReLU." Of course, they are real, potential values that can be assigned. In Table 3, it can be concluded that the results are not perfect. Therefore, we need to adjust them using backpropagation.

Where in the Eq. (11),

$$w'_{31} = w_{31} + \eta * f'_3(x) * (exp - real) * f_1(x) \tag{11}$$

we have measured the weight between the first neuron and final neuron, updated by adding an adequate modifier. If we continue to measure all weights as shown in the following Eq. (12), we receive the following:

$$w'_{11} = w_{11} + \eta * f'_3(x) * (exp - real) * f'_1(x) * w_{13} * input1 \tag{12}$$

Now, we can clearly see that the problem of "dying ReLU" affects neuron one Table 4. Therefore, this neuron will never evolve, and we cannot achieve XOR gate behavior. While values for this example chosen, are not impossible, nor disturbing one's prohibition.

Table 4 Post training weights value in sample XOR network

w11	w12	b1
0.49	0.49	−1.05
w21	w22	b2
0.94	0.94	−1.05
w31	w32	b3
0.99	0.94	0.44

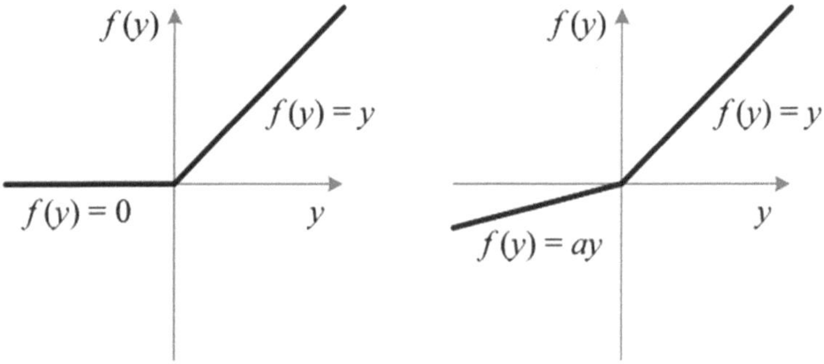

Fig. 9 ReLU compared to leaky ReLU function

4.4 Leaky Rectified Linear Unit (ReLU)

To keep away the dead nodes while using ReLU neurons, we need to reduce the possibility of returning 0 value as output. To achieve this, a small multiplier α, e.g., 0.01, can be introduced for the values below 0 [36]. This factor causes the inclination of the negative part of a function, therefore preventing proclivity for death Fig. 9.

$$f(x) = \begin{cases} \alpha * x, & \text{for } x \leqslant 0 \\ x, & \text{for } x > 0 \end{cases}$$

Comparing Leaky (also known as Parametric) ReLUs to standard ReLU layer gives no activation value for negative values. Thus, one can take advantage to save some energy in a hardware implementation, like the minimized Internet of Things devices. Still, it does not make a difference when the network is run on a GPU. Figure 9 pictures a leaky ReLU function and its derivative.

4.5 Maxout

For leaky ReLU, it might be hard to choose the parameter for distorting the negative part arbitrarily. Because of that, a more generalized approach has been proposed by

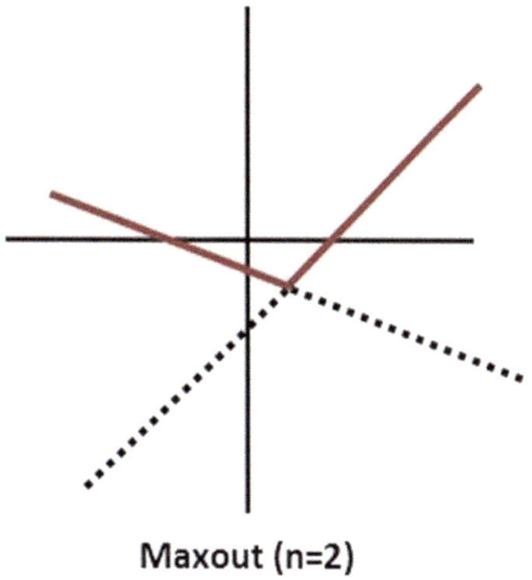

Fig. 10 The maxout activation function $n = 2$

the researchers. The Maxout neuron computes the function for the maximum out of two independent weight sets [37].

$$f(x) = \max(\overrightarrow{w_1} * \overrightarrow{x} + b_1, \overrightarrow{w_2} * \overrightarrow{x} + b_2) \quad (13)$$

However, for every single neuron, Maxout doubles the number of parameters, unlike the ReLU neurons. Moreover, all the benefits of ReLU units, such as no saturation and the linear regime of operations, are embraced by the Maxout neuron. While it can provide the best results among all ReLU-like functions, it comprises the highest cost of training; thus, should be used as the last resort [38]. This "last resort" solution can be recognized by the phone speech where the Maxout function is successfully applied [39]. Figure 10 shows the Maxout when $n = 2$.

4.6 *Softplus Activation Function*

Another alternative to dead ReLU is softplus. It was first described as a better alternative to $sigmoid$ or $tanh$. The output produced by the softplus function ranges from 0 to $\int$, whereas the $tanh$ and $sigmoid$ produces an output with lower and upper limits. The softplus equation is shown in Eq. 14.

$$f(x) = \ln(1 + e^x) \quad (14)$$

One of the biggest pros for softplus is its smooth derivative used in backpropagation. The derivative of softplus function is equal to sigmoid function. Moreover, the local minima of greater or equal quality are attained by the network trained with

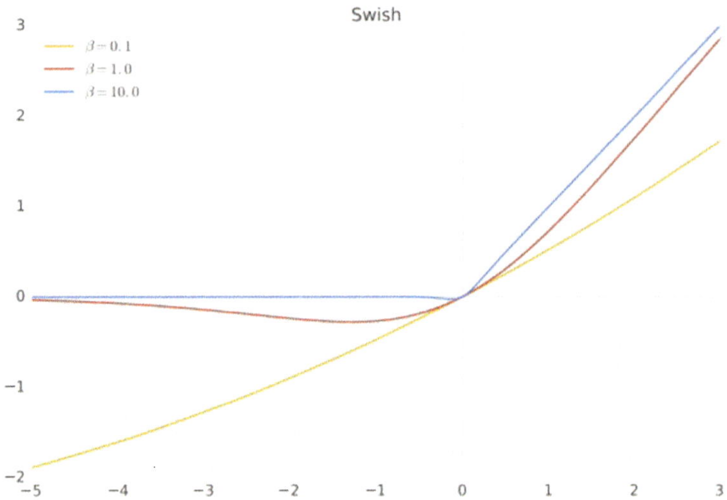

Fig. 11 Different values of β in Swish function

the rectifier activation function despite the hard threshold at zero [40]. However, the derivative of the softplus is equal to a sigmoid function. Figure 11 shows a softplus function.

4.7 Swish Activation Function

According to research by Google brain team [41], swish activation function is an alternative to ReLU. Though for both backpropagation and feed-forwarding, the cost is much higher for the computation of the function [42]. Although ReLU having many limitations, it plays a crucial role in deep learning studies. However, research indicates that ReLU is over-performed by the new activation function for the deep network. Swish activation function is represented by the Eq. 14.

$$f(x) = x * \sigma(x) = \frac{x}{1 + e^{-x}} \tag{15}$$

Moreover, some authors proposed [42] a non-zero β parameter for enabling a basic swish, the equation for this modification is as shown in Eq. 16.

$$f(x) = \beta x * \sigma(\beta x) = \frac{\beta x}{1 + e^{-\beta * x}} \tag{16}$$

The figure below shows E-swish producing different graphs using different values of beta Fig. 11.

A linear function will be attained if β is moved closer to zero; however, the function will look like ReLU if β gets closer to one. However, the beta used in this simulation is among the learned as well as a parameter.

The Swish function is also easily derivable Eq. 17.

$$f'(x) = f(x) + \sigma(x) - \sigma(x) * f(x) = \frac{e^{-x}(x+1)+1}{(1+e^{-x})^2} \tag{17}$$

This function can deal with vanishing gradient problem that sigmoid is unable to. Moreover, experiments prove that swish is doing better than the ReLU—so far the most efficient activation function of deep learning. It is a fact, the computation of the function has much higher cost than ReLU or even its siblings for both feed-forwarding and backpropagation [43].

5 Comparisons and Results Analysis

This section will compare various functioning and properties of the activation functions. It will involve numerous properties of the AFs discussed above summary, which will lead to an appropriate conclusion on the paper. Also, real-world applications of the AFs will be considered here other than theoretical analysis. The criteria for this analysis will be based on the speed of training and accuracy o f the classification. The table below shows the analysis of the various AFs with their equations and the range. All hidden nodes have the same activation function for each series of tests of given formula, with the exception of the final layer which always uses softmax AF. The network has been implemented using keras and tensorflow frameworks for Python 3. All operations were done on a single GPU unit: Nvidia GeForce GTX 860M.

The experiment was performed with dataset CIFAR-10 [44]. The CIFAR-10 dataset consists of 60,000 32 × 32 color images splitted into ten classes, with 6000 images per class. There are 50,000 training images and 10,000 test images.Training consists of 25 epochs. The final accuracy is not perfect, nonetheless aim is to compare efficiency of different types of activation functions in the same model, therefore for the experiment a simple network [45] with just two convolution layers has been implemented. In order to alleviate randomness, there are three series of each AF application and the final result is the average value for given AF.

Results are shown in three diagrams. First one Fig. 12 presents the most important parameter: rate how many images were correctly classified. It appears that ReLU and Leaky ReLU were the most successful, since all other networks completed task with less than 70% accuracy Table 5. This proves that that ReLU is still reliable and even "dying ReLU" is not able to decline overall performance.

Regarding time performance, ReLU is unquestionably leader. It took over 5% less time to train than second one—sigmoid and 15% less time than its sibling—Leaky ReLU (Fig. 13). In addition for ReLU-network, it took only 2.4 s to classify ten

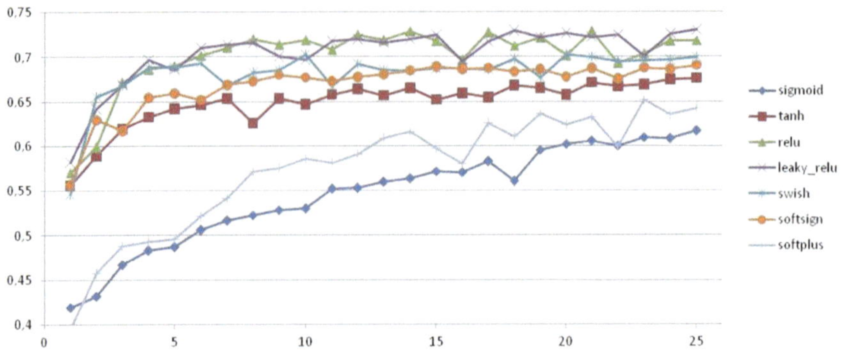

Fig. 12 Network accuracy over training for each activation function

Table 5 Final ratio of correctly classified images

	Sigmoid	tanh	Relu	Leaky_relu	Swish	Softsign	Softplus
Final accuracy	0.6166	0.6757	0.7179	0.7295	0.6989	0.6901	0.6598

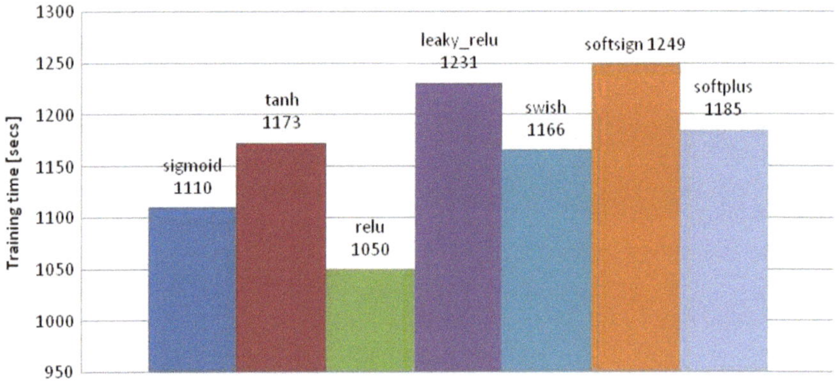

Fig. 13 Average training time for AF model in seconds

thousand images, while the second in row—softplus almost 2.5 s spent on this task (Fig. 14). This empirically proves superiority of basic ReLU activation function.

6 Conclusions

This paper shows that there is no ultimate answer for questions like "which activation function should I choose?" However, after this comprehensive summary of activation functions used in deep learning, we can make a few but certain recommendations

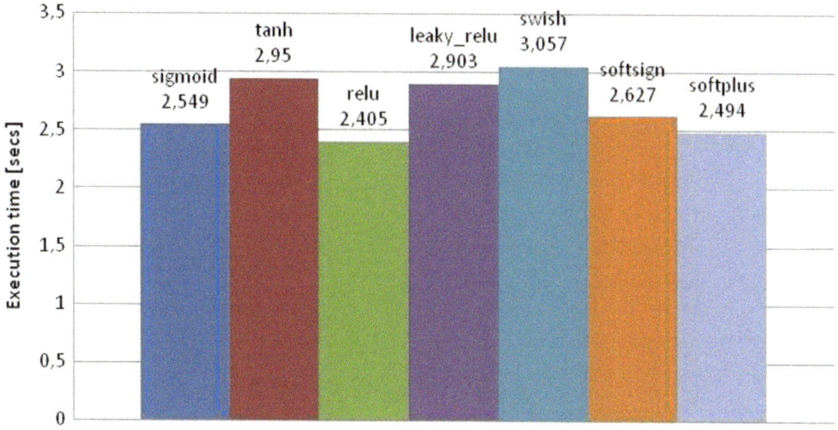

Fig. 14 Average time to classify 10,000 images with each AF

Table 6 Recommendation when to use which activation function in deep neural networks

Function	Comment	When to use?
Step	Does not work with backpropagation algorithm	Rather never
Sigmoid	Prone to the vanishing gradient function and zigzagging during training due to not being zero centered	Can fit into Boolean gates simulation
Tanh	Also prone to vanishing gradient	In recurrent neural network
Softsign		Rather never
ReLU	The most popular function for hidden layer. Although, under rare circumstances, prone to the "dying ReLU" problem	First to go choice
LReLU	Comes with all pros of ReLU, but due to not-zero output will never "die"	Use only if You expect "dying ReLU" problem
Maxout	Far more advanced activation function than ReLU, immune to "dying," but much more expensive in case of computation	Use as last resort
Softplus	Similar to ReLU, but a bit smoother near 0. Comes with comparable benefits as ReLU, but has more complex formula, therefore network will be slower	Rather never
Swish	Same as leaky ReLU, but according to above research it does not outperform ReLU. Might be more useful in networks with dozens of layers	Worth to give a try in very deep networks
SoftMax		For output layer in classification networks
OpenMax		For output layer in classification with open classes possibility

based on the provided theory as shown in the table above. Thus, this paper summarizes comprehensively the activation functions applied in deep learning (DL) (Table 6), highlighting the current trends in the application of these functions, which is most essential and has never been published in any literature. It started by a presentation of a brief introduction on the activation function and deep learning, which was then followed by analysis of different AFs and a discussion of some applications fields of these functions can be used based on systems and architectures in the development of deep neural networks. Moreover, the activation functions can adjust to the process better or worse since they have the ability to improve the learning of the certain data patterns. Therefore, the functions develop over the years and still there is a need to do a proper research before applying any of these functions in deep learning for each project separately and this paper is only a guidance.

References

1. Deng, L.: A tutorial survey of architectures, algorithms, and applications for deep learning. APSIPA Trans. Signal Inf. Process. **3**, e2 (2014)
2. Hertz, J.A.: Introduction to the theory of neural computation. CRC Press (2018)
3. LeCun, Y., Boser, B., Denker, J.S., Henderson, D., Howard, R.E., Hubbard, W., Jackel, L.D.: Backpropagation applied to handwritten zip code recognition. Neural Comput. **1**, 541–551 (1989)
4. He, K., Zhang, X., Ren, S., Sun, J.: Deep residual learning for image recognition. Comput. Vis. Pattern Recognit. (CVPR) **7** (2015)
5. Krizhevsky, A., Sutskever, I., Hinton, G.E.: Imagenet classification with deep convolutional neural networks. In: Proceedings of the 25th International Conference on Neural Information Processing Systems, vol. 1, pp. 1097–1105 (2012). NIPS'12, Curran Associates Inc., USA
6. Simonyan, K., Zisserman, A.: Very deep convolutional networks for large-scale image recognition (2014). arXiv:1409.1556
7. Szegedy, C., Liu, W., Jia, Y., Sermanet, P., Reed, S., Anguelov, D., Erhan, D., Vanhoucke, V., Rabinovich, A.: Going deeper with convolutions. In: Computer Vision and Pattern Recognition (CVPR), pp. 1–17
8. Piczak, K.J.: Recognizing bird species in audio recordings using deep convolutional neural networks. In: CLEF (Working Notes), pp. 534–543
9. Yakopcic, C., Westberg, S., Van Esesn, B., Alom, M.Z., Taha, T.M., Asari, V.K.: The history began from alexnet: a comprehensive survey on deep learning approaches (2018)
10. Huang, G., Sun, Y., Liu, Z., Sedra, D., Weinberger, K.Q.: Deep networks with stochastic depth. In: Leibe, B., Matas, J., Sebe, N., Welling, M. (eds.), ECCV (4), volume 9908 of *Lecture Notes in Computer Science*, pp. 646–661. Springer (2016)
11. Robbins, H., Monro, S.: A stochastic approximation method. Ann. Math. Statist. **22**, 400–407 (1951)
12. Nielsen, M.A.: Neural Networks and Deep Learning, Determination Press (2015)
13. Byrd, R.H., Hansen, S.L., Nocedal, J., Singer, Y.: A stochastic quasi-newton method for large-scale optimization. SIAM J. Opt. **26**, 1008–1031 (2016)
14. Banerjee, A., Dubey, A., Menon, A., Nanda, S., Nandi, G.C.: Speaker recognition using deep belief networks (2018). arXiv:1805.08865
15. Hecht-Nielsen, R.: Theory of the backpropagation neural network. In: Wechsler, H. (ed.), Neural Networks for Perception, pp. 65–93. Academic Press (1992)
16. LeCun, Y., Bottou L., Orr, G.B., Müller, K.R.: Efficient BackProp, pp. 9–50. Springer, Berlin, Heidelberg

17. He, K., Zhang, X., Ren, S., Sun, J.: Delving deep into rectifiers: surpassing human-level performance on imagenet classification. In: Proceedings of the IEEE International Conference on Computer Vision, pp. 1026–1034
18. Hara, K., Kataoka, H., Satoh, Y.: Learning spatio-temporal features with 3d residual networks for action recognition. In: Proceedings of the IEEE International Conference on Computer Vision, pp. 3154–3160
19. Godfrey, L.B., Gashler, M.S.: A continuum among logarithmic, linear, and exponential functions, and its potential to improve generalization in neural networks. In: 7th International Conference on Knowledge Discovery and Information Retrieval, pp. 481–486
20. Neal, R.M.: Connectionist learning of belief networks. Artif. Intell. **56**, 71–113 (1992)
21. Karpathy, A.: Yes you should understand backprop. https://medium.com/@karpathy/yes-you-should-understand-backprop-e2f06eab496b (2016). Accessed 30 Nov 2018
22. Liu, W., Wang, Z., Liu, X., Zeng, N., Liu, Y., Alsaadi, F.E.: A survey of deep neural network architectures and their applications. Neurocomputing **234**, 11–26 (2017)
23. Mollahosseini, A., Chan, D., Mahoor, M.H.: Going deeper in facial expression recognition using deep neural networks. In: 2016 IEEE Winter Conference on Applications of Computer Vision (WACV), pp. 1–10. IEEE
24. Glorot, X., Bengio, Y.: Understanding the difficulty of training deep feedforward neural networks. In: Teh, Y.W., Titterington, M. (eds.), Proceedings of the Thirteenth International Conference on Artificial Intelligence and Statistics, volume 9 of Proceedings of Machine Learning Research, pp. 249–256. PMLR, Chia Laguna Resort, Sardinia, Italy (2010)
25. Elliott, D.L.: A better activation function for artificial neural networks, Maryland Publishing Unit (1998)
26. Turian, J., Bergstra, J., Bengio, Y.: Quadratic features and deep architectures for chunking. In: Proceedings of Human Language Technologies: The 2009 Annual Conference of the North American Chapter of the Association for Computational Linguistics, Companion Volume: Short Papers, NAACL-Short '09, Association for Computational Linguistics, pp. 245–248. Stroudsburg, PA, USA (2009)
27. Gibiansky, A., Arik, S.O., Kannan, A., Narang, S., Ping, W., Peng, K., Miller, J.: Deep voice 3: scaling text-to-speech with convolutional sequence learning. In: International Conference on Learning Representations, ICLR, pp. 1094–1099
28. Farzad, A., Mashayekhi, H., Hassanpour, H.: A comparative performance analysis of different activation functions in lstm networks for classification. Neural Comput. Appl. (2017)
29. Nielsen, M.A.: Neural networks and deep learning, Determination Press (2015)
30. Hahnloser, R., Sarpeshkar, R., Mahowald, M.A., Douglas, R., Sebastian Seung, H.: Digital selection and analogue amplification coexist in a cortex-inspired silicon circuit. Nature **405**, 947–951 (2000)
31. Hahnloser, R.H.R., Seung, H.S., Slotine, J.-J.: Permitted and forbidden sets in symmetric threshold-linear networks. Neural Comput. **15**, 621–638 (2003)
32. Ping, W., Peng, K., Gibiansky, A., Arik, S.O., Kannan, A., Narang, S., Raiman, J., Miller, J.: Deep voice 3: scaling text-to-speech with convolutional sequence learning (2017). arXiv:1710.07654
33. Chigozie Enyinna Nwankpa, A.G., Winifred Ijomah, S.M.: Activation functions: comparison of trends in practice and research for deep learning (2018)
34. Maas, A.L.: Rectifier nonlinearities improve neural network acoustic models. In: Proceedings 30th International Conference on Machine Learning, pp. 1–6
35. Bach, F.: Breaking the curse of dimensionality with convex neural networks. J. Mach. Learn. Res. **18**, 629–681 (2017)
36. He, K., Zhang, X., Ren, S., Sun, J.: Delving deep into rectifiers: surpassing human-level performance on imagenet classification. In: IEEE International Conference on Computer Vision (ICCV 2015), vol. 1502 (2015)
37. Goodfellow, I., Bengio, Y., Courville, A.: Deep Learning. MIT Press (2016)
38. Zhang, Y., Pezeshki, M., Brakel, P., Zhang, S., Laurent, C., Bengio, Y., Courville, A.: Towards end-to-end speech recognition with deep convolutional neural networks. In: Interspeech (2016), pp. 410–414

39. Tóth, L.: Phone recognition with hierarchical convolutional deep maxout networks. EURASIP J. Audio Speech Music Process. **25** (2015)
40. Glorot, X., Bordes, A., Bengio, Y.: Deep sparse rectifier neural networks. In: Gordon, G., Dunson, D., Dudík, M. (eds.) Proceedings of the Fourteenth International Conference on Artificial Intelligence and Statistics, volume 15 of Proceedings of Machine Learning Research, pp. 315–323. PMLR, Fort Lauderdale, FL, USA (2011)
41. Ramachandran, P., Zoph, B., Le, Q.V.: Searching for activation functions (2017). arXiv:1710.05941
42. Zoph, B.: Swish: a self-gated activation function (2017)
43. Sharma, J.: Experiments with swish activation function on mnist dataset, Medium Corporation (2017)
44. Krizhevsky, A., Nair, V., Hinton, G.: Cifar-10, Canadian Institute for Advanced Research (2015)
45. Szandała, T.: Benchmarking comparison of swish versus other activation functions on cifar-10 imageset. In: International Conference on Dependability and Complex Systems, pp. 498–505. Springer

The hDEBSA Global Optimization Method: A Comparative Study on CEC2014 Test Function and Application to Geotechnical Problem

Sukanta Nama, Apu Kumar Saha, and Arijit Saha

Abstract In geotechnical engineering, investigation of seismic earth pressure coefficient is a fundamental theme of study for retaining wall. During this study, the seismal active earth pressure coefficient on the rear of the wall supporting $c - \Phi$ backfill has been formulated by the help of limit equilibrium method. This type of problems is highly complex nonlinear optimization problems. Therefore, it will very be difficult to analyze the problem using classical optimization techniques. A hybrid algorithm called hDEBSA has been discussed in the present study which was proposed by joining the parts of DE with the segments of BSA calculation to investigation the pseudo-static seismic dynamic earth pressure coefficient. In hDEBSA, a modification of parameter of DE and BSA has been performed through self-adaption-based. The proficiency of the hDEBSA has been checked through CEC2014 test functions and applied to analyze the coefficient of seismic active earth pressure on the rear of the retaining wall supportive $c - \Phi$ backfill. The result obtained by this algorithm is compared with state-of-the-art other algorithms and are found to be in agreement. The achieved results of active earth pressure coefficient are in contrast with different results available found in the literature. Additionally, the impact of seismic parameters, soil and wall parameters on the earth pressure coefficient has been investigated.

Keywords Global optimization · Hybrid algorithm · $C - \Phi$ soil · Retaining wall · Active earth pressure

---

S. Nama
Department of Mathematics, Ramthakur College, Agartala, A.D. Nagar, West Tripura 799003, India
e-mail: sukanta1122@gmail.com

A. K. Saha (✉)
Department of Mathematics, National Institute of Technology Agartala, Tripura 799046, India
e-mail: apusaha_nita@yahoo.co.in

A. Saha
Department of Civil Engineering, National Institute of Technology Agartala, Tripura 799046, India
e-mail: sahaarijit20@gmail.com

A. K. Bhoi et al. (eds.), *Bio-inspired Neurocomputing*, Studies in Computational Intelligence 903, https://doi.org/10.1007/978-981-15-5495-7_12

1 Introduction

Global optimization method is efficient and robust tool for dealing with the complicated nonlinear problems of optimization. Based on different natural phenomena, many authors proposed different optimization algorithms. Some of the global optimization algorithms are genetic algorithm (GA) [1], differential evolution (DE) [2], particle swarm optimization (PSO) [3, 4], artificial bee colony (ABC) [5], harmony search (HS) [6, 7] algorithm, backtracking search optimization algorithm (BSA) [8], gravitational search algorithm (GSA) [9], etc. Since these algorithms do not guarantee to give the optimum solution in a finite amount of time, authors are trying constantly to improve the performance of the algorithms for enhancing their performances in respect of both solution quality and converging time. For this reason, many modified versions of these algorithms have been developed and still developing. For example, some of the improved versions can be seen in the references [10–18].

Recently, many authors attempted to enhance the overall performance of the algorithms by combining two algorithms or the component(s) of one algorithm with the component(s) of another algorithm. Such type of algorithms developed by combination of two or more algorithms is known as hybrid algorithms. Some of the hybrid algorithms can be seen in references [19–26] etc.

These improved as well as hybrid methods have been effectively incorporated to resolve different complicated problems of optimization on almost all branches of science and engineering. Such optimization algorithms are widely used in civil engineering to solve various problems with optimization. Some of the optimization problems are slope stability problems [27–30], foundation-related problem [31–34], retaining wall problem [16, 35, 36], etc.

One of the most critical optimization issues in civil and geotechnical engineering is retaining the wall problem. Limit equilibrium approaches are commonly used to derive optimization problem formulation instead of limits analysis and finite element analysis. For analysis of the algorithm's performance, one problem of optimization is considered which is the seismic active soil pressure of the wall [37, 38].

The principal contribution of this work is outlined here below:

(i) A new hybrid algorithm (called hDEBSA) has been studied which was proposed by the combination of DE and BSA algorithm.
(ii) Since the efficiency of the algorithm relies upon the control parameter, the output and convergence rate of the algorithm have been enhanced by using self-adjustment-based control parameters.
(iii) In the analysis of the retaining wall problem, the hDEBSA was applied.

The rest of the piece of the paper is composed as follows: Sect. 2 addresses the description of the basic algorithm; Sect. 3 shows the new hybrid algorithm; Sect. 4 describes the problem of geotechnical optimization in the real world; the efficiency and accuracy of the hybrid approach for CEC2014 to solve well-known optimization problems and geotechnical optimization problems in real world are shown in Sect. 5; and finally, Sect. 6 offers a description of this paper's findings and some possible directions for science.

2 Overview of Basic Algorithms

Since the heuristic global optimization algorithms are not widely used in geotechnical discipline, the following sub-sections provide a short portrayal of the DE and BSA.

2.1 *Overview of DE*

The difference evolution, which is based on the Darwin theory of evolution, is a stochastic evolutionary algorithm. There are two control parameter parameters, namely a weighting coefficient or scaling factor (F) which is utilized to produce new preliminary solutions and a crossover rate (CR), used to decide the amount of a preliminary solution ought to be embraced into the present solution, just as the DE conspire which is the significant thought behind DE for creating preliminary vectors to establish how much a trial solution should be incorporated into the current solution and a DE scheme that is the key idea behind DE for the generation of test vectors. DE has attempted to improve performance when performing the optimization process through rehashed cycles of mutation operator, crossover operator, and selection operator. The DE implementation process is as follows:

2.1.1 Mutation Operator

The mutation operation produces a mutant vector $v_{i,j}$ corresponding to the target vector $X_{i,j}$ in the existing population. The mutant vector $v_{i,j}$ is generated using some mutation operator (strategy) for every target vector $X_{i,j}$. During the literature review, some mutant vector generation strategy has been found. These are presented below:

$$\text{“DE/rand/1”} : v_{i,j} = X_{i,j}^{r_1} + F * \left(X_j^{r_2} - X_j^{r_3} \right) \tag{1}$$

$$\text{“DE/current-to-rand/1”} : v_{i,j} = X_j^{r_1} + K * \left(X_j^{r_1} - X_j^{r_3} \right) + F * \left(X_j^{r_4} - X_j^{r_5} \right) \tag{2}$$

$$\text{“DE/best/2”} : v_{i,j} = X_j^{\text{best}} + F * \left(X_j^{r_1} - X_j^{r_2} \right) + F * \left(X_j^{r_3} - X_j^{r_4} \right) \tag{3}$$

“DE/rand-to-best/1” or “DE/target-to-best/1”:

$$v_{i,j} = X_j^{r_1} + K * \left(X_j^{\text{best}} - X_j^{r_3} \right) + F * \left(X_j^{r_4} - X_j^{r_5} \right) \tag{4}$$

Here, the indices r_1, r_2, r_3, r_4 and r_5 are five different integer numbers that are chosen from the range [1, NP]. The weighting coefficient $F \in [0, 1]$; X_j^{best} represents the best individual vector according to the value of objective function and $K \in [0, 1]$. In this paper, the value of F is chosen as 0.5 for DE algorithm.

2.2 Crossover Operation

The DE binomial crossover process is used between target vector $X_{i,j}$ and mutant vector $v_{i,j}$ for producing the final form of trial vector $u_{i,j}$. The trial vector $u_{i,j} = (u_{i,1}, u_{i,2}, u_{i,3}, \ldots, u_{i,D})$ corresponding to target vector $X_{i,j}$ is produced with the help of Eq. (5).

$$u_{i,j} = \begin{cases} v_{i,j} & \text{if } \text{rand}_j(0, 1) < \text{CR} \\ X_{i,j} & \text{Otherwise} \end{cases} \tag{5}$$

Here, $i = 1, 2, 3, \ldots, NP$; $j = 1, 2, 3, \ldots, D$. For the DE algorithm, the crossover rate is considered as 0.9.

Once crossover operator is over, the selection operator is utilized in the target vector X_i and the trial vector $u_{i,j}$ to select the better one for the next generation by using Eq. (6).

$$X_{i,j} = \begin{cases} v_{i,j} & \text{if } f(u_{i,j}) < f(X_{i,j}) \\ X_{i,j} & \text{Otherwise} \end{cases} \tag{6}$$

The usage strategy of the DE is as per the following:

Stage 1. Generate algorithm control parameters and initial population;
Stage 2. Evaluate the objective function value for the beginning population set;
Stage 3. By selecting one mutation strategy from Eqs. (1–4), generate the mutant vectors for the population set.
Stage 4. Using crossover operator as given in Eq. (5), generate the final form of trial vectors.
Stage 5. Evaluate the fitness vector for the set of target vector.
Stage 6. Select individual from the target vector and trial vector according to Eq. (6).
Stage 7. If the end (stop) condition is not met, the individual component of the population will be restored with Stage 3. Else return the individual form the set of population with the best objective function value as the solution. In this paper, the maximum number of function calculation is considered as the termination criterion.

3 Overview of BSA

BSA has few control parameters and is easy to implement as it is a nature-based stochastic searcher technique. First of all, the beginning population is produced uniformly randomly within the search area. BSA measures the historical population "oldP" in order to identify the search route. The beginning historical population is determined with uniform arbitrary way inside the pursuit space. At the beginning, historical population "oldP" is redefined by using Eq. (7).

$$\text{if } a < b \text{ then, oldP} = P, \text{where}(a, b) \in \text{rand}(0, 1) \tag{7}$$

After calculating oldP, the position of the individuals in oldP should be changed randomly using Eq. (8).

$$\text{oldP} = \text{permuting (oldP)} \tag{8}$$

3.1 Mutation

During the iteration, BSA creates the underlying type of the preliminary population "Freak" by utilizing a unique mutation process that utilized just a single bearing individual for each target individual through Eq. (9).

$$M = P + F * (\text{oldP} - P) \tag{9}$$

Here, P represents the present population; oldP represents the historical population; and F is a scaling factor and it control the $(\text{oldP} - P)$. The value of the control parameter F is taken as $F = 3 * a$, here a $N(0, 1)$.

3.2 Crossover

After mutant activity is complete, the crossover cycle creates the ultimate form of the test population "T." Mutant is the initial value of the "T."

Algorithm 1: Crossover process of BSA

```
Define the mixrate; N (population Size), D (Dimension);
If a < b,     a, b ∈ rand(0,1)
  for i from 1 to N
    U = permuting (D);
    map (i, U(1: ceil(mixrate * rand * D))) = 1;
  end
else
for i from 1 to N
    map (i, randi(D)) = 1;
end
 M = P + (map.* F).* (oldP − P)
```

The BSA crossover strategy is non-uniform and more complex than the crossover approach of DE and GA. The crossover cycle of BSA is executed into two parts. The first step is to calculate a binary integer-evaluated matrix (map) $N \times D$, that indicates that the "*T*" individuals are to be manipulated with the relevant "*p*" individuals. Besides, utilizing the significant individual to the mutant individual, the applicable components (dimensions) of mutant individual are refreshed. Algorithm 1 is used to demonstrate the BSA crossover process. The mixing rate is assumed to be 1.

The usage methodology of BSA is as per the following:

Stage 1. Generate the original population, parameters of algorithm control;
Stage 2. Assess each individual's fitness in the population set;
Stage 3. At each iteration, generate the mutant population using Eq. (9).
Stage 4. Utilizing crossover process as given in Algorithm 1, create the ultimate form of the preliminary population.
Stage 5. Assess the fitness vector for every trial (preliminary) population (offspring).
Stage 6. Select individuals based on the fitness value from the target population to the preliminary population (offspring).
Stage 7. In the event that the halting measure is not fulfilled go to Stage 3, the person with the best value for fitness will return otherwise.

4 The hDEBSA Algorithm [25]

The hybrid metaheuristic has already been shown to deliver more effective action and versatility in dealing with real and large-scale problems. Many researchers have recently developed various hybrid methods for the metaheuristic optimization to solve problems of optimization in different fields.

The efficiency of any algorithm always relies upon the correct value of the parameters of the control algorithm. DE has two parameters of regulation F and CR. The exhibition of the DE algorithm is found in the literature to be sensitive to F and CR values [39]. The efficiency of the DE algorithm can also be enhanced by adjusting the values for control parameters F and CR in the course of the DE optimization process [40]. BSA's performance varies according to the F factor. Nonetheless, the lower value of F makes fine searching in small steps and causes slow conversion, and a larger F value speeds up the search, however it lessens the investigation capacity.

In this sense, a combination of the DE and BSA algorithms proposes the hybrid algorithm called hDEBSA [25]. The parameter control is also considered adaptive. The DE algorithm component will first be executed and then the BSA component will be executed in hDEBSA [25]. Just one worse individual is updated according to the probability when running the DE part algorithm. The probability value is determined by Eq. (10) [41].

$$p_i = \frac{r_i}{N} \tag{10}$$

Here, N is the size of the population and r_i is the ranking of every individual when the population is divided from the worst to the best. It could be noted that DE with ranking-based transformation operators presented in Eq. (10) is similar to the selection probability presented in [41, 42]. The following can be described for this selection strategy:

$$I_s = i, \text{if rand}(0, 1) > p_i, i = 1, 2, 3, \ldots N. \tag{11}$$

Here, I_s is selected by individual and optimized by DE.

The following is the detailed description of the parameters for the adaptive control and the hDEBSA algorithm.

4.1 *Scaling Factor/Weighting Coefficient* (**F**) *of DE Algorithm [25]*

A new set of trial vector is generated in the DE algorithm with the control parameter scaling factor (F). It was discovered that smaller value (smaller than 0.4) and greater value (more prominent than 1.0) of F are sporadically successful [2]. Certain researchers have also confirmed that the large F control value reduces the chance of stagnating in an optimal local value [43, 44]. The value of $F = 0.6$ or 0.5 has been found to be the correct initial value, while the value of Rönkkönen et al. [44] has been found to be $F = 0.9$. As indicated by Rönkkönen et al. [44], estimation of control parameter F is 0.4–0.95. Fluctuating the estimations of control parameter F during the enhancement procedure can improve the exhibition of the design algorithm [2], [40]. So, the change of scaling factor (F) can be characterized by Eq. (12) [25].

Instead of F, we have written as F_{DE} for scaling factor of DE to avoid confusion with the parameter F of BSA.

$$F_{\text{DE}} = F_{\text{DE}}^{\text{min}} + (F_{\text{DE}}^{\text{max}} - F_{\text{DE}}^{\text{min}}) * \frac{f_i^{\text{max}} - f_i^{textmin}}{f_0^{\text{max}} - f_0^{\text{min}}} \tag{12}$$

where $F_{\mathrm{DE}}^{\min} = 0.4$ and $F_{\mathrm{DE}}^{\max} = 0.95$; $f_0^{\max}$ and $f_0^{\min}$ are the highest and lowest fitness values, respectively, of the preliminary population; $f_i^{\max}$ and $f_i^{\min}$ are the highest and lowest fitness value, respectively, for the population of the ith iteration.

4.2 Crossover Rate (CR) of DE Algorithm [25]

The crossover rate for generating the final form of the test vector collection is added to the DE algorithm. It is utilized to decide the amount of a preliminary solution ought to be received into the present solution, just as the DE conspire which is the vital thought behind DE for producing preliminary vectors [45, 46]. Researchers have confirmed that the high CR value both accelerates convergence and reduces local search capability [43, 44, 47]. CR value = 0.1 is a correct choice of start, whereby the convergence speed of CR = 0.9 or 1.0 can be improved [2]. The right CR value can vary from 0.3 to 0.9 [43]. At the point when the estimation of CR is picked 1, the quantity of preliminary vector might be decreased drastically which may prompt fixed status [44, 48] and consequently, CR = 0.9 or 0.95 can be utilized rather than 1. The performance can be improved by changing the crossover rate CR values during the optimization process [40]. The adjustment of crossover rate (CR) can be characterized by Eq. (13) [25].

To avoid confusion, instead of CR, it is written as $\mathrm{CR}_{\mathrm{DE}}$.

$$\mathrm{CR}_{\mathrm{DE}} = \mathrm{CR}_{\mathrm{DE}}^{\max} - (\mathrm{CR}_{\mathrm{DE}}^{\max} - \mathrm{CR}_{\mathrm{DE}}^{\min}) * \frac{f_i^{\max} - f_i^{\min}}{f_0^{\max} - f_0^{\min}} \tag{13}$$

where $\mathrm{CR}_{\mathrm{DE}}^{\min} = 0.3$ and $\mathrm{CR}_{\mathrm{DE}}^{\max} = 0.9$; $f_0^{\max}$ and $f_0^{\min}$ are the highest and lowest fitness values, respectively, of the preliminary population; $f_i^{\max}$ and $f_i^{\min}$ are the highest and lowest fitness values, respectively, of the ith iteration population.

4.3 BSA Control Parameter [25]

In BSA, the control parameter (F) is used to control the amplitude of the search-direction matrix (oldPP). In the original BSA, the value of F was considered as $3{*}a, a \sim N(0, 1)$. And thus during the course of run, the fitness value is updated only with the proper choice of F. The lower F value allows for fine search in small steps, while the optimization algorithm works, however reducing the convergence speed, a larger F value speeds up the search but reduces the local search capacity. Seeing this circumstance, a self-adaption-based modification of BSA control parameter (F) is proposed. The modification of BSA control parameter (F) can be defined by Eq. (14) [25]. Instead of F, we have written F_{BSA} for clear understanding.

Initialization
D: Numberof variable in the optimization problem; NP: The set of population;
FEs_MAX: Set the maximum number of fitness evaluation;
Generate a set of individual of fix number of NP within the search space by uniform random initialization;
Evaluate the fitness value of each individual;
FEs = NP; % estimate the amount of function evaluations %;
% Main loop %
***While** Fes < FEs_MAX*
% DE mutation operator and Crossover operator %
Select one individual according to the rank. Calculate the new individual X_i^{new} *by DE mutation operator according to the Eqn. (1) with the associate parameter given in Eqn. (12).*
Then apply the crossover operator with the associate parameter given in Eqn. (13) to originate the trial individual's final form.
Compute the new individual's fitness value and upgrade if the value of the fitness is higher than the previous value.
FEs = FEs +1; % estimate the amount of function evaluations %;
***For** i = 1: NP*
% BSA mutation operator and Crossover operator %
Calculate the new individual X_i^{new} *by BSA mutation operator according to the Eqn. (9) with the associate parameter given in Eqn. (14). Then apply the crossover operator with the associate parameter given in Eqn. (14) to originate the trial individual's final form..*
Compute the new individual's fitness value and upgrade if the value of the fitness is higher than the previous value.
FEs = FEs +1; % estimate the amount of function evaluations %;
End
End While loop
***Output:** The best optimum individual's fitness value in the set of population;*

Fig. 1 Pseudo-code of the hDEBSA [42] algorithm

$$F_{\mathrm{BSA}} = F_{\mathrm{BSA}}^{\max} - \mathrm{rand}(0,1) * \left(F_{\mathrm{BSA}}^{\max} - F_{\mathrm{BSA}}^{\min}\right) + \mathrm{rand}(0,1) * \frac{f_i^{\max} - f_i^{\min}}{f_0^{\max} - f_0^{\min}} \tag{14}$$

where $F_{\mathrm{BSA}}^{\min} = 0.45$ and $F_{\mathrm{BSA}}^{\max} = 2.0$; $f_0^{\max}$ and $f_0^{\min}$ are the highest and lowest fitness value, respectively, of the preliminary population; $f_i^{\max}$ and $f_i^{\min}$ are the highest and lowest fitness values, respectively, of the ith iteration population.

Also, the value of mixrate is considered by Eq. (15) [42].

$$\mathrm{mixrate}_{\mathrm{BSA}} = 0.5 * (1 + \mathrm{rand}(0,1)) \tag{15}$$

So, the control parameter's value differs automatically during the execution of the suggested algorithm hDEBSA and improves algorithm performance. Figures 1 and 2 demonstrates the pseudo-code and flowchart of the hDEBSA algorithm.

5 Formulation of the Seismic Active Earth Pressure on Retaining Wall [37, 38]

Retaining walls are moderately rigid walls used to support sideways soil so that they can be maintained on different levels on different sides. Remaining walls are structures designed to manage soil at an irregular slant (usually an unhealthy, vertical or vertical slope). Remaining walls are utilized to bound soils between two unique heights regularly in regions of territory having unfortunate slants or in zones where

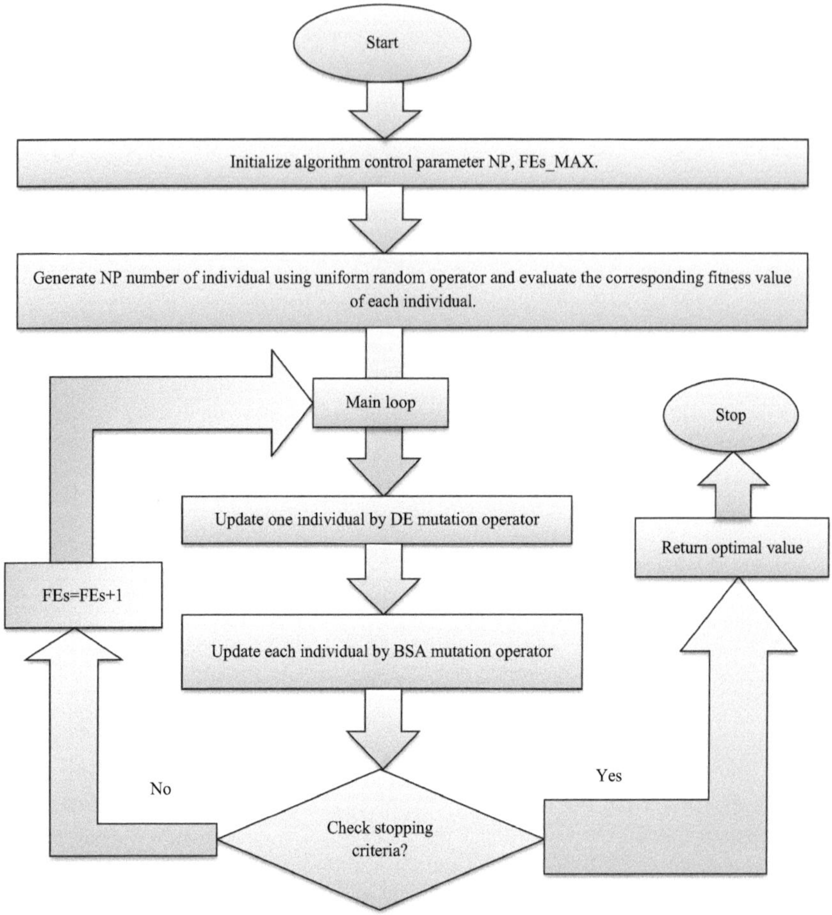

Fig. 2 Flowchart of the hDEBSA algorithm

the scene should be molded harshly and designed for progressively explicit purposes like hillside farming or roadway bridges.

A basic component in the seismic structure is the estimation of the dynamic earth pressure on the retaining wall by reverse (backfill) soil. Mononobe-Okabe analysis [49, 50] provided an arrangement for dynamic earth pressure to think of a single Φ soil input to the retaining wall. They extended out the Coulomb wedge theory by introducing seismic acceleration as inertia forces to discover the seismic dynamic earth impact. Additionally, Seed and Whitman [51] gave the response to seismic dynamic earth pressure take into account the backfill idea. More studies have been conducted by Ghosh et al. [52] to test the seismic dynamic earth pressure on retaining wall supporting $c - \varphi$ related inlay. Pressure because of weight, surcharge, and cohesion is independently optimized with their expository methodology and bunches of quantities of diagrams are presented to that. Subsequently, in [37, 38], an endeavor is made to discover the dynamic active earth force by enhancing the force to get a solitary wedge angle.

A schematic diagram of the equilibrium of the forces under active state of equilibrium of retaining wall is presented in Fig. 3. Here, "H" the height of retaining wall is bent in an angle "α" vertically, with a $c - \varphi$ to rear panel of the weight "β" the cohesion unit "c" the adhesion unit "c_a" and has a wall friction angle "δ," soil internal soil friction angle "Φ" and backfill inclination angle "" with the horizontal. The additional pressure q load per unit length works at the top of the backfill. Taking the seismic acceleration coefficients as k_h and k_v and considering a planner rupture surface BD formed an angle θ with the vertical forces acting on the wedge surface are shown in Fig. 3. The force acting on the retaining wall is "P_a" and the reaction is "R" when the equilibrium is involved. The following are geotechnical parameters and geometric elements.

$$\mathrm{AB} = H \sec\alpha \tag{16}$$

$$\mathrm{AD} = \frac{H \sec\alpha}{\cos(\theta + i)} \sin(\alpha + \theta) \tag{17}$$

$$\mathrm{BD} = \frac{H \sec\alpha}{\cos(\theta + i)} \cos(\alpha - i) \tag{18}$$

$$\Delta\mathrm{ABD} = \frac{H^2 \sec^2\alpha \sin(\alpha + \theta)\cos(\alpha - i)}{2\cos(\theta + i)} \tag{19}$$

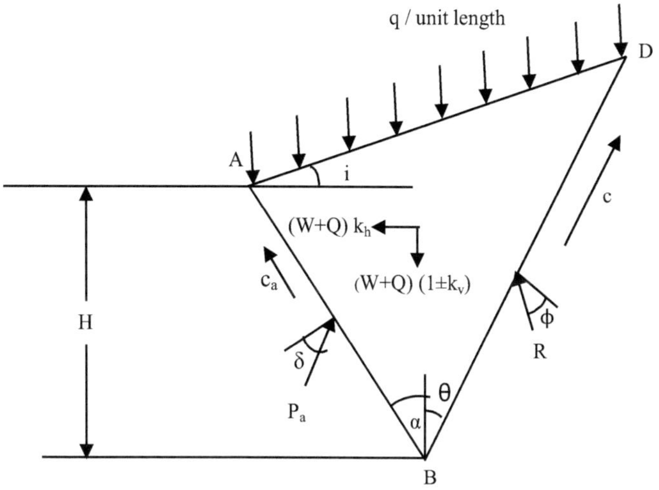

Fig. 3 Forces that act during an active state of balance in retaining the wall soil wedge system [38]

$$W = \frac{\lambda H^2 \sec^2\alpha \sin(\alpha+\theta)\cos(\alpha-i)}{2\cos(\theta+i)} \tag{20}$$

$$Q = q\frac{H \sec\alpha}{\cos(\theta+i)}\sin(\alpha+\theta) \tag{21}$$

The other forces are: Total cohesion

$$C = \frac{cH \sec\alpha}{\cos(\theta+i)}\cos(\alpha-i) \tag{22}$$

Total adhesion

$$C_a = c_a H \sec\alpha \tag{23}$$

Applying equilibrium conditions, i.e., $\sum V = 0$ and $\sum H = 0$, we get, respectively,

$$C_a\cos\alpha + C\cos\theta + R\sin(\theta+\varphi) + P\sin(\alpha+\delta) - (W+Q)(1\pm k_v) = 0 \tag{24}$$

$$-C_a\sin\alpha + C\sin\theta - R\cos(\theta+\varphi) + P\cos(\alpha+\delta) - (W+Q)k_h = 0 \tag{25}$$

Simplifying Eqs. (24) and (25) and putting the values of C, C_a, Q, and W, we get,

$$P_a = \left[\frac{\gamma H^2}{2} + qH\frac{\cos\alpha}{\cos(\alpha-i)}\right]\left\{\begin{array}{l}\left[\dfrac{\sin(\alpha+\theta)\cos(\alpha-i)\{(1\pm k_v)\cos(\theta+\varphi)+k_h\sin(\theta+\varphi)\}}{\cos(\theta+i)\cos^2\alpha\sin(\alpha+\delta+\varphi+\theta)}\right] \\ -m\dfrac{\sec\alpha\cos(\alpha-i)\cos\varphi}{\cos(\theta+i)\sin(\alpha+\delta+\theta+\varphi)} - n\dfrac{\sec\alpha\cos(\alpha+\theta+\varphi)}{\sin(\alpha+\delta+\theta+\varphi)}\end{array}\right\} \tag{26}$$

where

$$m = \frac{2c}{\gamma H + 2q\frac{\cos\alpha}{\cos(\alpha-i)}} \tag{27}$$

$$n = \frac{2c_a}{\gamma H + 2q\frac{\cos\alpha}{\cos(\alpha-i)}} \tag{28}$$

Equation (26) can be written as

$$P_a = \frac{\gamma H + 2q\frac{\cos\alpha}{\cos(\alpha-i)}}{2}K_{ae} \tag{29}$$

where

$$K_{ae} = \frac{\frac{\sin(\alpha+\theta)\cos(\alpha-i)}{\cos(\theta+i)\cos^2\alpha}\left[(1 \pm k_v)\cos(\theta+\varphi) + k_h \sin(\theta+\varphi)\right] - m\frac{\cos(\alpha-i)\cos\varphi}{\cos(\theta+i)\cos\alpha} - n\frac{\cos(\alpha+\theta+\varphi)}{\cos\alpha}}{\sin(\alpha+\delta+\varphi+\theta)} \quad (30)$$

Here, it is clear that for a specific retaining wall inlay framework, every one of the terms with the exception θ is constant. Optimizing the value of "K_{ae}" with respect to θ to get the active earth pressure coefficient (K_{ae}). The range of θ is taken from 0 to 90°.

6 Result and Discussion

Detailed results of the CEC2014 test functions and seismic active earth pressure for $c - \varphi$ backfill retentions have been provided in this segment.

6.1 Result and Discussion on CEC2014 Test Functions

The execution of the suggested hDEBSA is studied on CEC2014 special session on unconstrained continuous optimization problems. It has thirty test functions with varying difficulty levels. Such research functions comprise of four unique types of functions, such as

Unimodal functions F1–F3; simple multimodal functions F4–F16; hybrid functions F17–F22; composition functions F23–F30.

A comprehensive explanation of these thirty test functions can be seen in [53]. The present study takes into account the 10 and 30 dimensions (D) of the test functions. The algorithm is run 51 times with D * 10,000 fitness evaluation (FEs) and 50 population size. All the algorithms are runs in MATLAB R2010a. For every test function, the range of the domain space is considered as [−100, 100]. The best, worst, median, mean, and standard deviation of each test function are provided in order to determine the output of the algorithm. To evaluate statistics using the median result for all test functions, the Friedman rank test is applied to get the general rank of all algorithms.

Tables 1 and 2 show the performance result of DE [2], BSA [8] and hDEBSA on F1-F15 and F16-F30 CEC2014 special session on unconstrained continuous optimization problems with dimension 10. The results are shown with regard to best, worst, medium, mean, and standard deviation. Table 3 demonstrates the average rank obtained by the Friedman rank test for the median results of each algorithm's testing functions. The rank of hDEBSA is better than DE, and BSA is found in Table 3. So, hDEBSA is more effective than DE and BSA. It can be claimed.

The results of F1-F10, F11-F20, and F21-F30 CEC2014 test functions with dimensions 30 present in Tables 4, 5, and 6, respectively. The performance results are compared with CLPSO [10], UPSO [13], DNLPSO [14], DE-PSO [22], DESQI

Table 1 Experimental results of DE, BSA, and hDEBSA at dimension (*D*) 10 after reaching *D* * 10,000 FEs of F1-F15 CEC2014 benchmark functions over 51 runs and 50 population size (PS)

Function	Algorithms	Best	Worst	Median	Mean	Std
F1	DE	1.00E+02	1.00E+02	1.00E+02	1.00E+02	4.78E−08
	BSA	1.13E+02	2.22E+03	3.93E+02	5.47E+02	4.81E+02
	hDEBSA	1.00E+02	1.33E+03	1.05E+02	1.95E+02	2.48E+02
F2	DE	2.00E+02	2.00E+02	2.00E+02	2.00E+02	0.00E+00
	BSA	2.00E+02	2.00E+02	2.00E+02	2.00E+02	9.86E−07
	hDEBSA	2.00E+02	2.00E+02	2.00E+02	2.00E+02	0.00E+00
F3	DE	3.00E+02	3.00E+02	3.00E+02	3.00E+02	0.00E+00
	BSA	3.00E+02	3.00E+02	3.00E+02	3.00E+02	1.28E−09
	hDEBSA	3.00E+02	3.00E+02	3.00E+02	3.00E+02	0.00E+00
F4	DE	4.00E+02	4.07E+02	4.03E+02	4.03E+02	1.61E+00
	BSA	4.00E+02	4.04E+02	4.00E+02	4.00E+02	6.14E−01
	hDEBSA	4.00E+02	4.35E+02	4.04E+02	4.14E+02	1.65E+01
F5	DE	5.20E+02	5.20E+02	5.20E+02	5.20E+02	8.19E−02
	BSA	5.20E+02	5.20E+02	5.20E+02	5.20E+02	1.71E−02
	hDEBSA	5.04E+02	5.20E+02	5.20E+02	5.18E+02	4.63E+00
F6	DE	6.00E+02	6.04E+02	6.00E+02	6.00E+02	8.43E−01
	BSA	6.00E+02	6.03E+02	6.01E+02	6.01E+02	6.80E−01
	hDEBSA	6.00E+02	6.03E+02	6.01E+02	6.01E+02	9.31E−01
F7	DE	7.00E+02	7.00E+02	7.00E+02	7.00E+02	2.80E−02
	BSA	7.00E+02	7.00E+02	7.00E+02	7.00E+02	1.27E−02
	hDEBSA	7.00E+02	7.00E+02	7.00E+02	7.00E+02	2.67E−02
F8	DE	8.00E+02	8.20E+02	8.01E+02	8.02E+02	3.90E+00
	BSA	8.00E+02	8.00E+02	8.00E+02	8.00E+02	6.09E−10
	hDEBSA	8.00E+02	8.03E+02	8.01E+02	8.01E+02	8.44E−01
F9	DE	9.01E+02	9.20E+02	9.06E+02	9.07E+02	4.22E+00
	BSA	9.03E+02	9.10E+02	9.07E+02	9.06E+02	1.65E+00
	hDEBSA	9.03E+02	9.12E+02	9.09E+02	9.08E+02	1.91E+00
F10	DE	1.01E+03	2.22E+03	1.21E+03	1.31E+03	3.02E+02
	BSA	1.00E+03	1.01E+03	1.00E+03	1.00E+03	3.94E+00
	hDEBSA	1.00E+03	1.00E+03	1.00E+03	1.00E+03	1.73E−01
F11	DE	1.20E+03	2.76E+03	2.19E+03	2.12E+03	4.21E+02
	BSA	1.12E+03	1.35E+03	1.56E+03	1.34E+03	1.11E+02
	hDEBSA	1.13E+03	1.65E+03	1.27E+03	1.30E+03	1.33E+02
F12	DE	1.20E+03	1.20E+03	1.20E+03	1.20E+03	1.74E−01
	BSA	1.20E+03	1.20E+03	1.20E+03	1.20E+03	6.35E−02

(continued)

Table 1 (continued)

Function	Algorithms	Best	Worst	Median	Mean	Std
	hDEBSA	1.20E+03	1.20E+03	1.20E+03	1.20E+03	7.79E−02
F13	DE	1.30E+03	1.30E+03	1.30E+03	1.30E+03	4.04E−02
	BSA	1.30E+03	1.30E+03	1.30E+03	1.30E+03	3.86E−02
	hDEBSA	1.30E+03	1.30E+03	1.30E+03	1.30E+03	2.67E−02
F14	DE	1.40E+03	1.40E+03	1.40E+03	1.40E+03	4.91E−02
	BSA	1.40E+03	1.40E+03	1.40E+03	1.40E+03	4.31E−02
	hDEBSA	1.40E+03	1.40E+03	1.40E+03	1.40E+03	3.63E−02
F15	DE	1.50E+03	1.50E+03	1.50E+03	1.50E+03	4.93E−01
	BSA	1.50E+03	1.50E+03	1.50E+03	1.50E+03	1.45E−01
	hDEBSA	1.50E+03	1.50E+03	1.50E+03	1.50E+03	2.25E−01

[21], QPSO [20], PSOGSA [19] etc. Table 7 displays the mean rank for the median output of all test functions in every algorithm obtained by Friedman rank test. Table 7 displays the mean rank acquired by Friedman rank test concerning the median performance of all the test functions in every algorithm. Table 7 shows that the mean rank obtained by the Friedman rank test for hDEBSA is superior to the algorithm in comparison. Therefore, hDEBSA's efficiency is superior to the other algorithms.

Convergence analysis: To examine the convergence efficiency of hDEBSA, the objective function value of 51 runs for different fitness evaluation of some selected CEC 2014 benchmark functions have been plotted for 30 dimensions in Fig. 4. While Fig. 4 also represents the convergence characteristics on the execution of hDEBSA. In these figures, horizontal axis indicates the number of function evaluation ($D * 10{,}000$) and the vertical axis represents the objective function value of functions. The convergence of hDEBSA algorithms is apparent from this figure as being more efficient than former algorithms, i.e., reaches to the optima faster and avoids stagnation in most of the cases.

6.2 *Result and Discussion on Seismic Active Earth Pressure on Retaining Wall Supporting* $c - \varphi$ *Backfill*

The results of the seismic active earth pressure retaining wall supporting $c - \varphi$ backfill have been discussed in this segment. Also, this segment has provided a brief comparative study. The performance result, i.e., active earth pressure coefficients acquired by suggested algorithm are in contrast with other metaheuristic optimization algorithm. For determination of optimum output, hDEBSA is implemented with 1000 number of fitness evaluation and 50 population sizes.

Table 2 Experimental results of DE, BSA, and hDEBSA at dimension (*D*) 10 after reaching D*10000 FEs of F16-F30 CEC2014 benchmark functions over 51 runs and 50 population size (PS)

Function	Algorithms	Best	Worst	Median	Mean	Std
F16	DE	1.60E+03	1.60E+03	1.60E+03	1.60E+03	6.68E−01
	BSA	1.60E+03	1.60E+03	1.60E+03	1.60E+03	3.15E−01
	hDEBSA	1.60E+03	1.60E+03	1.60E+03	1.60E+03	2.33E−01
F17	DE	1.70E+03	1.74E+03	1.70E+03	1.71E+03	8.91E+00
	BSA	1.71E+03	1.83E+03	1.74E+03	1.75E+03	2.64E+01
	hDEBSA	1.70E+03	1.71E+03	1.70E+03	1.70E+03	4.87E+00
F18	DE	1.80E+03	1.80E+03	1.80E+03	1.80E+03	5.72E−01
	BSA	1.80E+03	1.80E+03	1.80E+03	1.80E+03	7.60E−01
	hDEBSA	1.80E+03	1.80E+03	1.80E+03	1.80E+03	7.55E−01
F19	DE	1.90E+03	1.90E+03	1.90E+03	1.90E+03	6.13E−02
	BSA	1.90E+03	1.90E+03	1.90E+03	1.90E+03	1.64E−01
	hDEBSA	1.90E+03	1.90E+03	1.90E+03	1.90E+03	8.79E−02
F20	DE	2.00E+03	2.00E+03	2.00E+03	2.00E+03	2.03E−01
	BSA	2.00E+03	2.00E+03	2.00E+03	2.00E+03	3.84E−01
	hDEBSA	2.00E+03	2.00E+03	2.00E+03	2.00E+03	2.74E−01
F21	DE	2.10E+03	2.12E+03	2.10E+03	2.10E+03	3.25E+00
	BSA	2.10E+03	2.11E+03	2.10E+03	2.10E+03	1.03E+00
	hDEBSA	2.10E+03	2.12E+03	2.10E+03	2.10E+03	2.31E+00
F22	DE	2.22E+03	2.22E+03	2.22E+03	2.22E+03	7.22E−01
	BSA	2.20E+03	2.21E+03	2.20E+03	2.20E+03	9.98E−01
	hDEBSA	2.20E+03	2.22E+03	2.20E+03	2.20E+03	2.33E+00
F23	DE	2.63E+03	2.63E+03	2.63E+03	2.63E+03	9.19E−13
	BSA	2.30E+03	2.63E+03	2.63E+03	2.62E+03	4.61E+01
	hDEBSA	2.63E+03	2.63E+03	2.63E+03	2.63E+03	9.19E−13
F24	DE	2.50E+03	2.52E+03	2.51E+03	2.51E+03	4.37E+00
	BSA	2.51E+03	2.52E+03	2.51E+03	2.51E+03	2.83E+00
	hDEBSA	2.51E+03	2.52E+03	2.52E+03	2.52E+03	3.11E+00
F25	DE	2.65E+03	2.70E+03	2.70E+03	2.70E+03	9.77E+00
	BSA	2.61E+03	2.70E+03	2.63E+03	2.63E+03	1.22E+01
	hDEBSA	2.61E+03	2.70E+03	2.63E+03	2.64E+03	2.84E+01
F26	DE	2.70E+03	2.70E+03	2.70E+03	2.70E+03	3.45E−02
	BSA	2.70E+03	2.70E+03	2.70E+03	2.70E+03	3.74E−02
	hDEBSA	2.70E+03	2.70E+03	2.70E+03	2.70E+03	3.28E−02
F27	DE	2.70E+03	3.21E+03	3.00E+03	2.98E+03	1.40E+02
	BSA	2.70E+03	3.03E+03	2.70E+03	2.78E+03	1.33E+02

(continued)

Table 2 (continued)

Function	Algorithms	Best	Worst	Median	Mean	Std
	hDEBSA	2.70E+03	3.10E+03	2.70E+03	2.75E+03	1.25E+02
F28	DE	3.11E+03	3.11E+03	3.11E+03	3.11E+03	4.71E−03
	BSA	3.16E+03	3.18E+03	3.16E+03	3.16E+03	5.43E+00
	hDEBSA	3.16E+03	3.17E+03	3.16E+03	3.16E+03	4.27E+00
F29	DE	3.10E+03	3.10E+03	3.10E+03	3.10E+03	4.08E−01
	BSA	3.07E+03	3.28E+03	3.12E+03	3.13E+03	3.58E+01
	hDEBSA	3.03E+03	3.13E+03	3.11E+03	3.09E+03	3.48E+01
F30	DE	3.21E+03	3.24E+03	3.22E+03	3.22E+03	2.61E+00
	BSA	3.48E+03	3.60E+03	3.51E+03	3.51E+03	2.37E+01
	hDEBSA	3.46E+03	3.60E+03	3.48E+03	3.49E+03	3.22E+01

Table 3 Ranks obtained by Friedman test on CEC2014 real-world benchmark functions with 10 dimensions of all the algorithms using their median performances

Algorithms	Mean rank
DE	2.05
BSA	2.13
hDEBSA	1.82

6.2.1 Parametric Study

The optimum output of pseudo-static seismic active earth pressure coefficient (K_{ae}) obtained by hDEBSA has been done in this section by considering different values of soil parameter. The values of soil parameter are considered as follows: $\alpha = -20°$, $0°$, $20°$; $\Phi = 20°, 30°, 40°$; $\delta = -\Phi/2, 0, \Phi/2, \Phi$; $i = 0°, 10°, 20°$; $m = 0.05, 0.10, 0.15, 0.20$; $n = 0.05, 0.10, 0.15, 0.20$; $k_h = 0, 0.1, 0.2$; $k_v = 0, \frac{k_h}{2}, k_h$;

Figure 5 represents the upshot of internal friction angle (ϕ) of soil on K_{ae} at $i = 10°$, $\alpha = 30°$, $m = 0.1$, $n = 0.05$, $k_v = k_h/2$, $\delta = \phi/2$ with respect to the horizontal seismic acceleration (k_h). From this figure, it very well may be seen that with increasing the value of internal friction angle (ϕ) of soil, the amount of K_{ae} will decrease. Also, it is seen that with the increases in the amount of k_h, the amount of K_{ae} will increase.

Figure 6 represents the effect of angle of wall frictions (δ) on K_{ae} at $i = 10°$, $\alpha = 30°$, $\phi = 30°$, $m = 0.1$, $n = 0.05$, $k_v = k_h/2$ with respect to the k_h. From this plot, it can be seen that increase in the value of angle of wall frictions (δ) starting 0–30° results in increase in the magnitude of K_{ae}. Also, it is observed that with the increase in the magnitude of k_h, the magnitude of K_{ae} increases.

Figure 7 represents the effect of seismic acceleration (k_v) on K_{ae} at $i = 10°$, $\alpha = 30°$, $\phi = 30°$, $\delta = \phi/2$, $m = 0.1$, $n = 0.05$ with respect to the horizontal seismic acceleration (k_h). From this plot, it is observed that with an increase in the value

Table 4 Experimental results of CLPSO, UPSO, DNLPSO, DE-PSO, DESQI, PSOGSA, QPSO, and hDEBSA at dimension (D) 30 after reaching D * 10,000 FEs of F1-F10 CEC2014 benchmark functions over 51 runs and 50 population size (PS)

Function		CLPSO	UPSO	DNLPSO	DE-PSO	DESQI	PSOGSA	QPSO	hDEBSA
F1	Best	6.47E+06	1.64E+05	6.65E+04	1.22E+06	1.56E+05	1.51E+05	1.87E+05	8.05E+04
	Worst	2.72E+07	2.97E+06	1.72E+07	7.42E+06	1.07E+06	2.22E+08	2.01E+06	2.06E+06
	Median	1.43E+07	7.21E+05	1.20E+06	3.08E+06	3.48E+05	2.61E+06	4.48E+05	8.57E+05
	Mean	1.47E+07	8.84E+05	2.47E+06	3.22E+06	3.79E+05	2.68E+07	7.01E+05	9.24E+05
	Std	4.15E+06	5.87E+05	3.17E+06	1.32E+06	1.51E+05	5.56E+07	5.36E+05	4.67E+05
F2	Best	2.00E+02	2.00E+02	2.00E+02	4.11E+06	2.00E+02	2.01E+02	2.00E+02	2.00E+02
	Worst	1.22E+03	2.62E+04	1.82E+09	1.12E+07	2.00E+02	3.49E+10	2.00E+02	2.00E+02
	Median	2.08E+02	2.76E+03	2.78E+02	7.24E+06	2.00E+02	6.81E+08	2.00E+02	2.00E+02
	Mean	2.58E+02	4.26E+03	4.68E+07	7.07E+06	2.00E+02	4.30E+09	2.00E+02	2.00E+02
	Std	1.57E+02	5.02E+03	2.63E+08	1.55E+06	4.68E−02	7.25E+09	1.64E−03	5.68E−15
F3	Best	3.08E+02	3.06E+02	3.05E+02	3.53E+02	3.00E+02	3.45E+02	3.00E+02	3.00E+02
	Worst	1.23E+03	6.08E+03	1.27E+05	5.05E+02	3.42E+02	1.47E+05	3.00E+02	3.00E+02
	Median	5.12E+02	7.75E+02	1.07E+03	4.16E+02	3.00E+02	9.88E+03	3.00E+02	3.00E+02
	Mean	5.61E+02	1.26E+03	8.75E+03	4.24E+02	3.02E+02	2.48E+04	3.00E+02	3.00E+02
	Std	2.24E+02	1.27E+03	2.30E+04	3.89E+01	6.14E+00	3.19E+04	2.96E−02	6.92E−14
F4	Best	4.22E+02	4.00E+02	4.09E+02	4.02E+02	4.00E+02	4.04E+02	4.05E+02	4.00E+02
	Worst	4.29E+02	4.78E+02	6.29E+02	6.30E+02	5.38E+02	4.47E+03	5.40E+02	5.33E+02
	Median	4.26E+02	4.21E+02	4.22E+02	5.19E+02	4.70E+02	5.36E+02	4.85E+02	4.67E+02
	Mean	4.26E+02	4.22E+02	4.37E+02	5.18E+02	4.59E+02	7.71E+02	4.98E+02	4.50E+02
	Std	1.38E+00	1.44E+01	4.08E+01	5.45E+01	4.22E+01	7.68E+02	2.88E+01	3.94E+01
F5	Best	5.20E+02	5.20E+02	5.20E+02	5.21E+02	5.21E+02	5.20E+02	5.20E+02	5.20E+02

(continued)

Table 4 (continued)

Function		CLPSO	UPSO	DNLPSO	DE-PSO	DESQI	PSOGSA	QPSO	hDEBSA
	Worst	5.21E+02	5.21E+02	5.21E+02	5.21E+02	5.21E+02	5.21E+02	5.21E+02	5.20E+02
	Median	5.20E+02	5.20E+02	5.21E+02	5.21E+02	5.21E+02	5.20E+02	5.21E+02	5.20E+02
	Mean	5.20E+02	5.20E+02	5.21E+02	5.21E+02	5.21E+02	5.20E+02	5.21E+02	5.20E+02
	Std	4.71E−02	1.52E−01	3.23E−01	1.11E−01	5.75E−02	2.38E−01	8.30E−02	3.48E−02
F6	Best	6.12E+02	6.18E+02	6.01E+02	6.06E+02	6.05E+02	6.14E+02	6.09E+02	6.00E+02
	Worst	6.16E+02	6.27E+02	6.19E+02	6.16E+02	6.13E+02	6.30E+02	6.21E+02	6.14E+02
	Median	6.14E+02	6.24E+02	6.06E+02	6.11E+02	6.09E+02	6.23E+02	6.17E+02	6.11E+02
	Mean	6.14E+02	6.24E+02	6.07E+02	6.11E+02	6.09E+02	6.23E+02	6.17E+02	6.10E+02
	Std	1.06E+00	1.92E+00	5.00E+00	2.35E+00	1.76E+00	3.66E+00	2.36E+00	2.92E+00
F7	Best	7.00E+02	7.00E+02	7.00E+02	7.01E+02	7.00E+02	7.00E+02	7.00E+02	7.00E+02
	Worst	7.00E+02	7.00E+02	7.00E+02	7.01E+02	7.00E+02	1.08E+03	7.00E+02	7.00E+02
	Median	7.00E+02	7.00E+02	7.00E+02	7.01E+02	7.00E+02	7.18E+02	7.00E+02	7.00E+02
	Mean	7.00E+02	7.00E+02	7.00E+02	7.01E+02	7.00E+02	7.47E+02	7.00E+02	7.00E+02
	Std	2.52E−05	4.10E−03	8.89E−01	1.78E−02	1.32E−02	6.78E+01	9.58E−03	1.19E−02
F8	Best	8.00E+02	8.41E+02	8.15E+02	8.25E+02	8.06E+02	8.86E+02	8.28E+02	8.00E+02
	Worst	8.00E+02	9.18E+02	9.61E+02	8.76E+02	8.28E+02	1.10E+03	8.87E+02	8.01E+02
	Median	8.00E+02	8.74E+02	8.39E+02	8.46E+02	8.16E+02	9.65E+02	8.54E+02	8.00E+02
	Mean	8.00E+02	8.76E+02	8.44E+02	8.48E+02	8.17E+02	9.72E+02	8.57E+02	8.00E+02
	Std	3.56E−12	1.52E+01	2.66E+01	1.17E+01	5.36E+00	4.85E+01	1.27E+01	1.95E−01
F9	Best	9.38E+02	9.60E+02	9.20E+02	9.30E+02	9.16E+02	1.02E+03	9.68E+02	9.28E+02
	Worst	9.73E+02	1.04E+03	1.07E+03	9.94E+02	9.62E+02	1.29E+03	1.05E+03	9.62E+02

(continued)

Table 4 (continued)

Function		CLPSO	UPSO	DNLPSO	DE-PSO	DESQI	PSOGSA	QPSO	hDEBSA
	Median	9.58E+02	9.91E+02	9.42E+02	9.56E+02	9.35E+02	1.13E+03	9.89E+02	9.40E+02
	Mean	9.57E+02	9.94E+02	9.52E+02	9.58E+02	9.36E+02	1.14E+03	9.92E+02	9.41E+02
	Std	7.72E+00	1.60E+01	3.21E+01	1.53E+01	8.81E+00	6.16E+01	1.53E+01	7.99E+00
F10	Best	1.00E+03	2.60E+03	1.42E+03	1.78E+03	1.00E+03	2.69E+03	1.24E+03	1.00E+03
	Worst	1.01E+03	4.24E+03	6.27E+03	7.37E+03	3.80E+03	6.83E+03	7.22E+03	1.00E+03
	Median	1.01E+03	3.48E+03	2.57E+03	4.26E+03	1.13E+03	4.56E+03	2.21E+03	1.00E+03
	Mean	1.01E+03	3.52E+03	2.79E+03	4.40E+03	1.61E+03	4.64E+03	2.30E+03	1.00E+03
	Std	2.58E+00	4.01E+02	1.10E+03	1.84E+03	8.14E+02	7.91E+02	8.24E+02	2.60E−01

Table 5 Experimental results of CLPSO, UPSO, DNLPSO, DE-PSO, DESQI, PSOGSA, QPSO, and hDEBSA at dimension (D) 30 after reaching D * 10,000 FEs of F10-F20 CEC2014 benchmark functions over 51 runs and 50 population size (PS)

Function		CLPSO	UPSO	DNLPSO	DE-PSO	DESQI	PSOGSA	QPSO	hDEBSA
F11	Best	3.08E+03	3.29E+03	3.26E+03	6.17E+03	1.80E+03	3.67E+03	3.02E+03	2.65E+03
	Worst	4.21E+03	4.83E+03	7.89E+03	8.21E+03	8.08E+03	7.22E+03	7.38E+03	3.81E+03
	Median	3.71E+03	4.22E+03	4.81E+03	7.34E+03	7.02E+03	5.17E+03	4.45E+03	3.34E+03
	Mean	3.69E+03	4.18E+03	5.05E+03	7.26E+03	6.24E+03	5.22E+03	4.53E+03	3.32E+03
	Std	2.80E+02	3.66E+02	1.33E+03	5.31E+02	1.92E+03	7.38E+02	9.20E+02	2.50E+02
F12	Best	1.20E+03	1.20E+03	1.20E+03	1.20E+03	1.20E+03	1.20E+03	1.20E+03	1.20E+03
	Worst	1.20E+03	1.20E+03	1.20E+03	1.20E+03	1.20E+03	1.20E+03	1.20E+03	1.20E+03
	Median	1.20E+03	1.20E+03	1.20E+03	1.20E+03	1.20E+03	1.20E+03	1.20E+03	1.20E+03
	Mean	1.20E+03	1.20E+03	1.20E+03	1.20E+03	1.20E+03	1.20E+03	1.20E+03	1.20E+03
	Std	7.90E−02	7.55E−02	1.08E+00	4.62E−01	3.14E−01	1.32E−01	4.02E−01	4.56E−02
F13	Best	1.30E+03	1.30E+03	1.30E+03	1.30E+03	1.30E+03	1.30E+03	1.30E+03	1.30E+03
	Worst	1.30E+03	1.30E+03	1.30E+03	1.30E+03	1.30E+03	1.31E+03	1.30E+03	1.30E+03
	Median	1.30E+03	1.30E+03	1.30E+03	1.30E+03	1.30E+03	1.30E+03	1.30E+03	1.30E+03
	Mean	1.30E+03	1.30E+03	1.30E+03	1.30E+03	1.30E+03	1.30E+03	1.30E+03	1.30E+03
	Std	4.92E−02	6.64E−02	1.42E−01	6.43E−02	4.77E−02	1.53E+00	4.94E−02	4.69E−02
F14	Best	1.40E+03	1.40E+03	1.40E+03	1.40E+03	1.40E+03	1.40E+03	1.40E+03	1.40E+03
	Worst	1.40E+03	1.40E+03	1.40E+03	1.40E+03	1.40E+03	1.48E+03	1.40E+03	1.40E+03
	Median	1.40E+03	1.40E+03	1.40E+03	1.40E+03	1.40E+03	1.40E+03	1.40E+03	1.40E+03
	Mean	1.40E+03	1.40E+03	1.40E+03	1.40E+03	1.40E+03	1.41E+03	1.40E+03	1.40E+03
	Std	3.12E−02	3.93E−02	5.25E+00	3.92E−02	3.90E−02	1.88E+01	3.36E−02	2.77E−02
F15	Best	1.51E+03	1.50E+03	1.50E+03	1.51E+03	1.50E+03	1.51E+03	1.50E+03	1.50E+03

(continued)

Table 5 (continued)

Function		CLPSO	UPSO	DNLPSO	DE-PSO	DESQI	PSOGSA	QPSO	hDEBSA
	Worst	1.51E+03	1.51E+03	5.40E+03	1.52E+03	1.51E+03	1.25E+05	1.52E+03	1.51E+03
	Median	1.51E+03	1.51E+03	1.51E+03	1.51E+03	1.51E+03	1.52E+03	1.51E+03	1.51E+03
	Mean	1.51E+03	1.51E+03	1.59E+03	1.51E+03	1.51E+03	6.12E+03	1.51E+03	1.51E+03
	Std	9.49E−01	1.79E+00	5.45E+02	1.91E+00	2.70E+00	1.76E+04	4.29E+00	6.80E−01
F16	Best	1.61E+03	1.61E+03	1.61E+03	1.61E+03	1.61E+03	1.61E+03	1.61E+03	1.61E+03
	Worst	1.61E+03	1.61E+03	1.61E+03	1.61E+03	1.61E+03	1.61E+03	1.61E+03	1.61E+03
	Median	1.61E+03	1.61E+03	1.61E+03	1.61E+03	1.61E+03	1.61E+03	1.61E+03	1.61E+03
	Mean	1.61E+03	1.61E+03	1.61E+03	1.61E+03	1.61E+03	1.61E+03	1.61E+03	1.61E+03
	Std	2.82E−01	3.72E−01	7.66E−01	4.35E−01	3.97E−01	4.75E−01	6.06E−01	4.07E−01
F17	Best	3.55E+05	4.24E+04	2.21E+04	3.09E+04	2.20E+04	9.94E+03	2.41E+04	1.37E+04
	Worst	1.59E+06	3.97E+05	5.50E+05	7.87E+05	6.32E+05	6.64E+06	4.54E+05	5.98E+05
	Median	8.99E+05	1.31E+05	1.67E+05	1.63E+05	1.94E+05	6.40E+04	7.01E+04	1.34E+05
	Mean	9.60E+05	1.49E+05	2.17E+05	2.18E+05	2.10E+05	6.40E+05	9.06E+04	1.66E+05
	Std	3.72E+05	7.80E+04	1.45E+05	1.68E+05	1.31E+05	1.30E+06	7.03E+04	1.33E+05
F18	Best	1.84E+03	1.95E+03	1.90E+03	2.13E+03	1.81E+03	2.11E+03	1.81E+03	1.81E+03
	Worst	2.32E+03	8.13E+03	1.50E+05	1.30E+04	3.71E+03	1.09E+09	3.47E+03	5.26E+03
	Median	1.92E+03	2.75E+03	3.96E+03	3.08E+03	2.12E+03	1.26E+04	1.95E+03	2.05E+03
	Mean	1.93E+03	3.20E+03	1.31E+04	3.74E+03	2.28E+03	4.45E+07	2.09E+03	2.36E+03
	Std	7.42E+01	1.28E+03	2.47E+04	1.90E+03	5.34E+02	1.76E+08	3.51E+02	7.54E+02
F19	Best	1.91E+03	1.91E+03	1.90E+03	1.91E+03	1.90E+03	1.91E+03	1.90E+03	1.90E+03
	Worst	1.91E+03	1.92E+03	1.99E+03	1.98E+03	1.96E+03	2.13E+03	1.91E+03	1.91E+03

(continued)

Table 5 (continued)

Function		CLPSO	UPSO	DNLPSO	DE-PSO	DESQI	PSOGSA	QPSO	hDEBSA
	Median	1.91E+03	1.91E+03	1.91E+03	1.91E+03	1.90E+03	1.95E+03	1.91E+03	1.90E+03
	Mean	1.91E+03	1.91E+03	1.91E+03	1.92E+03	1.91E+03	1.97E+03	1.91E+03	1.90E+03
	Std	1.33E+00	2.31E+00	1.18E+01	2.10E+01	1.16E+01	6.29E+01	1.74E+00	9.04E−01
F20	Best	2.76E+03	2.56E+03	2.32E+03	2.13E+03	2.01E+03	2.78E+03	2.19E+03	2.01E+03
	Worst	1.16E+04	1.49E+04	3.91E+04	2.36E+03	1.13E+04	1.61E+05	3.98E+03	3.33E+03
	Median	5.06E+03	7.36E+03	3.59E+03	2.20E+03	2.41E+03	8.13E+03	2.56E+03	2.07E+03
	Mean	5.70E+03	7.56E+03	6.99E+03	2.20E+03	3.35E+03	1.77E+04	2.63E+03	2.15E+03
	Std	1.95E+03	3.10E+03	7.90E+03	4.17E+01	2.21E+03	2.79E+04	3.20E+02	2.35E+02

Table 6 Experimental results of CLPSO, UPSO, DNLPSO, DE-PSO, DESQI, PSOGSA, QPSO, and hDEBSA at dimension (D) 30 after reaching D * 10,000 FEs of F21-F30 CEC2014 benchmark functions over 51 runs and 50 population size (PS)

Function		CLPSO	UPSO	DNLPSO	DE-PSO	DESQI	PSOGSA	QPSO	hDEBSA
F21	Best	1.38E+04	7.23E+03	9.01E+03	7.81E+03	1.62E+04	5.16E+03	1.03E+04	2.94E+03
	Worst	2.93E+05	3.23E+05	3.14E+06	1.01E+05	4.27E+05	2.70E+07	6.98E+04	9.70E+04
	Median	1.17E+05	4.21E+04	7.71E+04	3.40E+04	1.25E+05	4.13E+04	3.81E+04	1.30E+04
	Mean	1.27E+05	5.49E+04	2.14E+05	3.59E+04	1.41E+05	1.33E+06	3.79E+04	1.87E+04
	Std	7.04E+04	5.10E+04	4.52E+05	1.92E+04	9.93E+04	5.31E+06	1.15E+04	2.02E+04
F22	Best	2.23E+03	2.27E+03	2.24E+03	2.23E+03	2.32E+03	2.44E+03	2.35E+03	2.22E+03
	Worst	2.52E+03	2.72E+03	3.08E+03	2.67E+03	2.63E+03	3.76E+03	3.14E+03	2.58E+03
	Median	2.38E+03	2.49E+03	2.51E+03	2.36E+03	2.35E+03	3.13E+03	2.77E+03	2.35E+03
	Mean	2.37E+03	2.48E+03	2.53E+03	2.37E+03	2.40E+03	3.09E+03	2.76E+03	2.35E+03
	Std	6.83E+01	1.06E+02	1.89E+02	8.99E+01	7.27E+01	2.79E+02	1.81E+02	7.97E+01
F23	Best	2.61E+03	2.61E+03	2.61E+03	2.62E+03	2.62E+03	2.62E+03	2.62E+03	2.62E+03
	Worst	2.61E+03	2.61E+03	2.64E+03	2.62E+03	2.62E+03	2.78E+03	2.62E+03	2.62E+03
	Median	2.61E+03	2.61E+03	2.61E+03	2.62E+03	2.62E+03	2.63E+03	2.62E+03	2.62E+03
	Mean	2.61E+03	2.61E+03	2.62E+03	2.62E+03	2.62E+03	2.64E+03	2.62E+03	2.62E+03
	Std	8.18E−04	4.59E−13	5.99E+00	6.23E−01	2.54E−11	3.49E+01	9.66E−06	1.39E−12
F24	Best	2.60E+03	2.62E+03	2.62E+03	2.62E+03	2.62E+03	2.62E+03	2.60E+03	2.62E+03
	Worst	2.63E+03	2.64E+03	2.70E+03	2.65E+03	2.63E+03	2.73E+03	2.62E+03	2.64E+03
	Median	2.63E+03	2.63E+03	2.64E+03	2.64E+03	2.62E+03	2.66E+03	2.62E+03	2.63E+03
	Mean	2.62E+03	2.63E+03	2.64E+03	2.64E+03	2.62E+03	2.66E+03	2.62E+03	2.63E+03
	Std	4.17E+00	4.51E+00	1.23E+01	7.29E+00	1.49E+00	2.61E+01	1.02E+01	4.48E+00
F25	Best	2.70E+03	2.70E+03	2.70E+03	2.71E+03	2.70E+03	2.70E+03	2.70E+03	2.70E+03

(continued)

Table 6 (continued)

Function		CLPSO	UPSO	DNLPSO	DE-PSO	DESQI	PSOGSA	QPSO	hDEBSA
	Worst	2.70E+03	2.70E+03	2.76E+03	2.72E+03	2.71E+03	2.72E+03	2.71E+03	2.71E+03
	Median	2.70E+03	2.70E+03	2.70E+03	2.71E+03	2.71E+03	2.71E+03	2.70E+03	2.71E+03
	Mean	2.70E+03	2.70E+03	2.70E+03	2.71E+03	2.71E+03	2.71E+03	2.70E+03	2.71E+03
	Std	6.39E−01	1.79E−01	9.13E+00	1.81E+00	3.09E+00	4.01E+00	3.17E+00	7.42E−01
F26	Best	2.70E+03	2.70E+03	2.70E+03	2.70E+03	2.70E+03	2.70E+03	2.70E+03	2.70E+03
	Worst	2.70E+03	2.70E+03	2.94E+03	2.80E+03	2.80E+03	2.80E+03	2.80E+03	2.80E+03
	Median	2.70E+03	2.70E+03	2.70E+03	2.70E+03	2.80E+03	2.70E+03	2.80E+03	2.70E+03
	Mean	2.70E+03	2.70E+03	2.72E+03	2.71E+03	2.76E+03	2.73E+03	2.79E+03	2.70E+03
	Std	5.09E−02	6.97E−02	6.25E+01	2.71E+01	4.87E+01	4.39E+01	2.67E+01	1.40E+01
F27	Best	3.11E+03	3.15E+03	3.03E+03	3.10E+03	3.10E+03	3.10E+03	3.10E+03	3.02E+03
	Worst	3.15E+03	3.78E+03	3.61E+03	3.37E+03	3.32E+03	3.99E+03	3.48E+03	3.14E+03
	Median	3.12E+03	3.66E+03	3.14E+03	3.10E+03	3.18E+03	3.73E+03	3.34E+03	3.08E+03
	Mean	3.12E+03	3.62E+03	3.18E+03	3.14E+03	3.18E+03	3.69E+03	3.32E+03	3.07E+03
	Std	8.93E+00	1.26E+02	1.23E+02	7.74E+01	5.92E+01	2.19E+02	8.83E+01	3.10E+01
F28	Best	3.20E+03	3.20E+03	3.18E+03	3.64E+03	3.10E+03	3.89E+03	3.10E+03	3.60E+03
	Worst	3.23E+03	3.26E+03	3.34E+03	3.95E+03	3.84E+03	5.62E+03	7.55E+03	3.71E+03
	Median	3.21E+03	3.22E+03	3.20E+03	3.80E+03	3.64E+03	4.59E+03	4.29E+03	3.67E+03
	Mean	3.21E+03	3.22E+03	3.22E+03	3.81E+03	3.61E+03	4.62E+03	4.83E+03	3.66E+03
	Std	4.43E+00	1.53E+01	4.11E+01	6.13E+01	1.64E+02	4.52E+02	1.09E+03	3.02E+01
F29	Best	3.11E+03	3.11E+03	3.10E+03	5.79E+03	3.50E+03	4.10E+03	3.45E+03	3.62E+03
	Worst	3.11E+03	3.12E+03	3.28E+03	1.66E+04	4.54E+03	2.95E+07	4.59E+03	4.95E+03

(continued)

Table 6 (continued)

Function		CLPSO	UPSO	DNLPSO	DE-PSO	DESQI	PSOGSA	QPSO	hDEBSA
	Median	3.11E+03	3.11E+03	3.11E+03	8.95E+03	3.79E+03	4.00E+05	3.63E+03	3.99E+03
	Mean	3.11E+03	3.11E+03	3.11E+03	9.67E+03	3.83E+03	4.70E+06	3.70E+03	4.01E+03
	Std	1.65E+00	2.50E+00	2.60E+01	2.58E+03	2.21E+02	6.94E+06	2.32E+02	2.05E+02
F30	Best	3.30E+03	3.33E+03	3.32E+03	4.58E+03	3.64E+03	5.31E+03	4.51E+03	4.04E+03
	Worst	3.62E+03	3.91E+03	4.79E+03	1.13E+04	6.25E+03	1.91E+05	6.81E+03	6.51E+03
	Median	3.42E+03	3.59E+03	3.93E+03	7.04E+03	5.23E+03	1.38E+04	5.48E+03	4.96E+03
	Mean	3.43E+03	3.61E+03	3.94E+03	7.23E+03	5.25E+03	2.97E+04	5.55E+03	5.05E+03
	Std	7.76E+01	1.32E+02	2.95E+02	1.47E+03	6.03E+02	4.23E+04	4.44E+02	5.13E+02

Table 7 Ranks obtained by Friedman test on CEC2014 real-world benchmark functions with 30 dimensions of all the algorithms using their median performances

Algorithms	CLPSO	UPSO	DNLPSO	DE-PSO	DESQI	PSOGSA	QPSO	hDEBSA
Mean rank	3.82	4.37	4.28	5.17	4.10	6.62	4.43	3.22

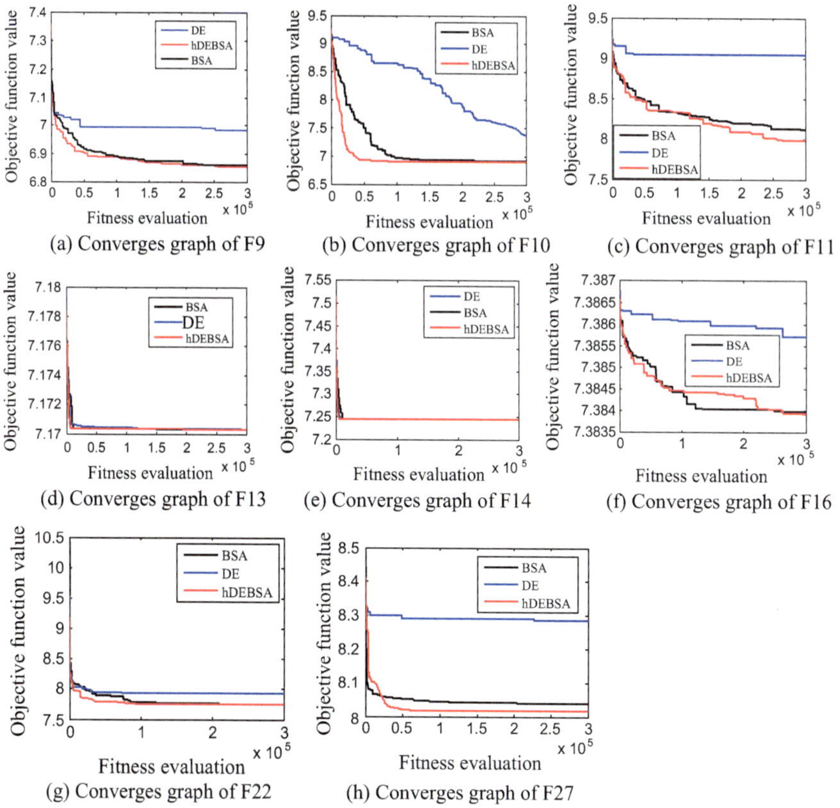

Fig. 4 Convergence graph of eight functions (F9, F10, F11, F13, F14, F16, F22 and F27) with $D = 30$ and 300,000 FEs over objective function value versus fitness evaluation

Fig. 5 Effect of angel of internal friction of soil (ϕ) on seismic active earth pressure coefficient (k_{ae}) at $\alpha = 30°$, $i = 10°$, $m = 0.1$, $n = 0.05$, $\delta = \phi/2$. $k_v = k_h/2$

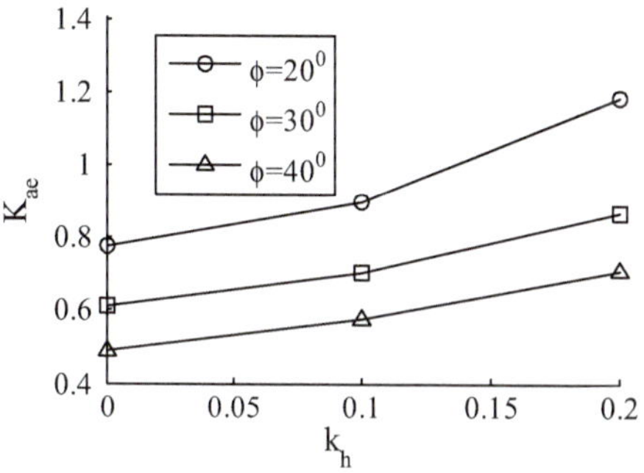

Fig. 6 Effect of angel of wall friction (δ) on seismic active earth pressure coefficient (K_{ae}) at $\alpha = 30°$, $\phi = 30°$, $i = 100$, $m = 0.1$, $n = 0.05$, $k_v = k_h/2$

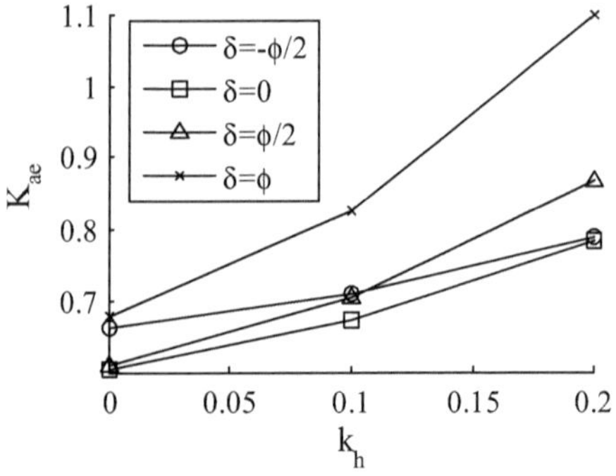

Fig. 7 Effect of seismic acceleration k_v on seismic active earth pressure coefficient (K_{ae}) at $\alpha = 30°$, $\phi = 30°$, $i = 10°$, $m = 0.1$, $n = 0.05$, $\delta = \phi/2$

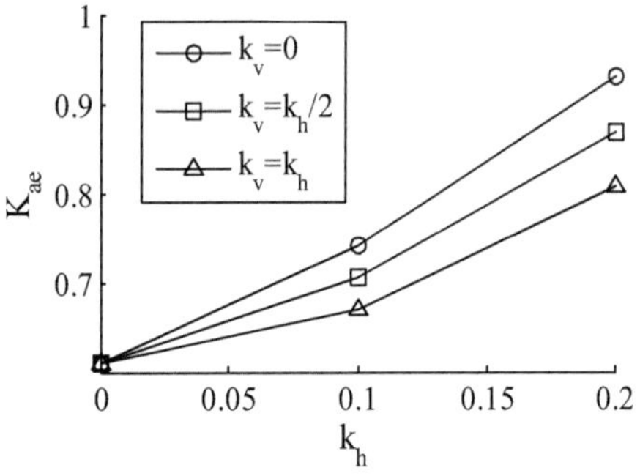

of seismic acceleration (k_v), the amount of K_{ae} decreases. Also, it is observed that through the growth in the amount of k_h, the amount of K_{ae} will also increase.

Figure 8 represents the effect of inclined angle of retaining wall (α) on K_{ae} at $i = 10°$, $\phi = 30°$, $\delta = \phi/2$, $m = 0.1$, $n = 0.05$, $k_v = k_h/2$ with respect to the k_h. From this figure, it can be perceived that with the growth in the value of inclined angel of retaining wall (α) from $-20°$ to $20°$, the magnitude of K_{ae} increases. Also, it is observed that with the increment in the amount of k_h, the amount of K_{ae} increases.

Fig. 8 Effect of angel α on seismic active earth pressure coefficient (K_{ae}) at $\phi = 300$, $\delta = \phi/2$, $i = 10°$, $m = 0.1$, $n = 0.05$, $k_v = k_h/2$

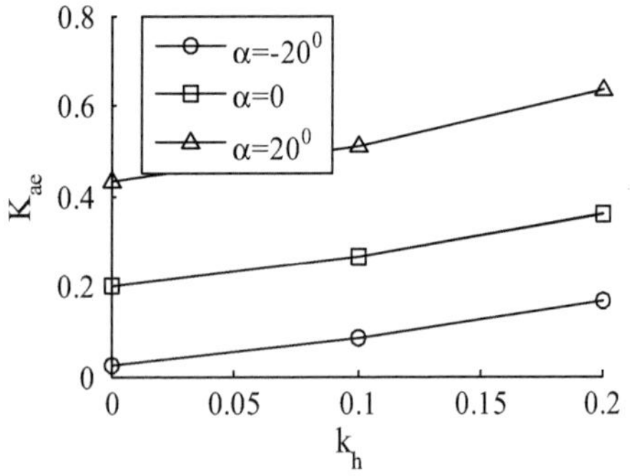

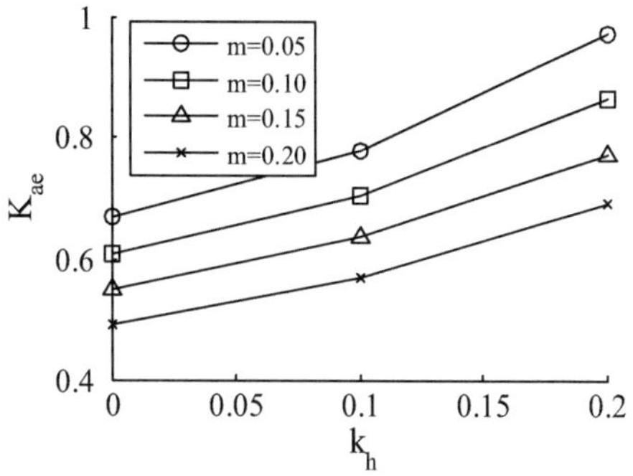

Fig. 9 Effect of cohesion parameter m on seismic active earth pressure coefficient (K_{ae}) at $\alpha = 30°$, $\phi = 30°$, $\delta = \phi/2$, $i = 10°$, $n = 0.05$

Figure 9 represents the effect of the magnitude of cohesion parameter (m) on K_{ae} at $i = 10°$, $\alpha = 30°$, $\phi = 30°$, $\delta = \phi/2$, $n = 0.05$, $k_v = k_h/2$ with respect to the k_h. From this plot, it is observed that with the increment in the amount of cohesion parameter (m), the amount of K_{ae} decreases. Also, it is observed that with the increment in the amount of value of k_h, the amount of K_{ae} increases. Figure 10 represents the effect of backfill inclination angle (i) on K_{ae} at $\alpha = 30°$, $\phi = 30°$, $\delta = \phi/2$, $m = 0.1$, $n = 0.05$, $k_v = k_h/2$ with respect to the k_h. From this figure, it can be observed that when the value of backfill inclination angel (i) increases from 0° to 20°, the magnitude of K_{ae} upturns. Also, it is observed that with the increment in the magnitude of k_h, the amount of K_{ae} increases.

Figure 11 represents the effect of the magnitude of adhesion parameter (n) on K_{ae} at $i = 10°$, $\alpha = 30°$, $\phi = 30°$, $\delta = \phi/2$, $m = 0.1$, $k_v = k_h/2$ with respect to the k_h. From this plot, it is observed that when the amount of adhesion parameter (m) increases, the amount of K_{ae} also increases. Also, it is observed that the increment in the amount of k_h results in the increment in the amount of K_{ae}.

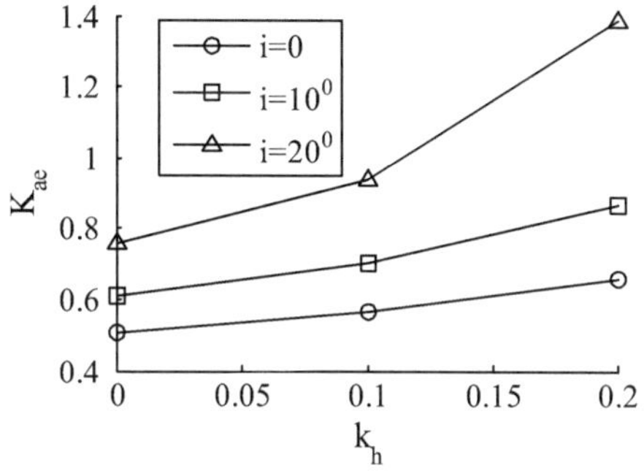

Fig. 10 Effect of backfill inclination angle i on seismic active earth pressure coefficient (K_{ae}) at $\alpha = 30°$, $\phi = 30°$, $\delta = \phi/2$, $m = 0.10$, $n = 0.05$

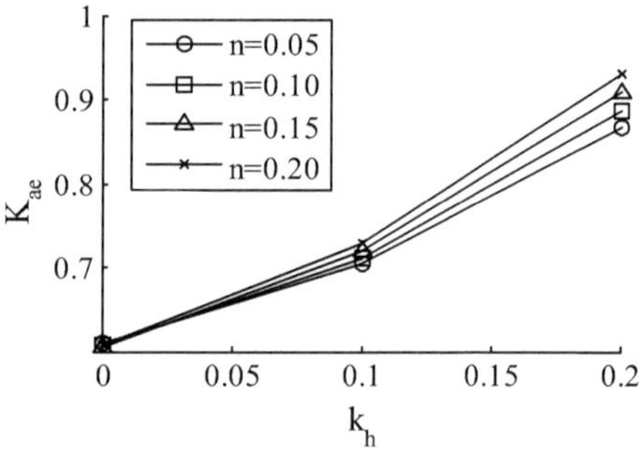

Fig. 11 Effect of adhesion parameter n on seismic active earth pressure coefficient (K_{ae}) at $\alpha = 30°$, $\phi = 30°$, $\delta = \phi/2$, $i = 100$, $m = 0.1$

6.2.2 Comparative Study

Table 8 demonstrates the performance result of PSO [4], HS [7], BSA [8], DE [2], ABC [5], and hDEBSA of the active earth pressure coefficient (K_{ae}) at $m = 0.1$, $k_h = 0.2$, $k_v = k_h/2$, $n = 0.05$, $\delta = \Phi/2$. Table 9 shows the comparison results of active earth pressure coefficient (K_{ae}) determined by hDEBSA with PSO, HS, BSA, DE, ABC at $m = 0.1$, $k_h = 0.2$, $k_v = k_h/2$, $n = 0.05$, $\delta = 2\Phi/3$. Table 10 depicts

Table 8 Comparison result of total active earth force on the back of a retaining wall with PSO, HS, BSA, DE, ABC, and hDEBSA for m = 0.1, n = 0.05, $k_h = 0.2$, $k_v = k_h/2$, $\delta = \phi/2$

i	ϕ	α	PSO	HS	BSA	DE	ABC	hDEBSA
0	20	−20	0.2888	0.2888	0.2888	0.2888	0.2888	0.2888
		0	0.43562	0.4341	0.43562	0.43562	0.43559	0.43562
		20	0.63484	0.63427	0.63484	0.63484	0.63437	0.63484
	30	0	0.29109	0.29089	0.29109	0.29109	0.28245	0.29109
		20	0.50043	0.50015	0.50043	0.50043	0.50042	0.50043
	40	0	0.19369	0.19197	0.19369	0.19369	0.19351	0.19369
		20	0.40411	0.4037	0.40411	0.40411	0.40411	0.40411
		30	0.57155	0.57049	0.57155	0.57155	0.57148	0.57155
10	20	−20	0.40243	0.39905	0.40243	0.40243	0.40216	0.40243
		0	0.60369	0.60365	0.60369	0.60369	0.59787	0.60369
		20	0.91477	0.91456	0.91477	0.91477	0.91432	0.91477
	30	−20	0.16727	0.16688	0.16727	0.16727	0.16227	0.16727
		0	0.35789	0.35788	0.35789	0.35789	0.35789	0.35789
		20	0.63582	0.63559	0.63582	0.63582	0.63517	0.63582
20	30	−20	0.24254	0.24254	0.24254	0.24254	0.24027	0.24254
		0	0.51489	0.51488	0.51489	0.51489	0.51307	0.51489
		20	0.96311	0.96099	0.96311	0.96311	0.96285	0.96311
	40	0	0.27909	0.2784	0.27909	0.27909	0.27814	0.27909
		20	0.62251	0.62055	0.62251	0.62251	0.62238	0.62251

Table 9 Comparison result of total active earth force on the back of a retaining wall of hDEBSA with PSO, HS, BSA, DE, ABC at $m = 0.1, n = 0.05, k_h = 0.2, k_v = k_h/2, \delta = 2\phi/3$

i	ϕ	α	PSO	HS	BSA	DE	ABC	hDEBSA
10	23	0	0.50821	0.50817	0.50821	0.50821	0.50578	0.50821
		30	1.0986	1.097	1.0986	1.0986	1.0978	1.0986
	40	0	0.23179	0.23168	0.23179	0.23179	0.2301	0.23179
		30	0.78484	0.78483	0.78484	0.78484	0.7763	0.78484
20	33	0	0.42674	0.42663	0.42674	0.42674	0.42672	0.42674
		30	1.2977	1.2968	1.2977	1.2977	1.2957	1.2977
	40	0	0.28788	0.28759	0.28788	0.28788	0.28736	0.28788
		30	1.0555	1.0536	1.0555	1.0555	1.0526	1.0555

Table 10 Comparison result of total active earth force on the back of a retaining wall of hDEBSA with Sharma and Ghosh [38, PSO, HS, BSA, DE, ABC at $m = 0.1, n = 0.1, k_h = 0.1, k_v = k_h/2, \delta = 2\phi/3$

i	ϕ	α	Sharma and Ghosh [38]	PSO	HS	BSA	DE	ABC	hDEBSA
10	20	0	0.305	0.41613	0.41612	0.41613	0.41613	0.416	0.41613
		30	0.83	0.92561	0.92441	0.92561	0.92561	0.9256	0.92561
	30	0	0.163	0.24676	0.24641	0.24676	0.24676	0.24675	0.24676
		30	0.671	0.74132	0.73844	0.74132	0.74132	0.74116	0.74132
	40	0	0.077	0.14415	0.14415	0.14415	0.14415	0.14406	0.14415
		30	0.574	0.62956	0.62749	0.62956	0.62956	0.62784	0.62956
20	30	0	0.215	0.32123	0.3199	0.32123	0.32123	0.3212	0.32123
		30	0.891	1.0067	1.0024	1.0067	1.0067	1.006	1.0067
	40	0	0.097	0.17423	0.17121	0.17423	0.17423	0.17171	0.17423
		30	0.701	0.7891	0.78578	0.7891	0.7891	0.78441	0.7891

the comparison result of total active earth pressure coefficient (K_{ae}) determined by hDEBSA with Sharma and Ghosh [38], PSO, HS, BSA, DE, ABC at $m = 0.1$, $n = 0.1$, $\delta = 2\Phi/3$, $k_h = 0.1$, $k_v = k_h/2$. From the observation of Tables 8 and 9, it may very well be seen that the exhibition result, i.e., K_{ae} of hDEBSA is better than HS and ABC. Also, from Table 10, it can be observed that the results achieved by hDEBSA are superior to Sharma and Ghosh [38], HS and ABC. Thus, the above research shows that hDEBSA works better in numerical results more satisfactorily than other comparable algorithms.

7 Conclusion

The efficiency of DE and BSA algorithm relies upon the correct estimation of standard algorithm-specific control parameters which are scaling factor, crossover rate, and mixrate. Additionally, during the advancement procedure, fluctuating the estimations of control parameter can improve the exhibition of an algorithm. This document provides the hybrid algorithm hDEBSA, which considers all control parameters as self-adapting and which is used to solve CEC2014's unconstrained test function and geotechnical optimization problem. This method has been used successfully. The obtained result has been compared with certain usual algorithm like DE, BSA, PSO, HS, and ABC; some improved variants like CLPSO, UPSO, and DNLPSO and also some hybrid variants like DE-PSO, DESQI, QPSO, and PSOGSA. Also, the performance result on geotechnical optimization problem of active earth pressure coefficient is also compared with Sharma and Ghosh [26]. From the contrast of the results acquired using the hDEBSA method with other methods in terms of numerical result shows the acceptability of the proposed method. The advantage of the hDEBSA algorithm lies in the information that all the usual specific control parameters are self-adaptive based. The parametric study, i.e., effect of the soil, wall, and seismic parameter on active earth pressure coefficient has also been investigated in this study. It is shown that the coefficient of seismic active earth pressure decreases with the increase in m, ϕ, n, whereas it increases with an increase in α, i, k_h, $k_{v,}$ etc. It may be recommended that the hDEBSA method can be utilized to find a solution to optimization problem of this kind.

References

1. Holland, J.H.: Adaptation in natural and artificial systems. University of Michigan Press (1975)
2. Storn, R., Price, K.: Differential evolution—a simple and efficient heuristic for global optimization over continuous spaces. J. Global Optim. **11**(4), 341–359 (1997)
3. Eberhart, R., Shi, Y.: Particle swarm optimization: developments, applications and resources in evolutionary computation. Proc. 2001 Congr. **81**, 81–86 (2001)
4. Shi, Y., Eberhart R.: A modified particle swarm optimizer. In: Evolutionary Computation Proceedings, IEEE World Congress on Computational Intelligence (1998)
5. Akay, B., Karaboga, D.: Artificial bee colony algorithm for large-scale problem and engineering design optimization. J. Intell. Manuf. **23**, 1001–1014 (2012)
6. Lee, K.S.: Geem ZW (2005) A new meta-heuristic algorithm for continuous engineering optimization: harmony search theory and practice. Comput. Methods. Appl. Mech. Eng. **194**, 3902–3933 (2005)
7. Mahdavi, M., Fesanghary, M., Damangir, E.: An improved harmony search algorithm for solving optimization problems. Appl. Math. Comput. **188**, 1567–1579 (2007)
8. Civicioglu, P.: Backtracking search optimization algorithm for numerical optimization problems. Appl. Math. Comput. **219**(15), 8121–8144 (2013)
9. Rashedi, E., Nezamabadi-pour, H., Saryazdi, S.: GSA: a gravitational search algorithm. Inf. Sci. **179**, 2232–2248 (2009)

10. Liang, J.J., Qin, A.K., Suganthan, P.N., Baskar, S.: Comprehensive learning particle swarm optimizer for global optimization of multimodal functions. IEEE Trans. Evol. Comput. **10**(3), 281 (2006)
11. van den Bergh, F., Engelbrecht, A.P.: A cooperative approach to particle swarm optimization. IEEE Trans. Evol. Comput. **8**, 225–239 (2004)
12. Mendes, R., Kennedy, J., Neves, J.: The fully informed particle swarm: simpler, may be better. IEEE Trans. Evol. Comput. **8**, 204–210 (2004)
13. Parsopoulos, K.E., Vrahatis, M.N.: UPSO—a unified particle swarm optimization scheme. Lect. Ser. Comput. Sci. **1**, 868–873 (2004)
14. Nasir, Md, Das, S., Maity, D., Sengupta, S., Halder, U., Suganthan, P.N.: A dynamic neighborhood learning based particle swarm optimizer for global numerical optimization. Inf. Sci. **209**, 16–36 (2012)
15. Nama, S., Saha, A.K., Ghosh, S.: Improved symbiotic organisms search algorithm for solving unconstrained function optimization. Decis. Sci. Lett. **5**(3), 361–380 (2016)
16. Nama, S., Saha, A.K., Ghosh, S.: Improved backtracking search algorithm for pseudo dynamic active earth pressure on retaining wall supporting $c - \Phi$ backfill. Appl. Soft Comput. **52**, 885–897 (2017)
17. Nama, S., Saha, A.K.: An ensemble symbiosis organisms search algorithm and its application to real world problems. Decis. Sci. Lett. **7**(2), 103–118 (2018)
18. Ezugwu, A.E., Els, R., Fonou-Dombeu, J.V., Naidoo, D., Pillay, K.: Parallel symbiotic organisms search algorithm. In: Misra, S. et al. (eds) Computational Science and Its Applications—ICCSA 2019. ICCSA 2019. Lecture Notes in Computer Science, vol. 11623. Springer, Cham (2019)
19. Mirjalili, S., Mohd Hashim, S.Z.: A new hybrid PSOGSA algorithm for function optimization. In: International Conference on Computer and Information Application (ICCIA 2010) (2010)
20. Pant, M., Thangaraj, R., Abraham, A.: A new PSO algorithm with crossover operator for global optimization problems. Innov. Hybr. Intell. Syst. ASC **44**, 215–222 (2007)
21. Zhang, L., Li, H., Jiao, Y.-C., Zhang, F.-S.: Hybrid Differential Evolution and the Simplified Quadratic Interpolation For Global Optimization, Copyright is held by the author/owner(s). GEC'09, June 12–14, 2009, Shanghai, China. ACM 978-1-60558-326-6/09/06 (2009)
22. Pant, M., Thangaraj, R.: DE-PSO: a new hybrid meta-heuristic for solving global optimization problems. New Mathem. Nat. Comput. **7**(3), 363–381 (2011)
23. Nama, S., Saha, A.K., Ghosh, S.: A new ensemble algorithm of differential evolution and backtracking search optimization algorithm with adaptive control parameter for function optimization. Int. J. Ind. Eng. Comput. **7**(2), 323–338 (2016)
24. Nama, S., Saha, A.K., Ghosh, S.: A hybrid symbiosis organisms search algorithm and its application to real world problems. Memet. Comput. **9**(3), 261–280 (2017)
25. Nama, S., Saha, A.K.: A new hybrid differential evolution algorithm with self-adaptation for function optimization. Appl. Intell. **48**(7), 1657–1671 (2018)
26. Nama, S., Saha, A.K.: A novel hybrid backtracking search optimization algorithm for continuous function optimization. Decis. Sci. Lett. **8**(2), 163–174 (2019)
27. Bolton, H.P.J., Heymann, G., Groenwold, A.: Global search for critical failure surface in slope stability analysis. Eng. Optim. **35**, 51–65 (2003)
28. Cheng, Y.M.: Global optimization analysis of slope stability by simulated annealing method with dynamic bounds and "dirac function". Eng. Optim. **39**(1), 17–32 (2007)
29. Das S.K.: Slope Stability Analysis Using Genetic Algorithm. EJGE paper 2005-0504, (2014)
30. Deb, K., Goyal, M.: Optimizing engineering designs using combined genetic search. In: Proceedings of Seventh International Conference on Genetic Algorithms, pp. 512–28 (1997)
31. Cheng, Y.M., Li, L., Chi, S.C.: Performance studies on six heuristic global optimization methods in the location of critical slip surface. Comput. Geotech. **34**, 462–484 (2007)
32. Cheng, Y.M., Li, L., Chi, S., Wei, W.B.: Particle swarm optimization algorithm for location of critical non-circular failure surface in two dimensional slope stability analysis. Comput. Geotech. **34**(2), 92–103 (2007)

33. Sengupta, A., Upadhyay, A.: Locating the critical failure surface in a slope stability analysis by genetic algorithm. Appl. Soft Comput. **9**, 387–392 (2009)
34. Zolfaghari, A.R., Heath, A.C., Mc Combie, P.F.: Simple genetic algorithm search for critical non-circular failure surface in slope stability analysis. Comput. Geotech. **32**, 139–152 (2005)
35. Ahmadi-Nedushan, B., Varaee, H.O: Optimal design of reinforced concrete retaining walls using a swarm intelligence technique. In: Topping, B.H.V., Tsompanakis, Y. (eds) Proceedings of the First International Conference on Soft Computing Technology in Civil, Structural and Environmental Engineering, Civil-Comp Press, Stirlingshire, Scotland (2009)
36. Ghazavi, M., Bazzazian Bonab, S.: Optimization of Reinforced Concrete Retaining Walls Using Ant Colony Method, ISGSR 2011—Vogt, © 2011 Bundesanstalt für Wasserbau ISBN 978-3-939230-01-4 (2011)
37. Chakraborty S., Das S., Gupta S., Ghosh S.: A critical review of IS: 1893 (Part 3) (2014), International Geotechnical Engineering Conference on Sustainability in Geotechnical Engineering Practices and Related Urban Issues, September, pp. 23–24, Mumbai, India (2016)
38. Sharma, R.P., Ghosh, S.: Pseudo static seismic active response of retaining wall supporting $c - \varphi$ backfill. Electr. J. Geotech. Eng. (EJGE) **15**, 533 (2010)
39. Smuc, T.: Sensitivity of differential evolution algorithm to value of control parameters. In: Proceedings of the International Conference on Artificial Intelligence, pp. 108–1093 (2002)
40. Smuc, T.: Improving convergence properties of the differential evolution algorithm. In: Proceedings of MENDEL 2002, 8th International Mendel Conference on Soft Computing, pp. 80–86 (2002)
41. Gong, W.Y., Cai, Z.H.: Differential evolution with ranking based mutation operators. IEEE Trans. Cybern. **43**(6), 2066–2081 (2013)
42. Wang, L., Zhong, Y., Yin, Y., Zhao, W., Wang, B., Xu, Y.: A hybrid backtracking search optimization algorithm with differential evolution. Mathem. Prob. Eng. vol. 2015, Article ID 769245. http://dx.doi.org/10.1155/2015/769245,(2015)
43. Gämperle, R., Müller, S.D., Koumoutsakos, P.: A parameter study for differential evolution. Adv. Intell. Syst. Fuzzy Syst. Evol. Comput. **10**, 293–298 (2002)
44. Ronkkonen, J., Kukkonen, S., Price, K.V.: Real-parameter optimization with differential evolution. Proc. IEEE CEC **1**, 506–513 (2005)
45. Zaharie, D.: Influence of crossover on the behavior of differential evolution algorithms. Appl. Soft Comput. **9**(3), 1126–1138 (2009)
46. Zhang, C., Ning, J., Lu, S., Ouyang, D., Ding, T.: A novel hybrid differential evolution and particle swarm optimization algorithm for unconstrained optimization. Oper. Res. Lett. **37**, 117–122 (2009)
47. Storn, R.: On the usage of differential evolution for function optimization. In: Biennial Conference of the North American Fuzzy Information Processing Society (NAFIPS), IEEE, Berkeley, pp. 519–523 (1996)
48. Lampinen, J., Zelinka, I.: On stagnation of the differential evolution algorithm. In: Proceedings of MENDEL 2000, 6th International Mendel Conference on Soft Computing, pp. 76–83 (2000)
49. Mononobe, N., Matsuo, H.: On the determination of earth pressure during earthquakes. Proc. World Eng. Conf. **9**, 176 (1929)
50. Okabe, S.: General theory of earth pressure. J. Jpn. Soc. Civ. Eng. 12(1) (1926)
51. Seed, H.B., Whitman, R.V.: Design of Earth Retaining Structures for Dynamic Loads, Lateral stresses in the ground and design of earth retaining structures, pp. 103–107. ASCE, New York (1970)
52. Ghosh, S., Dey, G.N., Datta, B.: pseudostatic analysis of rigid retaining wall for dynamic active earth pressure. In: 12th International Conference of International Association for Computer Methods and Advances in Geomechanics, Goa, India, pp. 4122–4131 (2008)
53. Liang, J.J., Qu, B.Y., Suganthan, P.N.: Problem definitions and evaluation criteria for the CEC 2014. Special Session and Competition on Single Objective Real Parameter Numerical Optimization. Technical Report 11, Computational Intelligence Laboratory, Zhengzhou University, Zhengzhou China and Technical Report, Nanyang Technological University, Singapore, December (2013)

Effects of Squats Exercise with EMS on Muscle Strength, Endurance, and Body Function

Hye-im Han, Yu-Jin Jeong, Ha-yeong Sin, Dong-Yeop Lee, Ji-Heon Hong, Jin-Seop Kim, and Jae-Ho Yu

Abstract **Background/Objectives**: The purpose of this experiment is to find out the degree of improvement by clarifying the effects of squats after attaching them to EMS through the functions of the body, such as muscle strength, jumping, and muscle endurance. **Methods**: The participants were divided into EMS squat group (ESG), squat group (SG), and control group (CoN) without squat exercise, and measured all subjects. Squat exercise was performed at 90° knee flexion with broad legs and straightened waist. Using ultrasonography, we measured the thickness of vastus intermedius (IM), rectus femoris (RF), vastus medialis (VM), vastus lateralis (VL), rectus abdominis (RA), external abdominal oblique (EO), internal abdominal oblique (IO), and transverse abdominis (TrA). And we measured jump, height, endurance, and agility. **Findings**: Squat movement applied with EMS improved jump, endurance, and agility, and increased the thickness of quadriceps muscle except VL, VM, and abdominal muscle, except TrA. However, we found that in endurance, there is not much difference from squat exercise. **Improvements**: There was a time constraint because the study was conducted for a short period, and the study was limited due to the occurrence of the elimination due to the strength of exercise.

H. Han · Y.-J. Jeong · H. Sin · D.-Y. Lee · J.-H. Hong · J.-S. Kim · J.-H. Yu (✉)
Department of Physical Therapy, Sunmoon University, Asan-SI, Chungnam, South Korea
e-mail: naresa@sunmoon.ac.kr

H. Han
e-mail: hanh456@naver.com

Y.-J. Jeong
e-mail: useful77755@naver.com

H. Sin
e-mail: hyhy1998@naver.com

D.-Y. Lee
e-mail: kan717@hanmail.net

J.-H. Hong
e-mail: hgh1020@sunmoon.ac.kr

J.-S. Kim
e-mail: skylove3373@hanmail.net

A. K. Bhoi et al. (eds.), *Bio-inspired Neurocomputing*, Studies in Computational Intelligence 903, https://doi.org/10.1007/978-981-15-5495-7_13

Keywords EMS · Squat · Ultrasonography · Jump · Endurance · Body function

1 Introduction

The muscular endurance and muscle strength of the lower body are important factors not only for athletes but also for the public [1]. Squat is the most efficient exercise to improve these factors. Squat exercise is a multi-joint movement that is preferred because it is easy to apply in everyday life and requires many large groups of muscles to function together [2]. And this exercise is often applied to muscle strengthening, repeatedly bending and stretching knee joints. It also improved the muscles and stamina of the lower extremities, and the ability to jump and run [3–5]. Various studies have been conducted on squats with these advantages. Among them, Watts et al. reported that with the application of EMS in the traditional squats, he showed many improvements in muscle strength, muscle strength, and jumping and running skills [5].

EMS are safe and reasonable, with minimal time-saving in cohort studies that cannot participate in existing exercise programs [6, 7]. EMS are also frequently used to relieve symptoms of many neurological disorders and improve muscle strength [8]. The combination of dynamic motion and EMS is very effective. It is easy to support weakness muscles, especially when EMS are applied in squats [9, 10]. In particular, strength, short distance, endurance, jumping, and body fat are important factors in the application of EMS. Recent studies have shown that squat motion using EMS improves maximum strength and power [5]. In addition, the core muscle stability exercise with an eight-week EMS has significantly improved endurance and abdominal muscle thickness compared to the control group [11]. Studies of jumping ability have shown that in EMS training for athletes, it improves their ability to jump and run [12].

EMS strongly activates anaerobic glycosidic, producing lactose and energy to acidify more cytoplasm than spontaneous contractions [9]. Also, it is applied to dynamic movements which can increase the ability to activate muscle length and muscle contraction [12]. These devices can also stimulate multiple muscle groups at the same time and multiple muscle groups such as agonist and antagonist [9]. As a result, it reduced fat and back pain and improved muscle mass and function [13]. And the application of EMS for two weeks improved the strength and endurance of the back extension muscle has a significant effect [14].

However, young adults were similar to the training effect in groups with EMS, especially those without EMS, except for a significant increase in leg extension. In addition, the low current strength and the EMS used for a short period failed to prevent muscle atrophy in critically ill patients. Wirtz et al. reported the cause of this threshold. Most muscle fibers were already active with high resistance [9]. Other opinion, Kemmler argued that the impact of the EMS group on endurance performance parameters was minimal [10].

Table 1 General characteristics of participants ($N = 39$)

Division	Subject
Age (year)	22.34 ± 3.85
Height (cm)	167.35 ± 24.73
Weight (kg)	68.42 ± 39.74

There have been many papers that have studied the effects of squats and EMS each so far. But few studies have been conducted and compared at the same time. And interventions that improve endurance were still debatable. Therefore, the purpose of this study is to study how much applied with EMS improves muscle strength, endurance, agility, and jump.

2 Materials and Methods

2.1 *Participants*

The study design was about randomized control and multiple intervention. In this study, 39 healthy female and male adults were attending Sun Moon University in Asan. Before taking part, all subjects were given sufficient explanations about the purpose and method of this study, and 39 participants conducted experiments in CoN. The ESG wearing EMS and did squat exercise, SG did squat exercise, and CoN did not apply any exercise. The exclusion in this study is history of ACL damage, knee damage, or surgical experience. The physical characteristics of the subjects are given in (Table 1). This study was conducted with the approval of the Sun Moon University Research by the Institutional Review Board at Sun moon University Research Ethnics Board (SM-201904-029-1).

2.2 *Experimental Procedures*

We attached the EMS and divided it into EMS squat group (ESG) attached to the EMS and squat group (SG) not attached to the EMS to experiment with improving muscle strength, muscle endurance, and agility during squat exercise. ESG wearing EMS, SG, and CoN without EMS had a squat three days a week for four weeks. Squatting ESG and SG should be performed at knee flexion 90° with legs spread wide and waist straightened to prevent bending (Fig. 1). ESG and SG breathe in before they go down to work out, energize the abdominal muscle, and breathe out when they stand up. They perform squat repeatedly 25 in one set for four times.

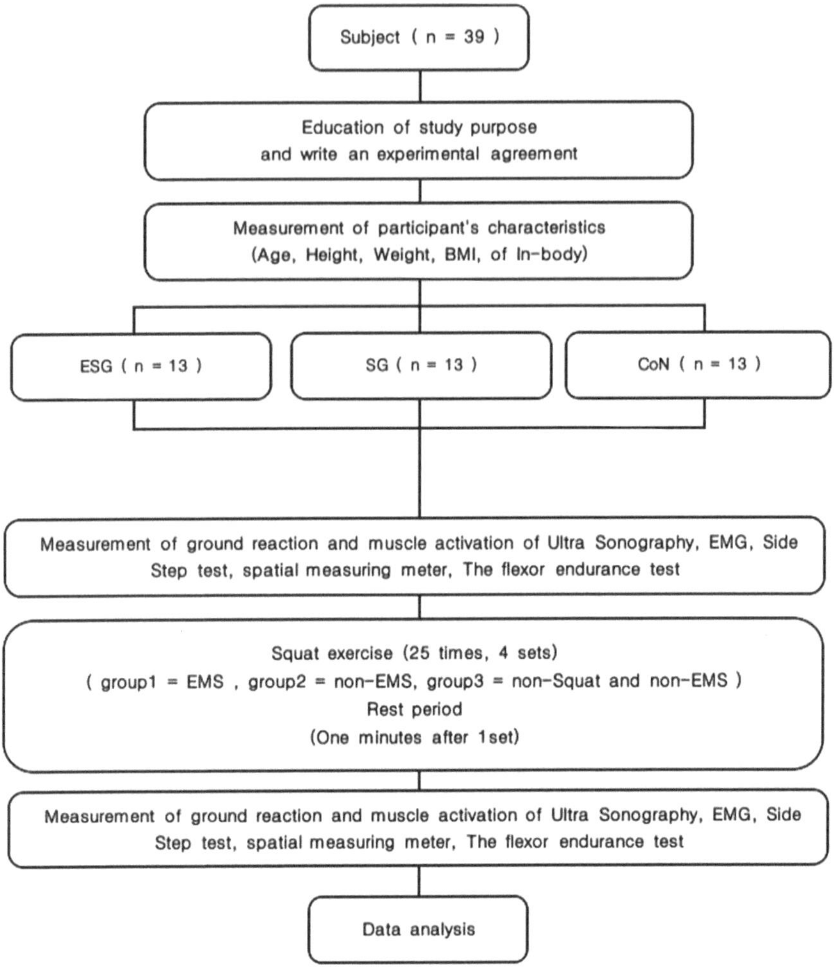

Fig. 1 Experiment protocol flowchart

2.2.1 Ultrasonography Measurement Muscle

See Table 2.

2.2.2 Squat Program

According to Nicolas et al., a similar addition was made after a six-week 10 RM back squat program with only redundant EMS (ESG) and 10 RM back squat motion (SG). Special adaptation for ESG was advanced in leg curl muscles, and SG and ESG applied SJ, CMJ, and pendulum sprint performance. No implementation was

Table 2 Measurement quadriceps and abdonimal muscle

Muscle	Location
IM	Deep midway between the ASIS and superior border of the patella above 10 cm
RF	Midway between the ASIS and superior border of the patella above 10 cm
VM	Medial way between the ASIS and superior border of the patella above 10 cm
VL	Lateral way between the ASIS and superior border of the patella above 10 cm
IO	The ASIS to the pubic symphysis and lines from one ASIS to the other
EO	Anterior end of the eighth rib
TrA	Measured in supine hook lying position with the hand under the head
RA	Measure it against both ribs and biceps (bone at the cendulum, double at the navel, double at the cendulum)

IM vastus intermedius, *RF* rectus femoris, *VM* vastus medialis, *VL* vastus lateralis, *IO* internal abdominal oblique, *EO* external abdominal oblique, *TrA* transverse abdominis, *RA* rectus abdominis

obtained from linear sprint performance. Nicolas et al. therefore, this study is a program that applied a four-week squat exercise program with EMS to the subject based on the above paper. A total of four sets of 25 squats are performed, and each set is given a one-minute break. Squat is a common sport among athletes and the public, especially used to increase the strength of the leg muscles, and it is required to squat and perform squats exercise at a moderate angle to minimize the stress on the joints and the possibility of injury to the waist and knees. The subjects should stand with their legs in the width of the pelvis and their hands behind their heads. The subjects shall squat to reach the maximum depth of the squat position with a bent hip point 90° and relaxed knee position, the squats shall be performed as low as possible without the heel and the maximum squared position shall be maintained for 5 s. Therefore, based on the above paper, this study applied a four-week squared motion program to the subject to compare before and after arbitration.

2.2.3 Measurement

Flexor endurance test

One of the physical tests, flexor endurance test (located right away and the legs do hip flexion 90° and knee flexion 90°, cross the arms and hold them on for 10 s. Repeat it six times a minute) to measure the muscle endurance.

Jump height

The standing vertical jump measured how high an individual can climb from a standstill. In this study, the height of the jump was measured by vertical jump after squat movement to check the enhancement of body function.

Side step

This is a kind of agility test. Draw three parallel lines at 120 cm intervals on the floor as shown. Stand between the legs with the centerline between the legs and jump sideways so that the right line is between the open legs, then back to the center line in the same way, and then back to the left line between the open legs. Then, jump back to the center line. As such, it is possible to calculate the number of 20 s by repeating the side steps as soon as possible, which alternate between the centerline and the left and right lines with the legs spread apart.

Ultrasonography
Ultrasonography is a diagnostic medical imaging technology that uses ultrasound to visualize muscles, tendons, many internal organs, their size, structure, and pathological damage in real-time single-layer images. In this study, we measure the thickness of IM, RF, VM, VL, RA, EO, IO, TrA through this device.

2.2.4 Data Analysis

SPSS version 17.0 for Windows was used for research data analysis. One-way ANOVA was used. Also, all statistical significance levels were set at $p < 0.05$. To calculate the physical characteristics of the subjects of mean and standard deviation of all variables, the descriptive statistic was used. Also to compare before and after exercise, it used the paired T-test.

3 Result

3.1 The Mean Quadriceps and Abdominal Muscle Each Group (After—Before) Differing Result After Four Weeks

Among the quadriceps muscle, there were significant differences in ESG, SG, and CoN in the vastus intermedius and rectus femoris, respectively. Vastus medialis and Vastus lateralis did not show a significant difference between ESG and SG. And this did not show a significant difference between ESG and CoN, SG and CoN. Among the abdominal muscle, there were significant differences between each group in rectus abdominis, external abdominis, and internal abdominis muscle. Transverse abdominis did not show a significant difference between ESG and SG. And this did not show a significant difference between ESG and CoN (Fig. 2; Table 3).

3.2 The Jump, Endurance, Agility Each Group Differing Result After Four Weeks

Sergent jump was significant differences between the ESG, SG, and CoN, respectively. Endurance did not show a significant difference between ESG and SG. And this

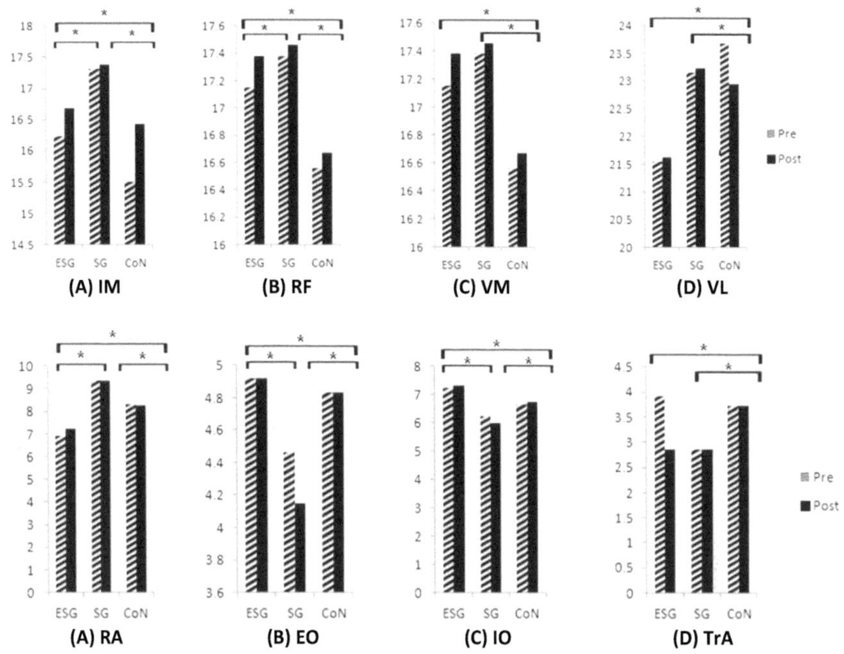

Fig. 2 Mean IM, RF, VM, VL muscle and RA, EO, IO, TrA muscle each group (after–before) differing result after four weeks. IM: vastus intermedius, RF: rectus femoris, VM: vastus medialis, VL: vastus lateralis. RA: rectus abdominis, EO: external abdominal oblique, IO: internal abdominal oblique, and TrA: transverse abdominis

did not show a significant difference between ESG and CoN, SG and CoN. Agility does not show a significant difference between ESG and SG. And this did not show a significant difference between ESG and CoN, SG, and CoN (Fig. 3; Table 3).

4 Discussion

In this study, the effect of applying EMS on squats was compared by dividing by strength, endurance, and physical function. Among the quadriceps muscle, there were significant differences in ESG, SG, and CoN in IM, RF. In the VM and VL, CoN had significant differences with ESG and SG, and there were no significant differences between ESG and SG. Among the abdominal muscle, a significant difference between each group was found in the RF, EO, IO, and the difference between the ESG and the ESG in the TRA. The jump showed significant differences in ESG, SG, and CoN, respectively. Endurance showed significant differences between ESG and SG for CoN, and no significant differences between ESG and SG. Agility showed a significant difference in CoN and no significant difference between ESG and SG.

Table 3 Jump, endurance, agility and muscle comparing before and after means and standard deviation

	ESG		*t*	SG		*t*	CoN		*t*	*F*
	Pre	Post		Pre	Post		Pre	Post		
IM	16.23 ± 1.83	16.69 ± 1.65	−3.207**	16.04 ± 1.79	16.22 ± 1.85	−1.000*	15.77 ± 2.12	15.85 ± 2.23	−0.562	2.016***
RF	17.15 ± 1.57	17.56 ± 1.6	−1.897*	17.02 ± 1.6	17.23 ± 2.22	−1.000*	17.38 ± 2.84	17.42 ± 3.09	−0.433	6.049***
VM	26.92 ± 2.62	27.38 ± 2.69	−3.207**	26.15 ± 4.39	26.60 ± 4.21	−1.760**	23.23 ± 1.83	23.15 ± 1.9	0.562	3.750***
VL	21.54 ± 1.85	21.62 ± 1.8	−1	23.15 ± 2.76	23.23 ± 2.92	−1	24.15 ± 5.82	24.15 ± 5.82	0	3.976***
RA	6.92 ± 3.47	7.23 ± 3.51	−2.309*	9.38 ± 1.04	9.38 ± 0.87	0	8.85 ± 1.77	8.77 ± 1.83	−1	8.835***
EO	4.92 ± 0.76	4.92 ± 0.86	0	4.46 ± 0.96	4.15 ± 0.8	1.298	5.00 ± 1	5.00 ± 1	0	1.694***
IO	7.23 ± 1.3	7.40 ± 1.25	−9.733***	6.23 ± 1.69	6.00 ± 1.73	0.562***	7.15 ± 1.86	7.23 ± 1.78	−1	2.552***
TrA	3.92 ± 0.64	2.85 ± 0.68	−1	2.85 ± 0.68	2.85 ± 0.68	0	3.85 ± 0.89	3.77 ± 0.92	1000	2.178***
Endurance	53.15 ± 10.54	57.00 ± 11.88	−5.839***	55.54 ± 7.68	57.62 ± 7.68	−3.566**	53.77 ± 7.71	52.85 ± 8.44	2.984*	20.070***
Jump	28.69 ± 4	30.85 ± 4.27	−6.791***	24.46 ± 3.9	25.15 ± 4.25	−2.920*	25.54 ± 4.96	25.15 ± 5.35	1.443	21.377***
Agility	39.38 ± 4.33	43.00 ± 4.12	−8.374***	39.38 ± 2.84	42.08 ± 3.96	−4.815***	38.46 ± 2.81	37.23 ± 2.2	2.421*	26.227***

*$p < 0.05$; **$p < 0.01$; ***$p < 0.001$
IM vastus intermedius, *RF* rectus femoris, *VM* vastus medialis, *VL* vastus lateralis
RA rectus abdominis, *EO* external abdominal oblique, *IO* internal abdominal oblique
TrA transverse abdominis

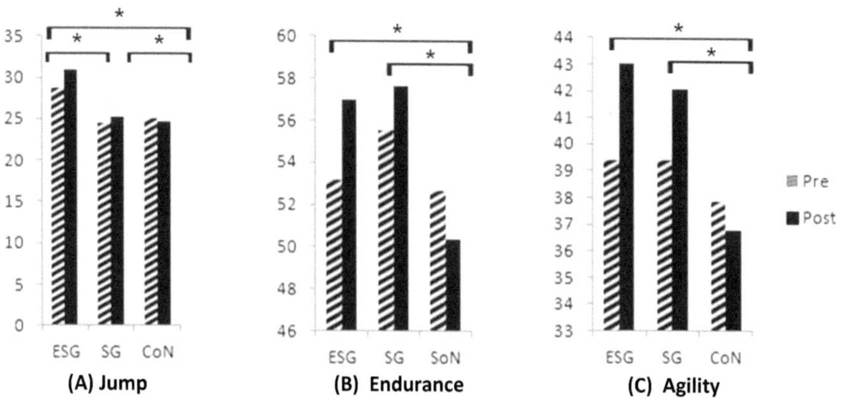

Fig. 3 The jump, endurance, agility of each group (after–before) differing result after four weeks

According to previous studies, the eight-week core stabilization exercise program improved the thickness of the abdominal muscle more than that of the control group [12] Jo et al. compared the effects on abdominal muscle thickness and lumbar stability with those of the squat exercise. Exercise lasted 30 min and six days a week and ultrasound was applied with a 12 MHz linear proved to measure changes in the body's muscle thickness and abdominis. In results, there were significant differences in thickness before and after exercise in the RF, EO, IO [15]. Participants also performed a set of squats 48 times a week for a total of two 12-week period, and the thigh muscle thickness increased, although the thigh circumference did not change. Thomas et al. described that the results of measuring the EMG value of the quadriceps muscle during the squat training conducted by home-based training showed that the peak value was higher compared to the other stretching tunneling [16]. Tsuzuku's study reported that in 85 healthy adult men and women for 12 weeks of squats and sit-ups, muscle strength and thickness of quadriceps muscle were significantly increased compared to control group [17]. Therefore, this study also showed a significant difference between the abdominal muscle excluding TRA and the quadriceps muscle excluding VM, VL.

Wirtz et al. found one of the reasons why the squats with EMS did not set the EMS at its maximum intensity for not showing any significant improvement in muscle groups [5]. The EMS strength of this study was the medium degree, it is believed that in order to achieve more significant results in all muscle groups, and it is recommended to re-evaluate the EMS after it has been applied. Carrie et al. evaluated the effectiveness of the core stability exercise program for the endurance and abdominal muscle thickness of the two groups by dividing the program consisting of seven core stabilization movements including squat [11]. Into experimental groups, which is conducted five days a week, and the control groups [11], that continued the previous individual exercise routine and do not implement the new exercise program [11], they examined the fore and after of the exercise and found that an eight-week core stabilization exercise program activates core muscles and increases core muscle endurance in a limited way [11]. Based on the results of the preceding study, this study was conducted to see if the squat movement applied with EMS affects the improvement of muscle endurance. As a result, we found no significant difference between ESG and SG endurance in the flexor endurance test. According to the preceding study, the overlapping effect of EMS in endurance for young healthy adults was similar to the effect of dynamic resistance training without EMS [11]. Therefore, in this study, it is recommended that additional core stabilization movements, including squat, are carried out together after EMS, since the effects of the attached EMS squat on the endurance improvement are not statistically significant.

Seichiro et al. demonstrated that high-strength squats were appropriate for athletes with a warm-up method, with a significant increase over the pre-recorded twist torque and jump height recorded after the squat movement and a warm-up method to improve jump performance. The squat movement at higher strength increased the height of the jump as well as the torque of the witch compared to the lower intensity movement [18]. Based on these studies, this study showed that the jumping heights of ESG and SG have increased significantly compared to the CoN. In addition, the application of EMS in the squat to improve jump height is considered effective, with the results in the ES being most clearly increased. Also, in Wirtz, 20 subjects were divided into ESG and SG for six weeks. The result was a significant increase in jump height and short-distance running capability in the ESG compared to the SG. Therefore, it was shown that training with EMS and strength training improved jumping and sprinting and the strength of the extensor muscle [5]. Shelly et al. reported that the squats training program for soccer players improved leg power, jump, and sprint performance [19]. Martinez LEJ also randomly assigned 98 young athletes to four groups for eight weeks to overlay the EMS, confirming that teenage athletes have significantly increased their squat-jump height [20]. Studies such as Lauturco showed that the speed of sprints has improved significantly in the jumping and half-squat movements of 22 competitors [21]. In studies such as Chaouachi et al., the squats training for NBA basketball players improved their short-range sprint performance and vertical jump capabilities [22]. In addition, Wang et al. studies showed a significant increase in the 40 m sprint speed and agility test applied to college rugby players [23]. The study of Julien et al. with 26 adult males in three groups resulted in a significant increase in short-range sprint and shuttle sprint performance compared to control groups [24]. Vandreka et al. performed three jump-squat exercises a week, dividing 62 experimenters into experimental and control groups for eight weeks. As a result, the experimental group discovered that the vertical jump performance was always better than the control group [25]. Therefore, the jump height in this study has also increased significantly in ESG and SG. It also showed that there was a significant difference between CoN and the ESG and SG.

This study has several limitations. The short period of mediation was not enough for a short period of four weeks, failing to study various age groups for college students and disproportionate sex ratios for each group. There were also deficiencies in the leadership of many subjects, such as the intensity of exercise and the application of regular exercise programs. In further studies, it is believed that it is advisable to supplement this point by controlling the personnel with sufficient arbitration period and applying other evaluation methods, such as muscle activity, to compensate for the deficiencies.

5 Conclusion

This study was conducted on 39 healthy adults in their 20s to recognize changes in the thickness of quadriceps muscle and abdominal muscle when overlapped with squat exercises, which is a technique widely used to alleviate symptoms of various muscle improved strength. A component pump was performed for the jump test, a flexor endurance test was performed for the endurance test, and a side step was performed for the agility test. Also, the quadriceps muscle and abdominal muscle thickness were measured with a sonography. The results of the study are as follows. Among the quadriceps muscle, there were significant differences in ESG, SG, and CoN in IM and RF. In VM and VL muscle CoN had significant differences with ESG and SG, and there were no significant differences between ESG and SG. Among the abdominal muscles, there were significant differences between each group in the TrA, EO, and IO; in the TrA, CoN showed significant differences between ESG and SG, and there was no significant difference between ESG and ESG. The jump showed significant differences in ESG, SG, and CoN, respectively. Endurance showed significant differences between ESG and SG for CoN, and no significant differences between ESG and SG. Agility showed a significant difference in CoN and no significant difference between ESG and SG. As a result of this study, the square movement overlaid with EMS has improved the jump and increased the thickness of the quadriceps muscle except for TrA and VM and VL, so we think it is effective for people who do squat exercise to attach the EMS.

References

1. Suqisaki, N., Kurokawa, S., Okada, J., Kanehisa, H.: Difference in the recruitment of hip and knee muscles between back squat and plyometric squat jump. PLoS ONE **9**(6), e101203 (2014). https://doi.org/10.1371/journal.pone.0101203
2. Cotter, J.A., Chaudhari, A.M., Jamison, S.T., Devor, S.T.: Knee joint kinetics in relation to commonly prescribed squat loads and depths. J. Str. Condition. Res. **27**(7), 1765–1774 (2013). https://doi.org/10.1519/JSC.0b013e3182773319
3. Bonnet, V., Mazza, C., Fraisde, P., Cappozzo, A.: Real-time estimate of body kinematics during a planar squat task using a single internal measurement unit. IEEE Trans. Biomed. Eng. **60**(7), 1920–1926 (2013). https://doi.org/10.1109/TBME.2013.2245131
4. Wu, H.W., Tsai, C.F., Liang, K.H., Chang, Y.W.: Effect of loading devices on muscle activation in squat and lunge. J. Sport Rehabil. **6**, 1–6 (2019). https://doi.org/10.1123/jsr.2018-0182
5. Nicolas, W., Christoph, Z., Ulrike, D., Heinz, K., Joachim, M.: Effects of loaded squat exercise with and without application of superimposed EMS on physical performance. J. Sports Sci. Med. **15**(1), 26–33 (2016)

6. Doheny, E.P., Caulfield, B.M., Minogue, C.M., Lowery, M.M.: The effect of subcutaneous fat thickness on the efficacy of transcutaneous electrical stimulation. IEEE Eng. Med. Biol. Soc. 5684–5687 (2008). https://doi.org/10.1109/iembs.2008.4650504
7. Kemmler, W., Von Stengel, S., Schwarz, J., Mayhew, J.L.: Effect of whole-body electromyostimulation on energy expenditure during exercise. J. Str. Condition. Res. **26**(1), 240–245 (2012). https://doi.org/10.1519/JSC.0b013e31821a3a11
8. Micke, F., Kleinoder, H., Dormann, U., Wirtz, N., Donath, L.: Effects of an eight-week superimposed submaximal dynamic whole-body electromyostimulation training on strength and power parameters of the leg muscles: a randomized controlled intervention study. Front. Phys. **5**(9):1719 (2018). https://doi.org/10.3389/fphys.2018.01719.eCollection
9. Wirtz, N., Wahl, P., Kleinoder, H., Wechsler, Achtzehn, S., Mester, J.: Acute metabolic, hormonal, and psychological responses to strength training with superimposed EMS at the beginning and the end of a 6 week training period. J. Musculoskelet Neur. Inter. **15**(4): 325–32 (2015)
10. Wolfgang, K., Anja, W., Sebastia, W., Mahdieh, S., Simon von, S., Andre, F., et al.: Efficacy and safety of low frequency whole-body electromyostimulation (WB-EMS) to improve health-related outcomes in non-athletic adults. a systematic review. Front. Phys. **9**:573 (2018). Doi: 10.3389/ fphys.2018.00573
11. Carrie, W.H., Aubrey, D.S., Colleen, F.H., Bridgette, D.G., Molly, F.P., Barri, L.S., et al.: The efficacy of an eight-week core stabilization program on core muscle function and endurance: a randomized trial. Int. J. Sports phys. Ther. **11**(4), 507–519 (2016)
12. Filipovic, A., Grau, M., Kleinoder, H., Zimmer, P., Hollmann, W., Bloch, W.: Effects of a whole-body electrostimulation program on strength, sprinting, jumping, and kicking capacity in elite soccer players. J. Sports Sci. Med. **15**(4), 639–648 (2016)
13. Park, J.H., Seo, K.S., Lee, S.U.: Effect of superimposed electromyostimulation on back extensor strengthening: A pilot study. J. Str. Condition. Res. **30**(9), 2470–2475 (2016). https://doi.org/10.1519/JSC.0000000000001360
14. Falavigna, L.F., Silva, M.G., Freitas, A.L., Silva, P.F., Paiva Junior, M.D., De Castro, C.M., et al.: Effects of electrical muscle stimulation early in the quadriceps and tibialis anterior muscle of critically ill patients. Physiother. Theory Pract. **30**(4), 223–228 (2014). https://doi.org/10.3109/09593985.2013.869773. Epub 2013 Dec 30
15. Cho, M.S.: The effects of modified wall squat exercises on average adults' deep abdominal muscle thickness and lumbar stability. J. Phys. Ther. Sci. **25**(6), 689–692 (2013)
16. Thomas, L.J., Markus, D.J., Lars, L.A., Henrik, H., Henrik, K., Thomas, B., et al.: Quadriceps muscle activity during commonly used strength training exercises shortly after total knee arthroplasty: implications for home-based exercise-selection. J. Experim. Orthopaedics **6**(1), 29 (2019). https://doi.org/10.1186/s40634-019-0193-5
17. Tsuzuku, S., Kajioka, T., Sakakibara, H., Shimaoka, K.: Slow movement resistance training using body weight improves muscle mass in the elderly: a randomized controlled trial. Scand. J. Med. Sci. Sports **28**(4), 1339–1344 (2018). https://doi.org/10.1111/sms.13039. Epub 2018 Jan 30
18. Fukutani, A., Takei, S., Hirata, K., Miyamoto, N., Kanehisa, H., Kawakami, Y.: Influence of the intensity of squat exercises on the subsequent jump performance. J. Str. Condition. Res. **28**(8), 2236–2243 (2014). https://doi.org/10.1519/JSC.0000000000000409
19. Chelly, M.S., Fathloun, M., Cherif, N., Amar, M.B., Tabka, Z., Van Praagh, E., et al.: The effects of modified wall squat exercises on average adults' deep abdominal muscle thickness and lumbar stability. J. Str. Condition. Res. **23**(8), 2241–2249 (2009). https://doi.org/10.1519/JSC.0b013e3181b86c40
20. Martinez, L.E.J., Benito, M.E., Hita, C.F., Lara, S.A., Martinez, A.A.: Effects of electrostimulation and plyometric training program combination on jump height in teenage athletes. J. Sports Sci. Med. **11**(4), 727–735 (2012)
21. Loturco, I., Pereira, L.A., Moraes, J.E., Kitamura, K., Cal Abad, C.C., Kobal, R., et al.: Jump-squat and half-squat exercises: selective influences on speed-power performance of elite rugby sevens players. PLoS ONE **12**(1), e0170627 (2017). https://doi.org/10.1371/journal.pone.0170627

22. Chaouachi, A., Brughelli, M., Chamari, K., Levin, G., Abdelkrim, N.B., Laurencelle, L., Castagna, C.: Lower limb maximal dynamic strength and agility determinants in elite basketball players. J. Str. Condition. Res. **23**(5), 1570–1577 (2019). https://doi.org/10.1519/JSC.0b013e3181a4e7f0
23. Wang, R., Hoffman, J.R., Tanigawa, S., Miramonti, A.A., La Monica, M.B., Beyer, K.S., et al.: Isometric mid-thigh pull correlates with strength, sprint, and agility performance in collegiate rugby union players. J. Str. Condition. Res. **30**(11), 3051–3056 (2016). https://doi.org/10.1519/JSC.0000000000001416
24. Jullien, H., Bisch, C., Largouet, N., Manouvrier, C., Carling, C.J., Amiard, V., et al.: Does a short period of lower limb strength training improve performance in field-based tests of running and agility in young professional soccer players? J. Str. Condition. Res. **22**(2), 404–411 (2008). https://doi.org/10.1519/JSC.0b013e31816601e5
25. Marian, V., Katarina, L., David, O., Matus, K., Simon, W.: Improved maximum strength, vertical jump and sprint performance after 8 weeks of jump squat training with individualized loads. J. Sports Sci. Med. **15**(3), 492–500 (2016)

Subscriber Location Prediction: A Neural Network Approach

Smita Parija and Santosh Das

Abstract Location forecasting of subscriber is critical in cellular network. Since many issues such as handoff, blocking probability, user prediction, etc., are greatly influenced by this location movement of mobile user. In this work, a comprehensive study of different artificial neural techniques and its wireless application, i.e., user forecasting, is presented. Three types of typical neural networks, namely back propagation (BP), Legendre (LN) and radial basis function (RBF), are extensively studied, investigated and analysed in the paper. The location user data used are the hourly mean mobile user data collected at sites in Chandrasekharpur, Bhubaneswar area. The performance is evaluated based on three metrics, namely training accurate, testing accuracy and processing time. The random user where the movement is frequent then conventional algorithms like multi-layer perceptron (MLP), radial basis function (RBF) do not outperform the Legendre neural network (LNN). For the best performance, the nonlinear neural network selected also depends on the type of collected mobile user data. The performance matrices such as processing time, training accuracy and testing time obtained from the simulation results that outperform, i.e., 0.006239 s, 84 and 81% than conventional neural algorithms. This indicates the implemented algorithm is a single robust and reliable that forecasts the location of a roaming user in a wireless network.

Keywords Cellular network · Location management · Polynomial Legendre neural network

S. Parija (✉)
Department of Electronics Engineering, CV Raman Global University, Bhubaneswar, India
e-mail: smita.parija@gmail.com

S. Das
Department of Electronics, National Institute of Technology, Rourkela, Odisha, India
e-mail: Das.santoshkumar@gmail.com

A. K. Bhoi et al. (eds.), *Bio-inspired Neurocomputing*, Studies in Computational Intelligence 903, https://doi.org/10.1007/978-981-15-5495-7_14

1 Introduction

The prime objective of the cellular network is to predict and identify the location in a cellular coverage area. In a mobile network, in order to forward a call appropriately and efficiently to the user, the mobile user location is required to be traced out cellular network [1]. For this, two key components are location search, i.e., paging, and location registration, i.e., location update. These two fundamental operations constitute mobility management technique. The 'paging' procedure is performed by the wireless network to locate the user, but location update procedure is done by the mobile device, and it transmits this signal to the network about user position [2]. The mobile user when moves in a network from one location to other location then there will be the updation in the location, Hence, a trade-off relation is between the user position update and network paging or network search technique. Based on the cell phone user location, hardly any number of calls required to be searched. For a frequent user as location is uncertain, less number of cells need to be tracked by mobile switching centre (MSC). Similarly, for a known location, the cost for paging increases [3].

Although the mobile terminal changes frequently its position, mobility management is necessary to keep connectivity with the network for better Quality of service (Qos) for the mobile user [4]. If the location of the mobile subscriber is known, Qos can be guaranteed. Good Qos can be ensured when mobile terminal location area is known to the network. In view of availability of mobile subscriber data, few methods have been proposed for forecasting the future location of the user in a wireless network. Based on the mobility in a network, there are three types of mobile user, i.e., uniform user whose location is on an average static, second type of user whose location is semi-random where the mobile terminal positions not known fully to the user, while the third type of user is random type whose location is not known to the network. For example, professional academic people or doctor belong to the first group whose current location is the parameter for forecasting the future position, the time is the performance parameter for second group mobile user in forecasting the future position. The above forecasting techniques fail to predict when mobile user is with variable movement pattern in different time [5]. In this case, nonlinear neural network algorithm (NNA) is implemented, and time as well as current position is taken as performance parameter to the input of neural network. This is employed in the last case of user where NNA is implemented to forecast the roaming user's consequent position. In any case, this conventional MLPT and RBF algorithm possesses many disadvantages such as slow processing time, and convergence period is very slow and oscillates to the local minimum. Even if it is used in difficult problems, still its MLPT contains a long training sequence that takes the long time to converge. As a result, LeNNA utilised in the prediction of position for the roaming user whose location is not fully known to the network [6].

The rest of the work in this paper is organised as follows. Sect. 2 explains about the forecasting of user location with mobile data. This forecasting issue resolved with the proposed NNA such as conventional multi-layer perceptron technique (MLPT), radial basis function (RBF) and Legendre neural network method (LeNNM) that analysis is discussed in Sect. 3. Simulation result and discussion is presented in Sect. 4. Section 5 concludes the paper.

2 Forecasting the Position of User with Mobile Data (UMD)

User mobile data is collected for simulation purpose where each cell in the coverage area is assumed to be hexagonal for convenient. Along with it, the location of a mobile user is identified by latitude (LAT) and longitude (LOG). This LAT-LOG file present in the network is updated with updation of roaming user. Longitude is taken in X-coordinate and latitude is given to Y-coordinate. For simulation, the parameters are time and location. Similarly, the next location is the predicted cell obtained from outputs with X- and Y-coordinate [7].

Many techniques have been adopted such as nonlinear back propagation (BP), network optimisation method and local search techniques, but all resulted with no satisfactory performance [8]. Hence, a low-cost methods need to be developed with scalability with the advanced neural network [9].

There is a link that needs to be maintained between the mobile terminate (MT) and the base transceiver (BT) in the wireless network. There is a frequent transmission between MT and BT, so that network can send the signal all time and become more responsive. On the other hand, for cost reduction, it is needed to convey the signal as sparingly as possible. A satisfactory trade-off performance which could be expected from a network is the objective of this work. For this, if forecasting of a location of the roaming user is possible in a network coverage to arrive at a subsequent cell in a coverage region, then cost of the network can be minimised at any moment. It can anticipate the next cell with adopting of many new nonlinear algorithms. This adaptive algorithm with better system model predicts the subsequent cell in an efficient manner of a roaming subscriber [10].

Data modelling is done which contain some nonlinearity then conventional linear models that fail to predict as only current time and present state are considered to forecast the user's position in a cell. On the other hand, due to random and nonlinearity nature in the data, artificial nonlinear adaptive algorithm neural networks performs better in both nonlinear and linear problems with help of the data being used in modelling. Both time and position are used as a parameter during modelling the system [11].

Figure 1 presents system model of conventional MLP deployed in forecasting the desired parameter, position. This is a supervised type network where desired output is known. Here, position is the desired parameter. Using collected data, the developed model maps between input and output. If the desired output is not known to the user, then the output is created by the developed system models. These MLP technique with perceptron is used to control the nonlinearity and is also helpful for adaptive identification that is sensed through collected data [7].

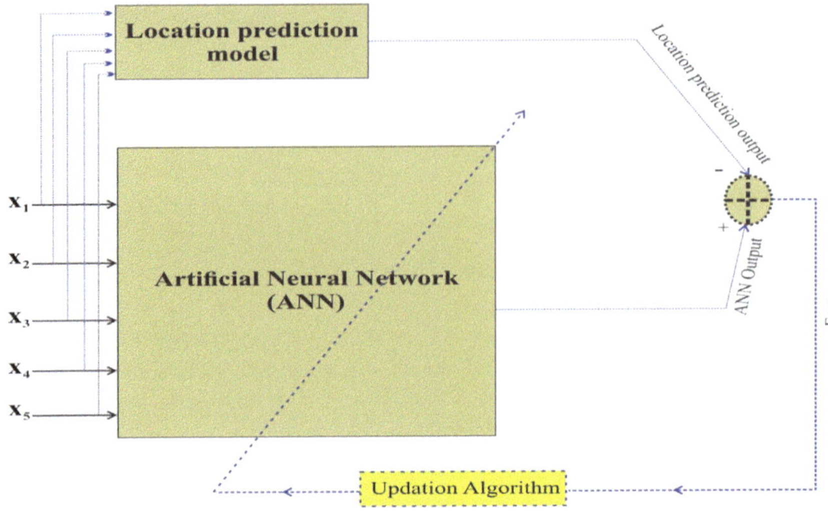

Fig. 1 Architecture of NNA

Here, Rumelhart et al. in 1986 developed and tried BP learning algorithm and neural network algorithm with more than one layer, i.e., MLP based as shown in Fig. 1. This MLP is a multi-layer and feed-forward network. The nodes are the processing units present in each layer. There are three different layers named input layer, hidden layer and output layer.

In general, one input layer, more than one hidden layer(s) and one output layer are present in multi-layer perceptron algorithm. Based on dimension, there is a mapping between input and output layer. But in this issue, one hidden layer is considered for location forecasting issue. The layers are connected to the network with connecting weights. Bias which is given with a weight of +1 signifies a one-to-one relation between the first input layer and the last output layer.

In neural network, there is a training phase and testing phase. In training phase, an error signal is generated by comparing the actual position, i.e., actual output and desired output which is my predicted position in subsequent cell. Again this error is minimised with the help of connecting weights of NNA through BP and tried to bring it close to perfections. These weights are updated adaptively in alterative manner until a minimum mean square is achieved.

The neural network pertaining to location management involves three inputs, namely time, two-dimensional Cartesian coordinates of the cells and two outputs, namely two-dimensional Cartesian coordinates of the expected cell in next interval of time. The learning power by using BP algorithm is not good enough; however, in conjunction with techniques such as momentum, resilient propagation and quick propagation, it can be improved significantly. These techniques coupled with normalised input data have been used in the current work. Location update region and paged region are assumed to be same, and one location area is confined to one cell [8].

3 Architecture of RBF Neural Network

Legendre polynomials were named after Adrien-Marie Legendre (1752–1833). On substituting these polynomials in place of the conventional polynomials in a neural network, one obtains Legendre neural network (LNN). LNN, unlike other neural algorithms such as trigonometric neural network, lays less computational burden and trains faster. The prominent features of LNN are (a) orthogonal polynomial expression and (b) spherical symmetry problems and angular dependence problems which involve LNN.

The architecture of LNN is shown in Fig. 2. It may be observed that in general, the Legendre polynomials require less number of multiplications and additions compared to other algorithms. The Legendre polynomials are a set of orthogonal polynomials which represent solution to the differential equation is the argument of the Legendre polynomials, whereas *n* designates their order.

The Legendre polynomials are represented as follows:

$$\left(1 - x^2\right)\frac{d^2 y}{dx^2} - 2x\frac{dy}{dx} + n(n + 1)y = 0 \tag{1}$$

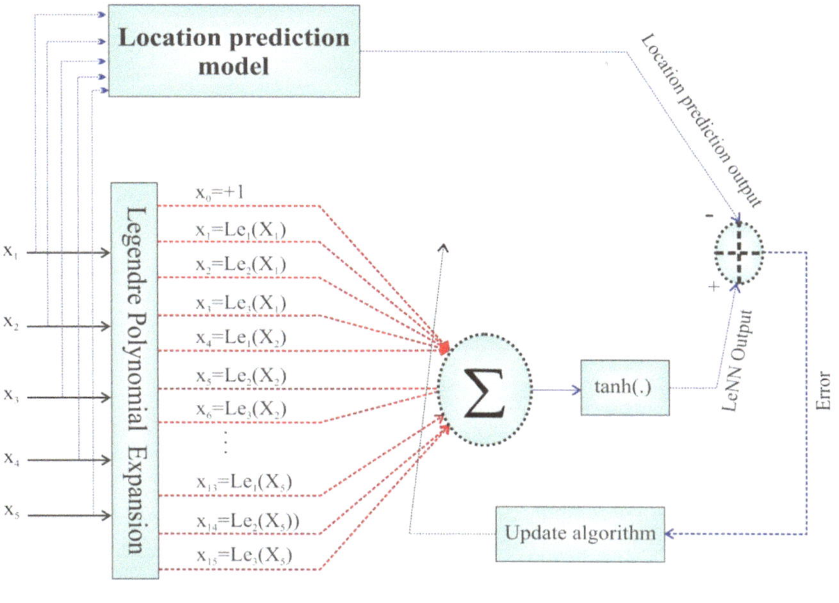

Fig. 2 Nonlinear architecture of LNN

For $n \geq 0$, $|x| \leq 1$

This is the Legendre polynomials of n order and value ranges between $1 < x <$ 1 1. They constitute second-order differential equation. The first several Legendre polynomials are listed below:

$$P_0(x) = 1 \tag{2}$$

$$P_1(x) = x \tag{3}$$

$$P_2(x) = \frac{1}{2}(3x^2 - 1) \tag{4}$$

$$P_3(x) = \frac{1}{2}(5x^3 - 3x) \tag{5}$$

$$P_4(x) = \frac{1}{8}(35x^4 - 30x^2 + 3) \tag{6}$$

$$P_5(x) = \frac{1}{8}(63x^5 - 70x^3 + 15x) \tag{7}$$

The recurrence formula used for generating higher order polynomials is as follows

$$P_{n+1}(x) = \frac{(2n+1)}{n+1} x P_n(x) - \frac{n}{n+1} P_{n-1}(x) \tag{8}$$

In case of high-order Legendre polynomials such as 2, 3 and 4, it needs three, five and six multiplications subsequently. For example, five multiplications in fourth order are required for evaluation. Hence, on an average, even in case of fourth order, the number of required addition is 1, but in high order, other neural algorithm like functional approximations requires more number of additions and multiplications in the trigonometric evaluation process because of involvement of complex power series expansion. This is the advantage of LNN over other neural algorithms.

4 Result and Discussion

Extensive simulation studies were carried out with several examples of nonlinear dynamic systems. Different neural algorithms such as NNA, RBF and LNN are implemented for forecasting the location using the performance parameter in cellular network to resolve location management issue. For simulation purpose, MATLAB software is used in the configuration Dell Inspiron 3584 15.6-in. FHD Laptop–Intel Core i3 7th Gen, 8 GB and 64-bit operating system. The neural network system models are implemented using mathematical analysis. [12].

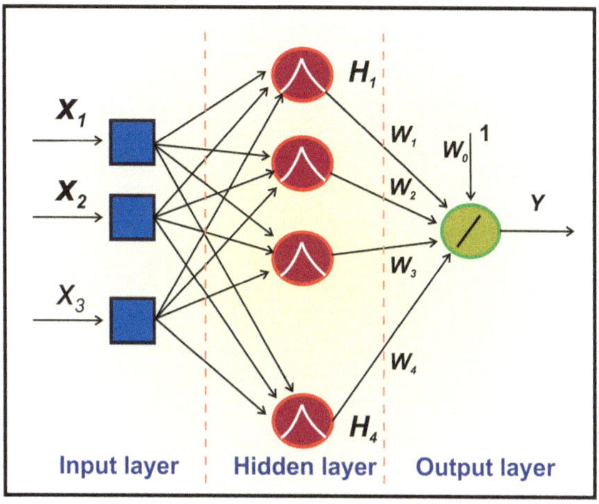

Fig. 3 Nonlinear architecture of RBF

The system is simulated taking the help of the architecture of NNA, RBF and LNN algorithm, and the proposed system is simulated. The performance analysis is carried out in terms of mean square error (MSE). MSE graph is obtained and shown in Fig. 4a–c. With similar analysis, the prediction error is MSE which is shown in Fig. 4c–e. For system model of NNA, ten hidden nodes were chosen, and epoch of 600 were selected to obtain a better result, but in case of RBF architecture, the nodes in single hidden layer are five, and epoch of 500 is chosen for prediction. This prediction result is shown in Fig. 5d–f, respectively (Fig. 5).

The performance analysis is based on prediction accuracy and processing time of different algorithms given in Table 1. The processing time of different neural predictors is in ms that uses 3000 iterations to train the NN. Prediction that the LNN predictor takes is 0.006239 s, i.e., minimum time to converge, whereas the NNA based predictor takes maximum training time of 116.547 s to converge. The training time of LNN and RBF based predictors has found to be 0.0023686 s and 0.039051 s, respectively. It is observed that NNA based equalizer takes maximum processing time of116.547 s, whereas for RBF, it is 11.933 s.

To study the convergence characteristics, it is observed that MSE of NNA is maximum 2.5412487 and RBF is 0.1639051, whereas in case of LNN, MSE is 0.016444 which outperforms when compared to other neural network predictor.

From Table 1, results were obtained and observed for analysis that the overall training accuracy for NNA is 78% but 74% in their testing accuracy. Same in the case of RBF training accuracy is 77% and testing accuracy is 75%, whereas in LNN training, both training and testing accuracy are 84% and 81%, respectively, after training the network.

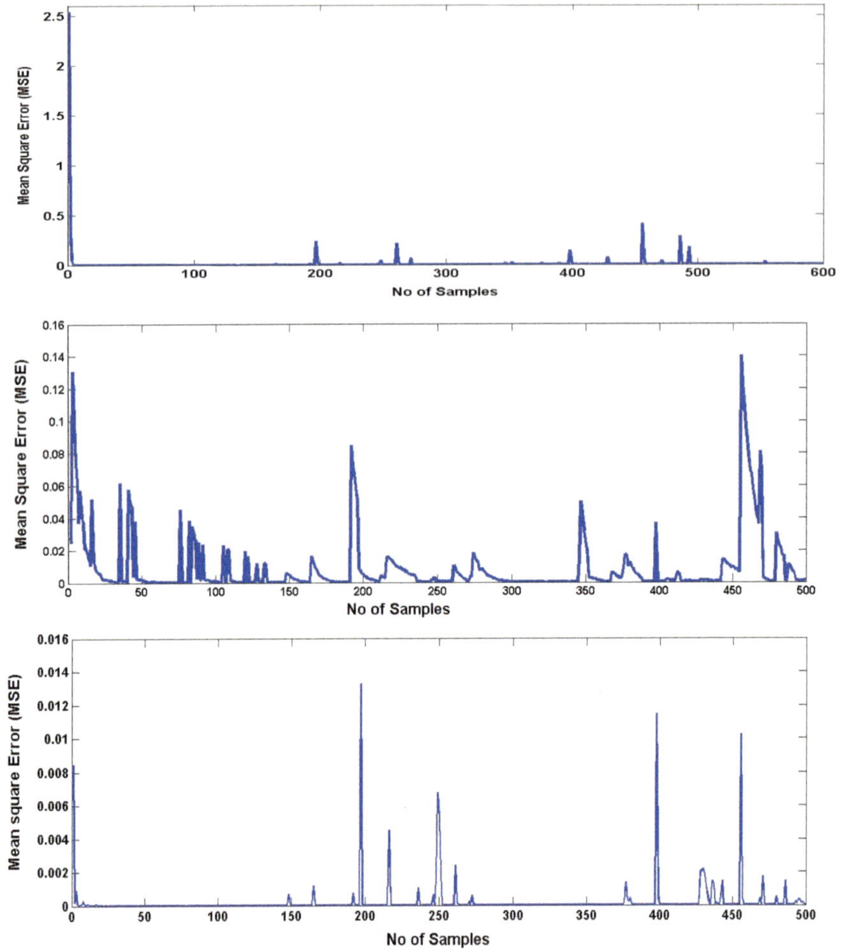

Fig. 4 Mean square error of NNA, RBF and LNN (hidden layer with epoch 600, 500 and 500, respectively)

5 Conclusion

In cellular networks, locating the roaming user quickly and effectively in an efficient manner is a critical task. In this work, conventional nonlinear NNA such as MLPT and RBFA fails to predict the prediction of mobile user due to slow convergence and processing time. For this, alternatively the LNNA efficiently forecasts the location of mobile roaming user. These algorithms are studied, investigated and compared with LNNA. It is observed that from simulation results that performance analysis of LNNA is better than conventional algorithms because of efficient convergence rate. The convergence learning rate is faster in LNNA than LNN MLPA as a single

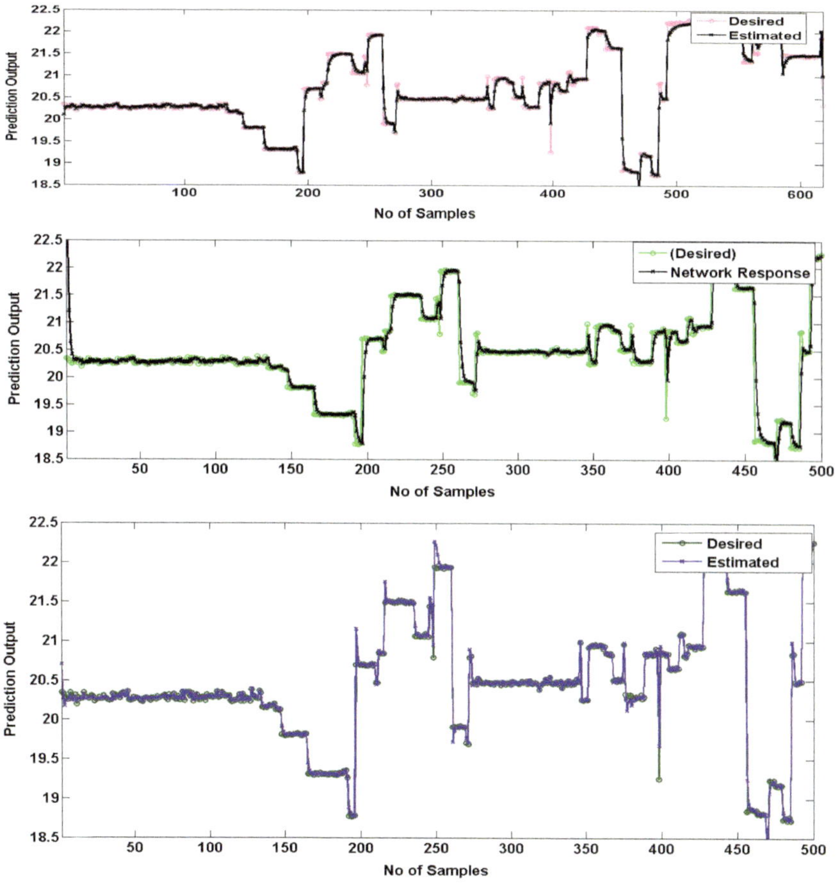

Fig. 5 Performance prediction analysis of NNA, RBF and LNN

Table 1 Results during simulation for computational time, training and testing accuracy

Methods	Processing time	Accuracy in training (%)	Accuracy in testing (%)
NNA	116.547 s	78	74
RBF	11.933	77	75
LNN	0.006239 s	84	81

nonlinear orthogonal functional expansion layer is present in LNN MLPA. From the simulation result, the computational complexity and processing time with reduced number of bias and weights obtained are less in LNNA than conventional NNA which is an advantage of Legendre approach. This proposed algorithm may be implemented for online application because of its satisfactory results.

References

1. Sidhu, B., Singh, H.: Location management in cellular networks. Proc. World Acad. Sci. Eng. Tech. **21**(January), 314–319 (2007)
2. Khashei, M., Bijari, M.: A novel hybridization of artificial neural networks and ARIMA models for time series forecasting. Appl. Soft Comput. **11**(2), 2664–2675 (2011)
3. Bilurkar, K.P., Rao, N., Krishna, G., Jain, R.: Application of neural network techniques for location predication in mobile networking. In: Neural Information Processing, 2002. ICONIP'02. In: Proceedings of the 9th International Conference on, vol. 5. IEEE, pp. 2157–2161
4. Dehuri, S., Roy, R., Cho, S.-B., Ghosh, A.: An improved swarm optimized functional link artificial neural network (iso-flann) for classification. J. Syst. Softw. **85**(6), 1333–1345 (2017)
5. Parija, S., Nanda, S.K., Sahu, P.K., Singh, S.S.: Novel intelligent soft computing techniques for location prediction in mobility management. In: 2013 Students Conference on Engineering and Systems SCES (2013)
6. Patra, J.C., Meher, P.K., Chakraborty, G.: Nonlinear channel equalization for wireless communication systems using Legendre neural networks. Sig. Process. **89**(11), 2251–2262 (2010)
7. Jain, R., Lin, Y.B., Lo, C.N., Mohan, S.: A caching strategy to reduce network impacts of PCS. IEEE J. Select. Areas Commun. **12**(8), 1434–1445 (1994)
8. Haykin, S.: Neural Networks, 2nd edn. Prentice-Hall, Upper Saddle River, NJ (1999)
9. Patra, J.C., Chin, W.C., Meher, P.K., Chakraborty, G.: Legendre-FLANN-based nonlinear channel equalization in wireless, IEEE International Conference on Systems, Man, and Cybernetics, pp. 1826–1831 (2008)
10. Patra, J.C., Bornand, C.: Nonlinear dynamic system identification using Legendre neural network. IEEE Int, Joint Conf (2010)
11. Mall, S., Chakraverty, S.: Application of Legendre Neural Network for solving ordinary differential equations. J. Appl. Softcomput. **43**(3), 347–356 (2017)
12. Navabi, S., Wang, C., Bursalioglu, O.Y.: Predicting wireless channel features using neural networks, 2018 IEEE International Conference on Communications (ICC) (2018). Electronic ISBN: 978-1-5386-3180-5. https://doi.org/10.1109/icc.2018.8422221

SVM Based Temporal Compression Techniques for Video Compression

Anupama S. Budhewar and Dharmpal D. Doye

Abstract Compression is an important role in multimedia communication. In this paper, the problems of compression, quality degradation, best matched video, and image are discussed. Block matching algorithms compress the image block by block, and it also operates in a temporal manner. Proposed method is a SVM based algorithm compression. It is the combination of intensity and motion estimation. The proposed technique's compression gains of 92% accuracy with respective to static JPEG-LS, as it is applied in a frame basis. It achieves a good computation complexity and reduces it. An experimental result shows the better performance than existing methods. The method requires less best match points compared to existing methods. The quality of compression is improved when compared to the previous methods.

Keywords Support vector machine (SVM) · Support vector decomposition (SVD) · Hue saturation intensity (HSI) · Region of interest (ROI) · Probability error (PE)

1 Introduction

Artificial intelligence gives the power to learn and improve from expertise automatically without expressly programmed. Machine learning is used to focus on the event that will access knowledge. A SVM is employed to permit labeling to a variable such as in a given video that it detects the action and recognizes that action cycling, driving, etc. Also, SVM is used to classify the objects from its absolute boundaries. SVD is the process to decompose the matrix. Similarly, the SVD is used to suppress the pixels which are not part of the ROI. SVD uses a Kronecker product in which

A. S. Budhewar (✉)
Department of Computer Science and Engineering, Shri Guru Gobind Singhji Institute of Engineering and Technology, Nanded, Maharashtra, India
e-mail: asbudhewar@sggs.ac.in

D. D. Doye
Electronics and Telecommunication Engineering, Shri Guru Gobind Singhji Institute of Engineering and Technology, Nanded, Nanded, Maharashtra, India
e-mail: dddoye@sggs.ac.in

A. K. Bhoi et al. (eds.), *Bio-inspired Neurocomputing*, Studies in Computational Intelligence 903, https://doi.org/10.1007/978-981-15-5495-7_15

every element of a given matrix gets multiplied with a complete matrix. Consider a_{11}, a_{12}, a_{21}, a_{22} are the elements of matrix A, and in Kronecker product, a_{11} gets multiplied with elements of matrix B. Similarly a_{12} gets multiplied with elements of matrix B and goes on. Due to this method, averaging the intensities is possible. Easily foreground and background are distinguished. This method removes noise in the image also. The accuracy of the method is more. To compress video and image, find out the motion containing pixels. Once the pixels are detected with motion, easily no motion pixels can be distinguished and removed from image and video.

2 Literature Survey

To find temporal similarities and motion dissimilarities in angular directions, Zhou et al. [1] proposed a method and also used spectral clustering to classify trajectories into two parts. High-intensity pixels are recognized as foreground and low-intensity pixels as background. It is easy to distinguish the motion containing pixels of an image. The method requires more time, and computation complexity is more for large frame videos. Budhewar et al. [2] represented HSI based method to distinguish foreground and background. HSI based method recognizes the foreground and background efficiently. HSI finds out the intensities of pixels in an image. The minimum and maximum intensity pixels are distinguished using the threshold factor. The method requires less time for images and videos. SVC is used to empower the encoding of a high-quality video bit-stream that contains at least one subset bit-streams that would themselves be able to be decoded with an intricacy and reproduction quality like that accomplished utilizing the current [3, 4]. The block size of the SVC system can be varied to match the motion estimation that is used to reduce temporal redundancy between frames. It has seven modes for inter-prediction in multiple combinations of 2×2, 4×4, 8×8, and 16×16 macro-block size. The large size images require more space to store and also bandwidth to transmit across the network. Higher order interpolation is readjusted in various directions to accelerate the rate distortion. The approach performs various types of refining and sampling methods and substitution methods. This helps to detect change, dissimilar frames are captured and stored in buffer, and similar frames are removed to gain good compression [1]. Rank of the matrix is used to find out the minimum intensities pixels which contain background only. It recognizes the minimum intensity pixels and the pixels which are not the part of fore-ground pixels [5, 6]. Clustering regions are used to recognize the spatially coherent clusters. These clusters contain the motion containing foreground objects [7]. Block structure is method to find incorporation between the spatial prior [8, 9] sparsity structure that does not require the block size. Whether the block size is maximum or minimum, it detects the foreground object efficiently [10, 11]. The method maintains the balance between sparsity and group clustering priority. But the main advantage of these methods is that it detects minimum foreground pixels. So, it requires less time for temporal compression. Satkin et al. [12] proposed a technique to naturally perceive the transient distinction of the activity important to improve the exactness of activity acknowledgment. Be that as it may, this technique that overlooked spatial editing along these lines may incorporate people performing superfluous activities during preparing. Right now, content is used to identify frontal area pixels. This strategy learning spatial or worldly degree of activity may neglect to recognize its accurate degree in a video, and for the most part, it leads to incidental video substance from either fleeting or spatial space. The recognition execution is accordingly corrupted because of the incorporation of insignificant video substances in the preparing stage.

3 Proposed Method

SVM and rank of the framework are used to discover frontal area pixels. There are numerous applications such as marking the information set. SVM forestalls the overfitting. Here, SVD is a decay of a network into three grids, e.g., S, U, V, where S is the inclining lattice of particular qualities and U and V are the left and right the position of a network which is a proportion of the interesting data put away in a grid. Higher the position, the more the data put away in a network. Eigen vectors of a grid are headings of greatest spread or change of information. In these days, clicking photographs through the advanced cell is an interest and enthusiasm; however, sparing these photographs of the Web is cerebral pain. To store the pictures with less memory, SVD assists with limiting the size of a picture in bytes to a satisfactory degree of value. SVD assists with cutting the picture into three frameworks dependent on the initial scarcely any solitary qualities and gets a compacted guess of the first picture. There are some packed images which are indistinguishable from the original by human eyes. At the time of decompression matrix, completion operation is carried out to fill the missing entries. In frameworks of pictures, a picture is adjacent to the estimations of most pixels that rely upon the pixels around them. In this way, a low-position lattice can be a decent estimate of these pictures. Eigen values of the Eigen vectors help to find directions of maximum variance of Eigen vectors. High rank matrices store maximum information, so the approach selects high rank features and discards low rank features to achieve compression. Here, in this approach, spectral clustering is also used, and it is grouping of similar objects. It is an unsupervised machine learning technique. K-means clustering is a simple and powerful algorithm to classify similar objects. But there are certain situations where it does not work more accurately. Like in image compression, there are some pixels which are not part of ROI, but they present in the compressed image, and some of them are part of ROI but not in the compressed image. Here, in this approach, to find the higher similar intensity pixels, affinity matrix is used. Affinity matrix is used to represent how similar one object is to another. Here, in this approach, affinity matrix is used to higher the high intensity pixels and lower the low intensity pixels. Affinity matrix first finds the occurrences of the intensities of the pixels. Then, it groups first similar intensity rows and columns. Affinity score is high if objects are very similar. Distance score is very small if objects are close to each other. Here, distance is inversely proportional to affinity. Existing methods finds searching points for fixed number of frames. Loss of information was more which finds specific actions only, and more noise detects more frames with no motion. To find the minimum intensity pixels, rank of the matrix is used. In this, rank will give us all nonzero matrix rows and columns using following equation.

$$\text{Affi}(ai, aj) = \text{Cost}(Q_k) * \text{Freq}(Q_P) \quad (1)$$

where Cost (Q_k) is the cost of the matrix element. Freq (Q_P) is a number of occurrences of that element. Affinity matrix gives the sum of all nonzero row elements. So, it gives all maximum intensity pixels. Affinity matrix finds clustered affinity matrix as follows (Fig. 1).

The affinity score is high if objects are very similar. The distance score is very small if objects are close to each other. Here, distance is inversely proportional to affinity. Existing methods find searching points for a fixed number of frames.

To find the minimum intensity pixels, rank of the matrix is used. This rank will give us all nonzero matrix rows and columns using the following equation.

$$\text{Rank}[A] \leq \min \text{int} \quad (2)$$

$$\text{Let } [A^{-1}] = \begin{bmatrix} a_{11} & a_{12} \\ a_{21} & a_{22} \end{bmatrix}^{-1} = \frac{1}{|A|}[r][(a_{22} - a_{12}) * (a_{21} - a_{11})] \quad (3)$$

We find the rank of matrix A mathematically as follows
Rank of matrix $Q = UU^{\mathrm{T}}$

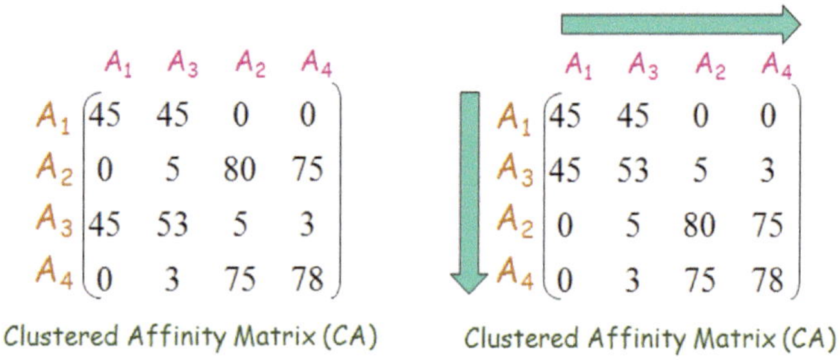

Fig. 1 Example of affinity matrix

$$A = \begin{bmatrix} 1 & 0 & 1 \\ 2 & 1 & 2 \\ 1 & 0 & -1 \end{bmatrix} \xrightarrow[R_3 - R_1]{R_2 - 2R_1} \begin{bmatrix} 1 & 0 & 1 \\ 0 & 1 & 0 \\ 0 & 0 & -2 \end{bmatrix} \xrightarrow{\frac{-1}{2} R_3} \begin{bmatrix} 1 & 0 & 1 \\ 0 & 1 & 0 \\ 0 & 0 & 1 \end{bmatrix} \xrightarrow{R_1 - R_3} \begin{bmatrix} 1 & 0 & 0 \\ 0 & 1 & 0 \\ 0 & 0 & 1 \end{bmatrix}$$

If we get identical matrix, then rank of the matrix will be the number of nonzero rows.

Spectral clustering is a technique with roots in graph theory, where the approach is used to identify communities of nodes in a graph based on the edges connecting them. The method is flexible and allows us to cluster non-graph data as well. SVM produces an accurate classification for new videos and also distinguishes between maximum and minimum intensity pixels. Here, rank of the matrix number will be treated as negative if minimum intensity pixels and above to that will be maximum intensity as follows

$$\min_{w,b}\left(\frac{1}{2}||\omega||^2 + c\sum_{i=i}^{m}\xi i\right) \tag{4}$$

where ω is a nonzero vector, c is a constant used to control the trade-off between margin maximization and error minimization [5], and $\xi i \geq 0$ is a function of the distance from the margin hyperplane to the points which has slack.

The rank of the matrix reduces Eigen noise. It is a nonzero row of a matrix having determinant zero. An affinity matrix is used to find the similarity between elements of a matrix.

Spectral clustering is applied to segment all boundaries into two clusters. Lower intensity clusters will be treated as background, and it is removed. The pixels with high intensity are getting higher values as it gets multiplied with occurrences, e.g., if any four rows of matrix have number of occurrences 5 or 6 times, its intensity value is 99, and it gets multiplied with frequency, so it gets boosted to $99 * 6 = 594$. And likewise, higher intensity pixels get higher values, and lower values become negative, i.e., values less than zero get discarded. The support vector machine uses labeling to variables of the training dataset. It is also used to find the absolute distance between boundaries and x elements. Singular vector decomposition is used to reduce the matrix to its constitutes. It is a major factor for the analysis of principal component decomposition and empirical orthogonal function.

It is a method of decomposing a matrix into three other matrices.

$$A = \mathrm{USV}^{\mathrm{T}} \tag{5}$$

where A is $m * n$ matrix, U is $m * n$ orthogonal matrix, S is $m * n$ diagonal matrix, and V is $m * n$ orthogonal matrix.

$$U^{\mathrm{T}}U = VV^{\mathrm{T}} = I \tag{6}$$

where I is an identity matrix. Here, rank and affinity matrix are used to find out minimum and maximum intensity pixels. To suppress noise and the pixels which are not part of ROI, SVD is used. SVD used Kronecker product in which the individual elements of the first matrix get multiplied with the next matrix, such as a_{11} is matrix B, a_{12} with matrix B, and so on. $R\min(m, n)$ is used to decompose the given matrix. It finds all minimum intensities pixels. So, temporal comparison requires less time. To detect the action, affinity matrix is used. Affinity matrix gives all maximum intensities pixels. The benefit of this method is that we get a sum of two matrices.

E.g., if A is 2×2 and B is also 2×2 matrix, then Kronecker method gives 4×4 matrices. However, normal matrix multiplication gives 2×2 as output matrix.

For even–odd number matrices, it shows index exceeds error. The proposed method finds minimum foreground pixels and reduces time to find dissimilar frames. Due to minimum foreground pixels, accuracy is improved by spatial compression as compared to other methods. Noise removal quality of an image is better than other methods.

4 Results

See Table 1; Figs. 2, 3, and 4.

Table 1 Comparative analysis of SVM with existing methods

S. no.	Image	Method 1 (HSI)	Method 2 (edge)	Method 3 (SVM + rank)	Detected FG pixels
1	News	76,800	485,604	48	76,800
2	Lena	262,144	262,144	107	262,144
3	Cricket	833,578	833,684	47	833,625
4	Parrot	168,710	169,701	130	168,840
5	Cartoon	709,241	712,360	329,524	1,038,765
6	Sports	828,321	41,913	24	2,485,035
7	News	688,080	476,709	4887	1,032,120
8	News2	3187	16,671	9415	1,438,920

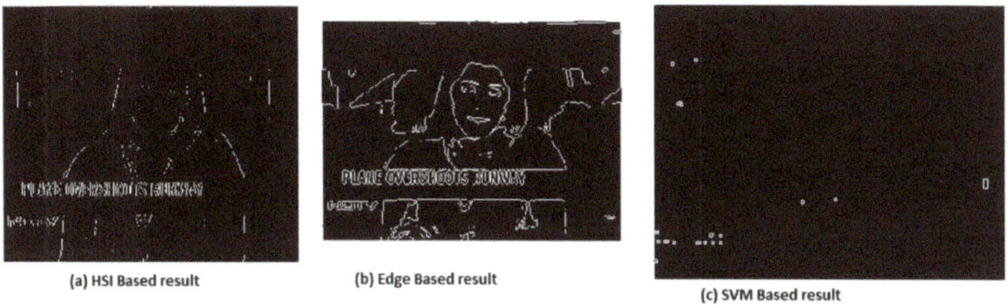

Fig. 2 Comparative analysis of SVM method with existing methods for the number of foreground pixels detected in **a** HSI based approach detected the pixels, **b** edge-based approach detected the pixels, and in **c** SVM based approach detected the pixels

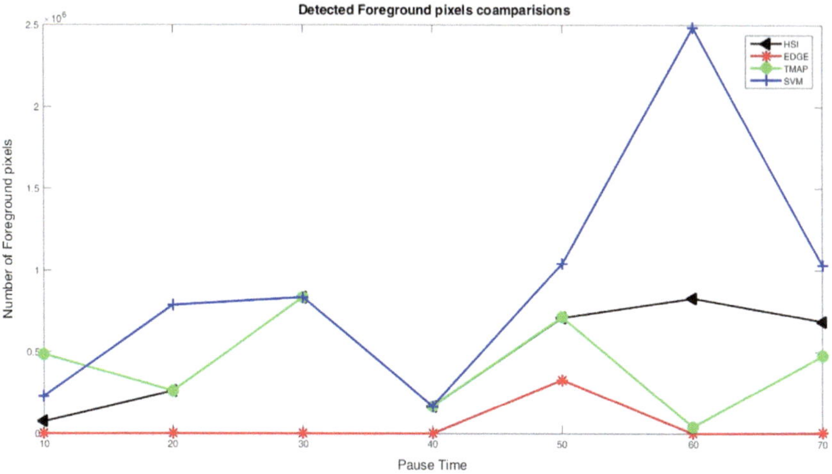

Fig. 3 Graphical analysis of SVM method with existing methods and the number of foreground pixels detected

Accuracy of proposed method is analyzed based on following parameters
PCC is percentage of correct classification

$$\text{Specificity} = \frac{\text{TN}}{\text{TN} + \text{FP}} * 100 \quad (7)$$

$$\text{Recall} = \frac{\text{TP}}{\text{TP} + \text{FN}} * 100 \quad (8)$$

$$\text{Precision} = \frac{\text{TP}}{\text{TP} + \text{FP}} * 100 \quad (9)$$

$$F_1 = \frac{2 * \text{Precision}}{\text{Precision} + \text{Recall}} \quad (10)$$

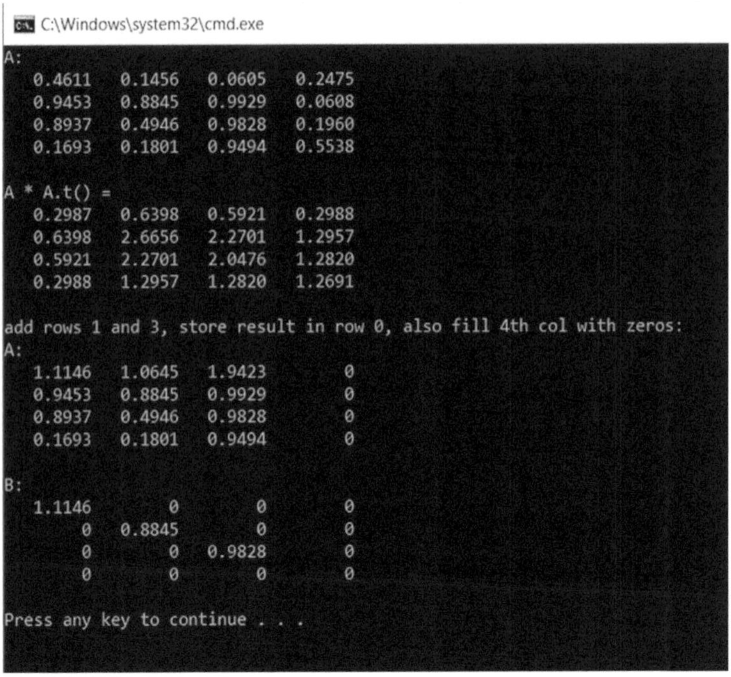

Fig. 4 Matrix values

See Table 2; Figs. 5, 6, 7, 8, 9, and 10.

Table 2 Comparative analysis of SVM with existing methods

S. no.	Image	PCC	Recall	Precision	F1	Specificity
1.	News	6.62	5.26	9.82	6.66	1
2.	Lena	6.64	5.23	9.87	6.66	1
3.	Cricket	6.62	5.31	9.8	6.66	1
4.	Parrot	6.65	5.02	9.9	6.66	1
5.	Cartoon	6.69	5.28	9.9	6.66	1
6.	Sports	6.68	5.31	9.2	6.66	1
7.	News	6.67	5.21	9.82	6.66	1
8.	News2	6.61	5.24	9.83	6.66	1

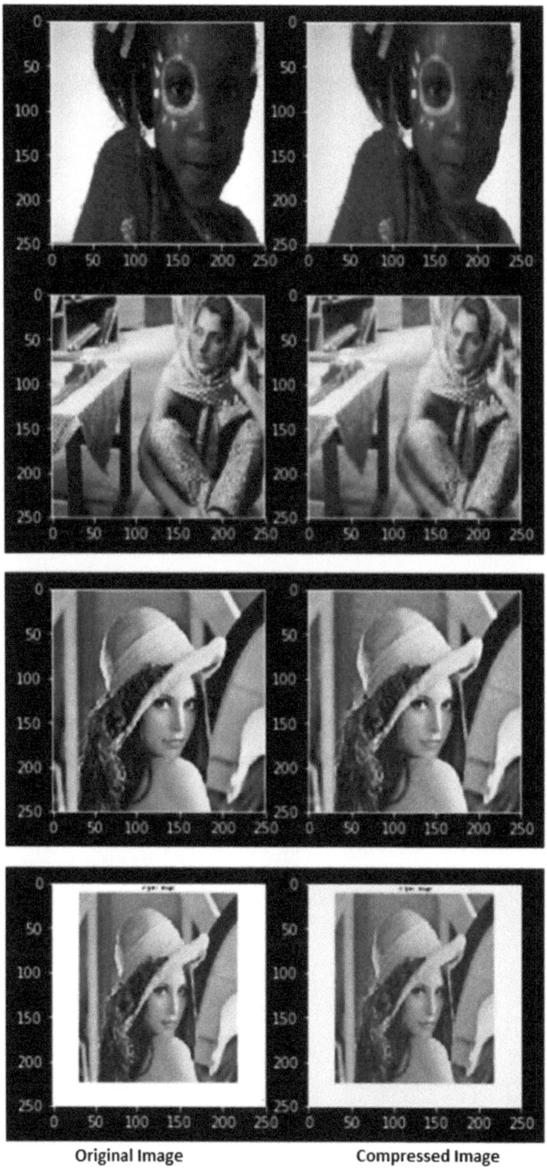

Fig. 5 Comparative results of compressed image with original image

Here, represented method incorporates five saliency maps generated from different stages to obtain the final saliency map. It includes $\mathbf{S_c}$ S_c^d, S_d^d, S_c^{3-D}, and S_d^{3-D} (Table 3).

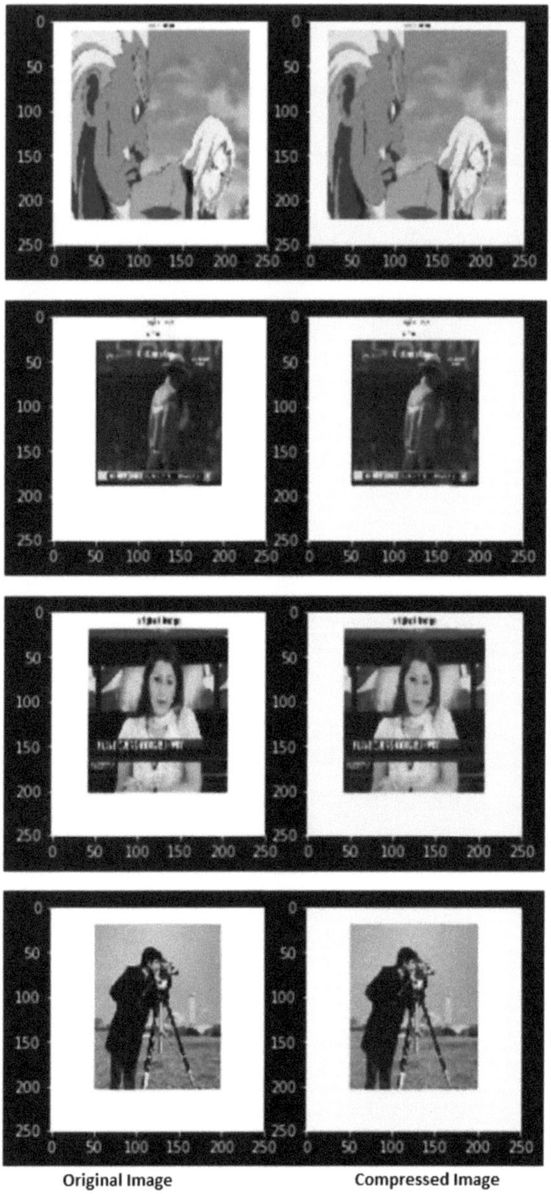

Fig. 6 Comparative results of compressed image with original image

Fig. 7 Comparative results of compressed image with original image

5 Conclusion

This paper represents the approach to detect foreground pixels using SVM. There are various block matching algorithms, among them FS searches 225 points for the best match, and other methods require less. The proposed method achieves good compression in less time, and it finds the best match in nine checking points only. Computation complexity and time complexity are less than the other methods. The proposed method finds averaging foreground pixels, so it requires less time for comparisons. The proposed method finds minimum foreground pixels and reduces time to find dissimilar frames. Due to minimum foreground pixels, accuracy is improved by spatial compression as compared to other methods. Noise removal quality of an image is better than other methods.

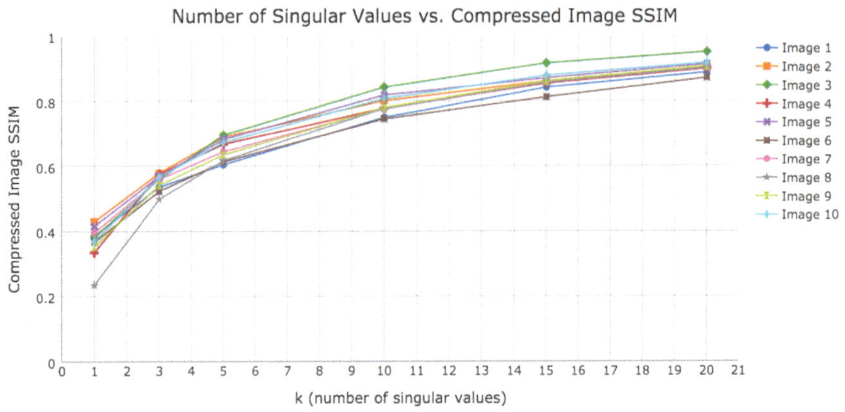

Fig. 8 Comparative results of compressed image with original image based on SSIM

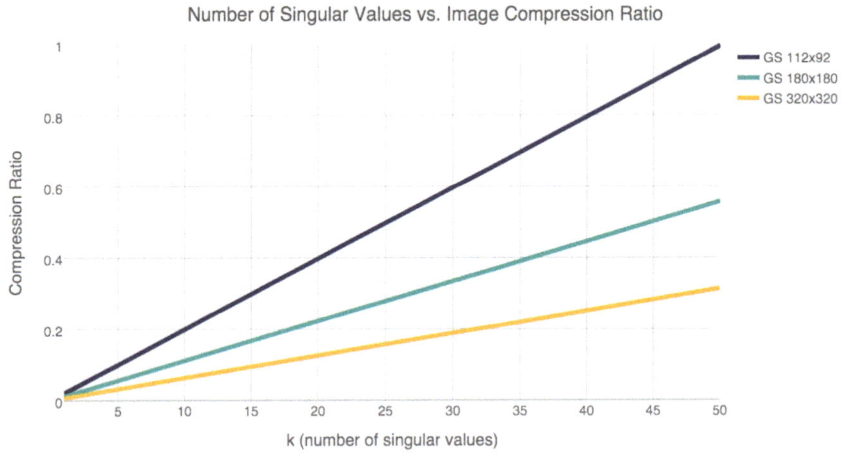

Fig. 9 Comparative results of compressed image with original image based on SSIM

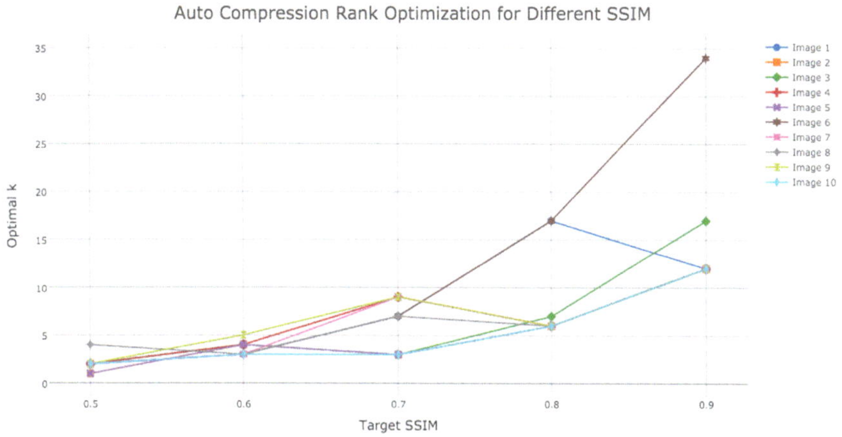

Fig. 10 Comparative results of compressed image with original image based on SSIM

Table 3 Comparative analysis of proposed method based on following parameters

Comp	S_c	S_c^d	$S_c^3 D$	S_d	S_d^d	S_c^{3-D}	S_d^{3-D}	Proposed method
Precision	0.606	0.667	0.701	0.475	0.588	0.666	0.694	0.982
Recall	0.786	0.733	0.697	0.567	0.652	0.708	0.717	0.523
F-measure	0.613	0.652	0.671	0.466	0.568	0.645	0.669	0.66

References

[1] Zhou, Z., Shi, F., Wu, W.: Learning spatial and temporal extents of human actions for action detection, IEEE Trans. Multim. 17:4 (2015)

[2] Budhewar, A.S., Doye, D.D.: Performance analysis of compression techniques for multimedia data. In: Mishra, M., Mishra, B., Patel, Y., Mishra R. (eds) Smart Techniques for a Smarter Planet. Studies in Fuzziness and Soft Computing, vol 374. Springer, Cham (2019)

[3] Budhewar, A.S., Doye, D.D.: Spatio-temporal compression of multimedia data using HSI and variable-length encoding. Int. J. Adv. Res. (2016)

[4] Budhewar, A.S., Thool, R.C.: Improving performance analysis of multimedia wireless sensor network: a survey. In: IEEE International Conference on Advances in Computing, Communication, and Informatics, 10–13 Aug 2015

[5] Candes, E., Li, X., Ma, Y., Wright, J.: Robust principal component analysis. J. ACM **58**(3), 1037 (2011)

[6] Cevher, V., Sankarnarayanan, A., Durate, F., Reddy, D., Baraniuk, G., Chellappa, R.: Compressive sensing for background subtraction. In: Proceedings of European Conference Computer Vision, pp. 155–168 (2008)

[7] Zhou, X., Yang, C., Yu, W.: Moving object detection by detecting continuous outliers in the low-rank representation. IEEE Trans. Patt. Anal. Mach. Intell. **35**(3), 597–610 (2013)

[8] Eldar, Y.C., Kuppinger, P., Boleskeri, H.: Block-sparse signals: uncertainty relations and efficient recovery. IEEE Trans. Signal Process **58**, 3042–3054 (2010)

[9] Stojnic, M., Parvaresh, F., Hassibi, B.: On the reconstruction of block sparse signals with an optimal number of measurements. IEEE Trans. Signal Process **57**(8), 3075–3085 (2009)

[10] Huang, J., Huang, X., Metaxas, D.: Learning with dynamic group sparsity. In Proceedings of International Conference on Computer Vision, pp. 64–71 (2009)
[11] Qiu, C., Vaswani, N.: ReProCS: a missing link between recursive robust PCA and recursive sparse recovery in large but correlated noise in arxiv preprint arxiv.org/abs/1106.3268 (2011)
[12] Satkin, S., Hebert, M.: Modeling the temporal extent of actions. In: Proceedings of European Conference Computer Vision, vol. 6311, pp. 536548 (2010)

A Bio-Inspired Chicken Swarm Optimization-Based Fuel Cell System for Electric Vehicle Applications

Neeraj Priyadarshi, Farooque Azam, Sandeep Singh Solanki, Amarjeet Kumar Sharma, Akash Kumar Bhoi, and Dhafer Almakhles

Abstract This research work demonstrates a bio-inspired chicken swarm optimization (CSO)-based maximum power point trackers (MPPT) for fuel cell-based electric vehicle applications. The natural clamp zero current switching-based bi-directed converter has been used which eliminates the requirement of passive snubber components which reduces the switching losses. The proposed model works on the hierarchal sequence of swarm chicken and its action which can be utilized for utilization of fuel cell fed electric vehicle. The proposed CSO-based MPPT provides optimal fuel cell power with high accuracy.

Keywords Bio-inspired · Chicken swarm optimization · Electric vehicle · Fuel cell · MPPT

N. Priyadarshi (✉) · A. K. Sharma
Department of Electrical Engineering, Birsa Institute of Technology (Trust), Ranchi 835217, India
e-mail: neerajrjd@gmail.com

A. K. Sharma
e-mail: maxeramar@gmail.com

F. Azam
School of Computing & Information Technology, REVA University, Bangalore 560064, India
e-mail: farooque53786@gmail.com

S. S. Solanki
Department of Electronics and Communications, Birla Institute of Technology, Mesra, Ranchi 835215, India
e-mail: sssolanki@bitmesra.ac.in

A. K. Bhoi
Department of Electrical and Electronics Engineering, Sikkim Manipal Institute of Technology, Sikkim Manipal University, Sikkim 737136, India
e-mail: akash730@gmail.com

D. Almakhles
Renewable Energy Lab, Department of Communications and Networks Engineering, College of Engineering, Prince Sultan University, Riyadh, Saudi Arabia
e-mail: dalmakhles@psu.edu.sa

A. K. Bhoi et al. (eds.), *Bio-inspired Neurocomputing*, Studies in Computational Intelligence 903, https://doi.org/10.1007/978-981-15-5495-7_16

1 Introduction

Due to devitalization of prevalent energy originators, the requirement of non-conventional energy originators is growing day by day [1–6]. Amid entire renewable energy sources, the fuel cell-based system has been used as a major promising technology for electric transportation purposes [7–11]. Fuel cell-based system provides low emission, high vehicle enforcement and low cost compared to other renewable energy originators. However, a maximum power point tracking (MPPT) is an imperative component to achieve high power from fuel cell-based system [12–15]. In this investigation effort, a peculiar bio-inspired chicken swarm optimization (CSO)-based MPPT algorithm is used for fuel cell electric vehicle applications. The three-ph electric motor with fuel cell-based system is a recent technology used for transportation electrification. In equated with classical battery operated vehicle, the fuel cell operated vehicle provides high driving range with no limitations. Figure 1 presents the complete schematic structure of fuel cell-based electric vehicle. Fuel stack system produces required electric current to drive the electric vehicle by chemical reactions of hydrogen and oxygen. Moreover, auxiliary power supply (battery) has been employed to provide better transient operation. Furthermore, bi-directed switched converter provides interfacing of lower voltage level to fickle fuel cell stack voltage level.

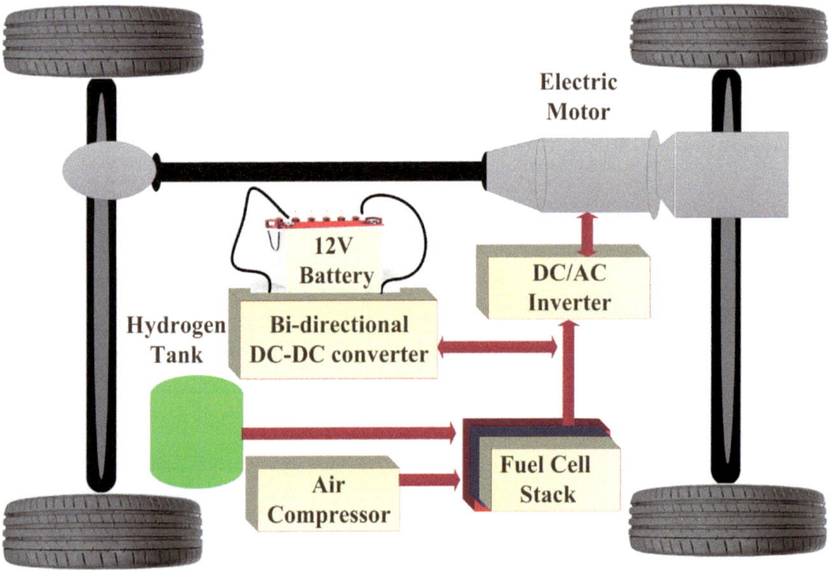

Fig. 1 Complete schematic of fuel cell-based electric vehicle

2 Proposed Chicken Swarm Optimization-Based MPPT for Fuel Cell System

Chicken swarm optimization is bio-inspired meta-heuristic algorithm invented in 2014 by Xianbing Meng. It works on hierarchical sequence of swarm chicken and its action. The rooster, hens and chicks are different sets in which swarm chicken is classified.

Figure 2 presents the hierarchical structure of swarm chicken in which rooster takes up dominant location with high food searched ability. The rooster also contends other chickens to occupy their neighborhood. The second location has been entertained by hens which supervenes the roosters for the food searching purposes. The third location is preoccupied by chicks which pursue their mom for investigation of food. Here, entire chicken counterpart with each other for the purpose of food investigation. Therefore, an objective function has been optimized to express the biological nature of chickens. Moreover, the mom–child relation is defined arbitrary. The categorization of hierarchical structure and mom–child relation can be amended subsequently in s time mark. The natural practice of hen follows rooster and chicks to go behind of hen for food searching purposes have been utilized effectively in the entire process of the algorithm. It is also noted that there is a completion among entire chickens during food searching process. Initialization and updation parameters are the important process of CSO algorithm. The swarm size and specifications of CSO algorithm has been initialized first with s time steps. The fitness function of population has been estimated first, and then, hierarchical structure has been formed. Figure 3 presents relation between swarm chicken. Mathematically, set theory relation has been developed as

$$\mathrm{MK} \subset \mathrm{HK} \tag{1}$$

where

MK Set of mom hens
RK Set of roosters
HK Set of hens

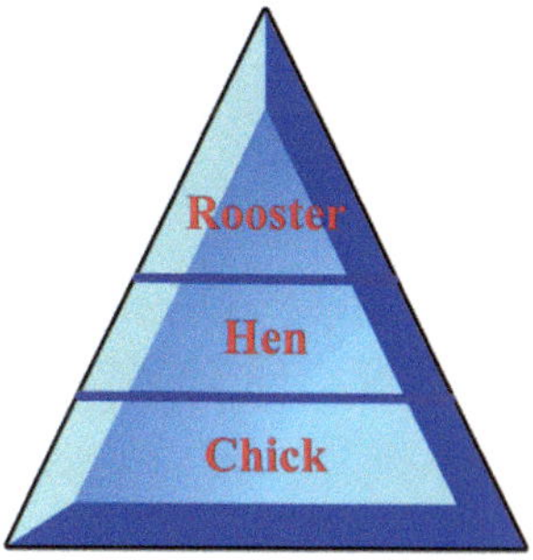

Fig. 2 Hierarchical structure of swarm chicken

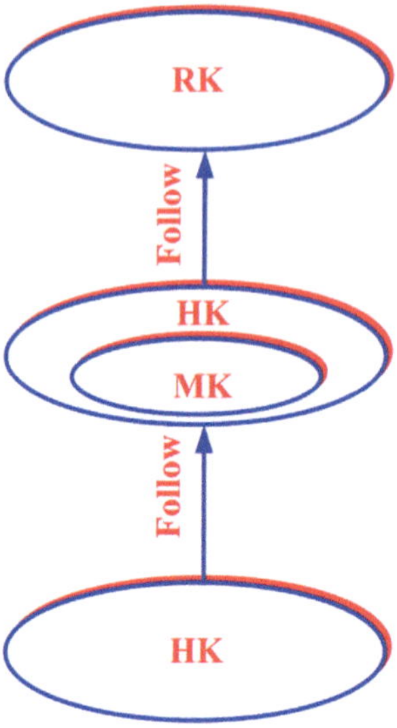

Fig. 3 Relation between swarm chicken

CK Set of chicks.

It is noted that the searching ability of food varies in the batch. The fitness function has been updated in the category as per the ability of searching food. Mathematically, the roosters food searched ability can be expressed as

$$X_{m,n}^{T+1} = X_{m,n}^{T} \times \left(1 + \mathrm{Rand}\left(0, \sigma_s^2\right)\right) \tag{2}$$

if $\mathrm{Fit}_m \leq \mathrm{Fit}_k$

$$\sigma_s^2 = 1 \tag{3}$$

else

$$\sigma_s = \exp\left(\frac{\mathrm{Fit}_m - \mathrm{Fit}_k}{|\mathrm{Fit}_m| + \in_s}\right) \tag{4}$$

where

$\mathrm{Rand}\left(0, \sigma_s^2\right)$	Gaussian distribution function
σ_s^2	Standard deviation
Fit	Fitness function
K	Rooster random index
$\in_s$	Small constant to prevent zero division error.

3 Simulation Results and Discussions

Figure 4 describes the complete process of CSO-based algorithm for MPPT. Figures 5, 6 and 7 present the equivalent circuitry of proposed fuel cell, fuel cell stack modeling and proposed fuel cell-based electric vehicle simulated model, respectively.

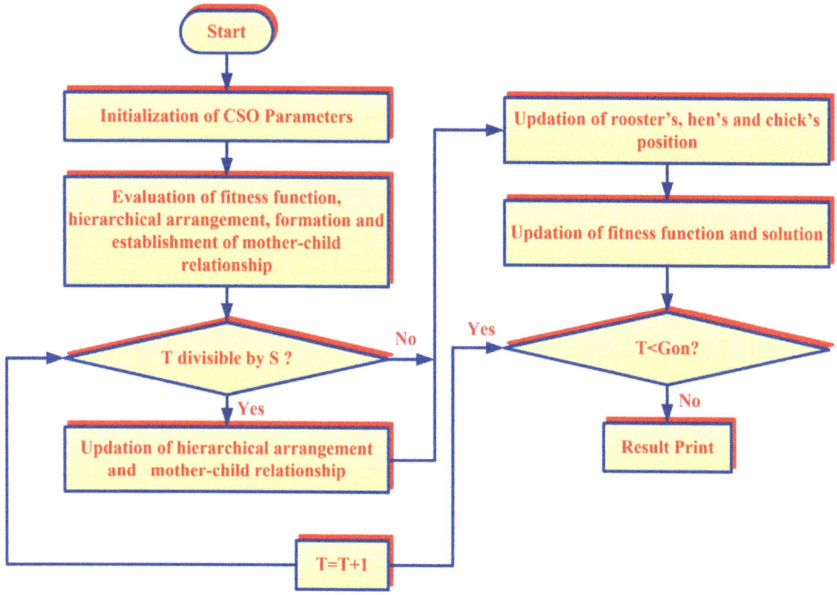

Fig. 4 Flowchart of CSO-based algorithm

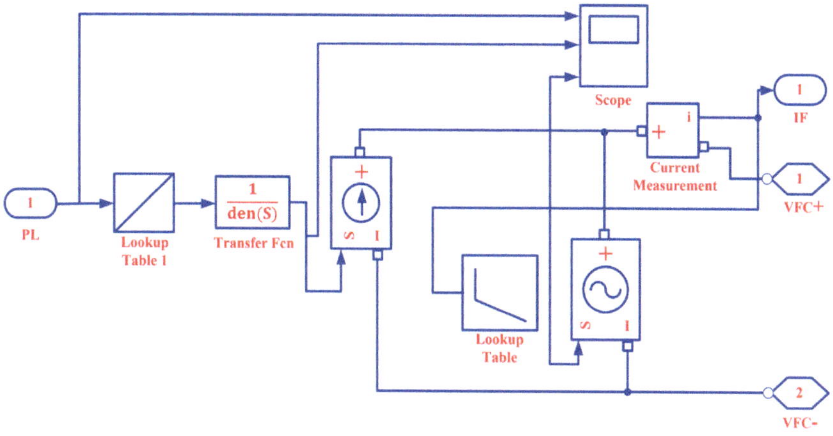

Fig. 5 Equivalent circuitry of proposed fuel cell

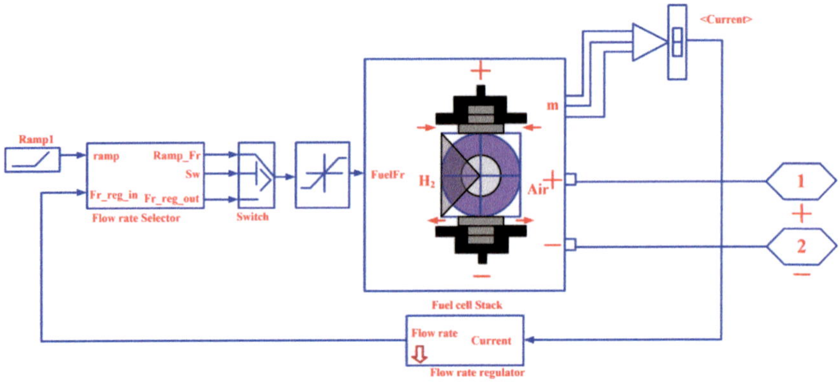

Fig. 6 Fuel cell stack modeling

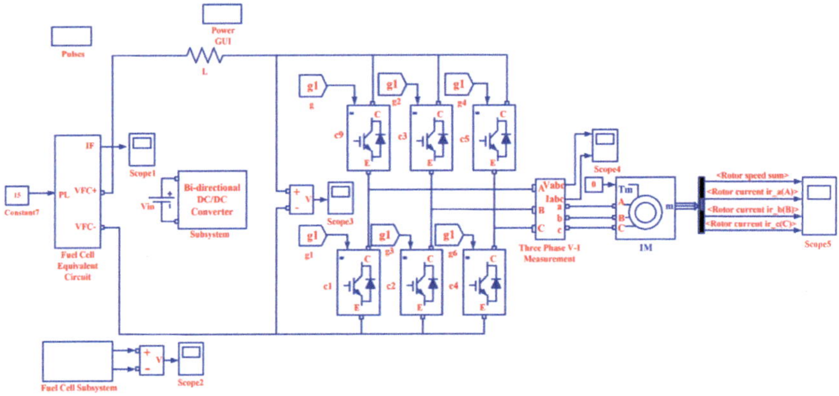

Fig. 7 Proposed fuel cell-based electric vehicle simulated model

Furthermore, several chickens in the batch have habit to take the food of others, and mathematically, updated equations can be expressed as

$$X_{m,n}^{T+1} = X_{m,n}^{T} + U_1 \times \text{Rand} \times \left(X_{ra1,n}^{T} - X_{m,n}^{T}\right) + U_2 \times \text{Rand} \times \left(X_{ra2,n}^{T} - X_{m,n}^{T}\right) \quad (5)$$

$$U_1 = \exp\left(\frac{\text{Fit}_m - \text{Fit}_{\text{ra1}}}{|\text{Fit}_m| + \in_s}\right) \quad (6)$$

$$U_1 = \exp(\text{Fit}_{\text{ra2}} - \text{Fit}_m) \quad (7)$$

where

$$\begin{aligned} \mathrm{ra1} &\in [1, N] \equiv \text{Rooster index} \\ \mathrm{ra2} &\in [1, N] = \text{Rooster/hen index} \\ \mathrm{ra1} &\neq \mathrm{ra2} \end{aligned} \tag{8}$$

The natural habit of chicks to pursue mom's food searching ability can be expressed mathematically as

$$X_{m,n}^{T+1} = X_{m,n}^{T} + \mathrm{GM} \times \left(X_{0,n}^{T} - X_{m,n}^{T}\right) \tag{9}$$

where
$X_{0,n}^{T}$ = Position for mth mom
GM = Parameter which identifies chick follows to their mom's.

$$\mathrm{GM} \in [0, 2]$$

Simulation waveforms of supply current and current flowing through inductors, respectively, are shown in Fig. 8. The phasor shift between inductor currents is found 180°. Therefore, supply current contains lower ripple factor due to ripple cancelation of inductor currents.

Figure 9 reveals that the phasor shift between output voltages of capacitors C_{01}, C_{02}, C_{03} and C_{04} of each cell is of 180° to each. Furthermore, cancelation

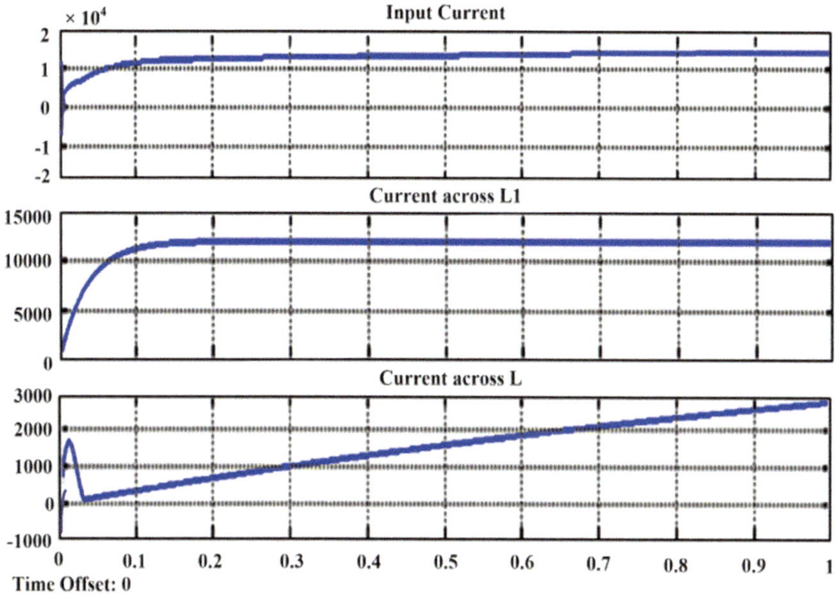

Fig. 8 Simulation feedback of input current and current through inductors

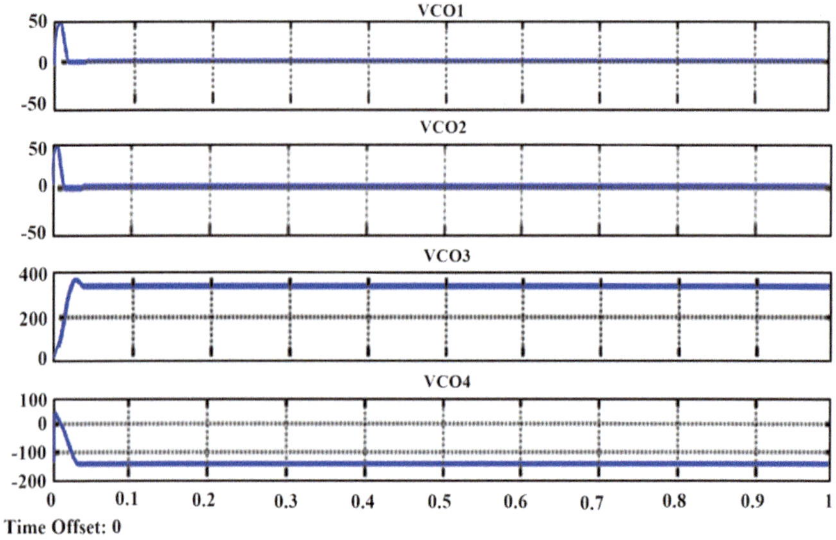

Fig. 9 Simulated response of voltage across output capacitors

methodology consequences low ripple in output voltage and mitigates the filter demand.

Figure 10 depicts the simulation waveform of current flowing through series inductors which flows more than constant values due to conduction of feedback diodes

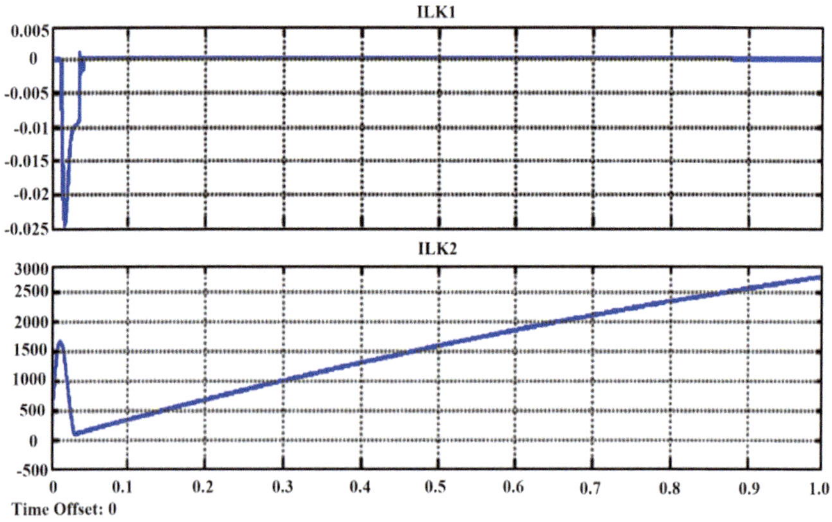

Fig. 10 Simulated response of inductor currents flowing in series

and makes ZCS turn-off. And the voltages of switches get natural commutation at voltage magnitude $V_0/4n$.

Figure 11 presents simulation feedback of primary switch current and secondary switch current. The above waveform of Fig. 12 shows the three-ph voltage/current

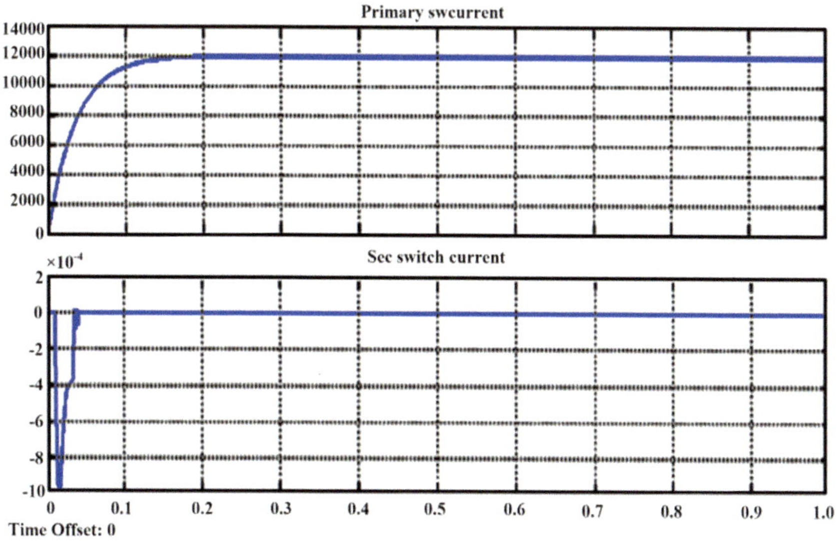

Fig. 11 Simulation feedback of primary switch current and secondary switch current

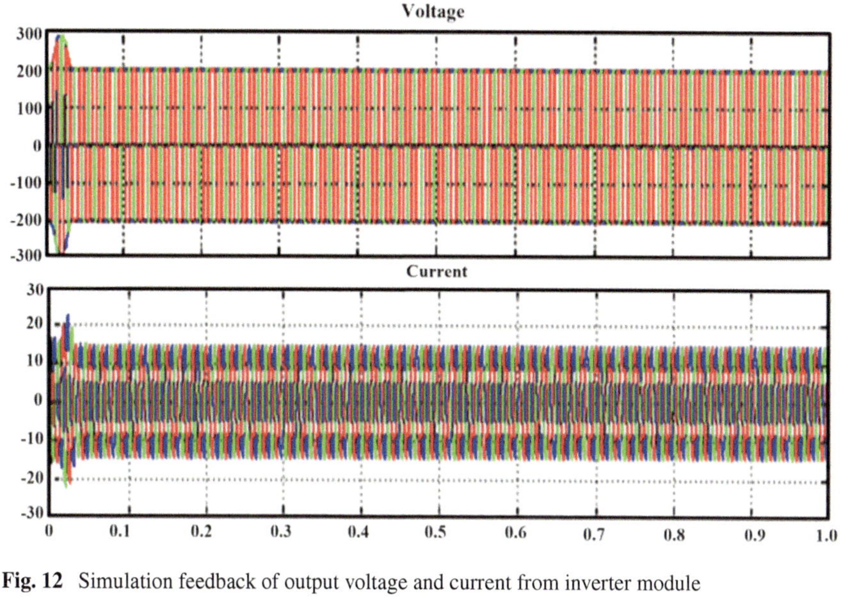

Fig. 12 Simulation feedback of output voltage and current from inverter module

of inverter. Both voltage as well as current are constant, and moreover, we get the constant outcomes voltage is 150–300 V.

As shown in Fig. 13 waveforms are available on scope 5, it shows the three phase induction motor rotor speed linearly increasing and in all the three phase rotor currents (i_a, i_b, i_c) are sinusoidal and free from harmonics. Above all simulation results obtaining justify that bi-directed power converter has output voltage is 150–300 V. Therefore, by using soft-switching technique, the simulation results show the outcomes voltage of bi-directed converter has maintained constant. The electrical motor will provide smooth and accurate control as the fuel cell supply is provided consistently.

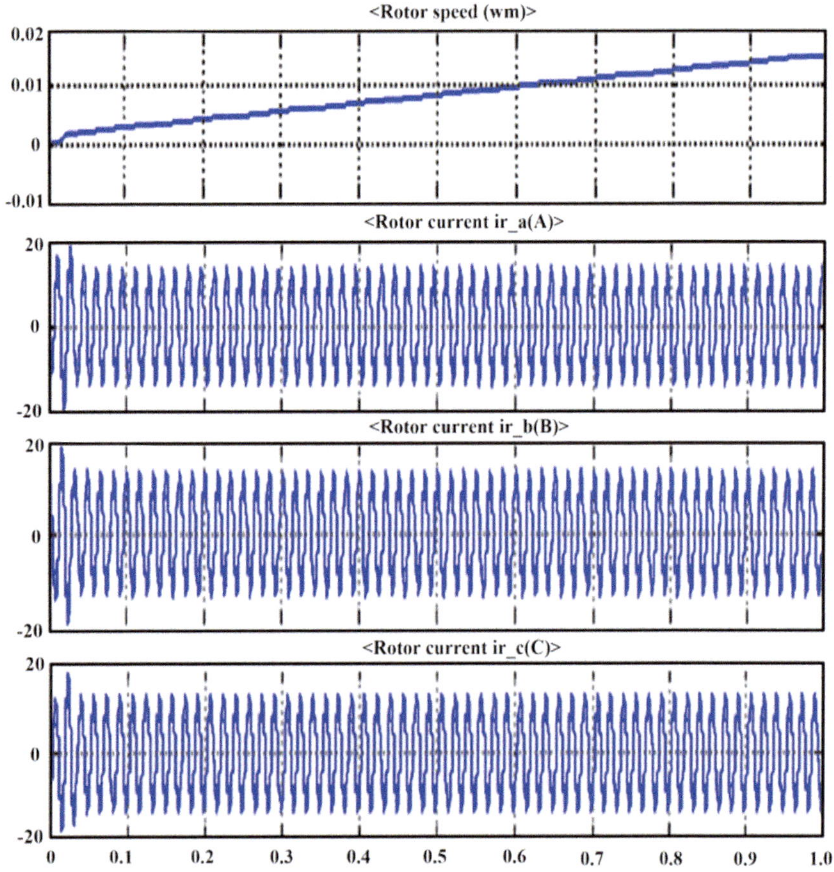

Fig. 13 Simulated response of three phase induction motor rotor speed and rotor current

4 Conclusion

This research paper described the bio-inspired chicken swarm optimization positioned peak point trackers for fuel cell electric vehicle utilization. Moreover, the bi-directed snubberless zero switching converter has been implemented which minimizes the switching losses and does not depend on loading conditions. The major advantages of the proposed control such as high switched frequency, robust design and lower switching losses have been achieved for fuel cell-based electric vehicle applications. Modulation technique has been employed to control the proposed converter which is simpler and easier to implement for high power applications.

References

1. Priyadarshi, N., Padmanaban, S., Maroti, P.K., Sharma. A.: An extensive practical investigation of FPSO-based MPPT for grid integrated PV system under variable operating conditions with anti-islanding protection. IEEE Syst. J. 1–11 (2018)
2. Priyadarshi, N., Padmanaban, S., Bhaskar, M.S., Blaabjerg, F., Sharma, A.: A fuzzy SVPWM based inverter control realization of grid integrated PV-wind system with FPSO MPPT algorithm for a grid-connected PV/wind power generation system: hardware implementation. IET Electr. Power Appl. 1–12 (2018)
3. Priyadarshi, N., Anand, A., Sharma, A.K., Azam, F., Singh, V.K., Sinha, R.K.: An experimental implementation and testing of GA based maximum power point tracking for PV system under varying ambient conditions using dSPACE DS 1104 controller. Int. J. Renew. Energy Res. **7**(1), 255–265 (2017)
4. Priyadarshi, N., Kumar, V., Yadav, K and Vardia, M.: An experimental study on zeta buck-boost converter for application in PV system. In: Handbook of Distributed Generation. Springer, https://doi.org/10.1007/978-3-319-51343-0_13
5. Priyadarshi N., Sharma A.K., Priyam S.: An Experimental realization of grid-connected PV system with MPPT using dSPACE DS 1104 control board. In: Advances in Smart Grid and Renewable Energy. Lecture Notes in Electrical Engineering, vol. 435. Springer, Singapore (2018)
6. Priyadarshi N., Sharma A.K., Priyam S.: Practical realization of an improved photovoltaic grid integration with MPPT. Int. J. Renew. Energy Res. **7**(4) (2017)
7. Lindahl, P.A., Shaw, S.R., Leeb, S.B.: Fuel cell stack emulation for cell and hardware-in-the-loop testing. IEEE Trans. Instr. Meas. **67**(9), 2143–2152 (2018)
8. Prajof, P., Agarwal, V.: Novel solar PV-fuel cell fed dual-input-dual-output dc-dc converter for dc microgrid applications. In: Proceedings of IEEE 42nd Photovoltaic Specialist Conference IEEE-PVSC'15, New Orieans, USA, pp. 1–6 (2015)
9. Padmanaban, S., Priyadarshi, N., Bhaskar, M.S., Holm-Nielsen, J.B., Hossain, E., Azam, F.: A hybrid photovoltaic-fuel cell for grid integration with jaya-based maximum power point tracking: experimental performance evaluation. IEEE Access **7**, 82978–82990 (2019)
10. Priyadarshi, N., Sharma, A.K., Azam, F.: A hybrid firefly-asymmetrical fuzzy logic controller based MPPT for PV-wind-fuel grid integration. Int. J. Renew. Energy Res. **7**(4) (2017)
11. Giustiniani, A., Petrone, G., Spagnuolo, G., Vitelli, M.: Low-frequency current oscillations and maximum power point tracking in grid-connected fuel-cell-based systems. IEEE Trans. Ind. Elec. **57**(6), 2042–2053(2010)
12. Rslan, H., Ahmed, M., Orabi, M., Youssef, M.: Development of grid connected power conditioner system compatible with fuel cell applications. In: Proceedings of 2nd IEEE Interantional Symposium on Power Electronics for Distributed Generation Systems, pp. 935–941 (2010)

13. Blavi, H.V., Moncusí, J.M.B., Marroyo, L., Guinjoan, F., Barrado, J.A., Salamero, L.M.: Adapting a low voltage PEM fuel-cell to domestic grid-connected PV system. In: Proceeding of 35th Annual Conference of IEEE Industrial Electronics, pp. 160–165 (2009)
14. Vasant, L.G., Pawar, V.R.: Solar-wind hybrid energy system using MPPT. In: Proceedings of International Conference on Intelligent Computing and Control Systems (ICICCS), pp. 595–597 (2017)
15. Wang, M.H., Huang, -M.L., Jiang, -W.J., Liou, -K.J.: Maximum power point tracking control method for proton exchange membrane fuel cell. IET Renew. Power Gener. **10**(7), 1–8 (2016)

The Effects of Stress and Organizational Commitment on Turnover Intention of Workers: The Moderated Mediation Effect of Organizational Communication

Chang Seek Lee, Ha Young Jang, and Eun Kyung Ryu

Abstract **Background/Objectives**: The purpose of this study was to investigate the moderated mediation effect of organizational communication on organizational commitment between worker stress and turnover intention. **Methods/Statistical analysis**: For this study, 400 workers were purposively sampled from H, N and T cities of Korea and data were collected through a survey. The collected data were analyzed by using difference verification, reliability, correlation and moderated mediation effect analyses. **Findings**: First, as a result of verifying the difference of turnover intention according to the general characteristics, there was a significant difference according to position. Second, there was a positive correlation between stress and turnover intention and organizational and organizational communication. Third, organizational communication showed moderated mediation effect on the mediating effect of organizational commitment between workers' stress and turnover intention. **Improvements/Applications**: Based on these findings, we discussed ways to reduce turnover intention due to workers' stress.

Keywords Stress · Organizational commitment · Turnover intention · Organizational communication · Moderated mediation effect

C. S. Lee
Department of Health, Counseling and Welfare, Hanseo University, Seosan-si 31962, South Korea
e-mail: lee1246@hanmail.net

H. Y. Jang · E. K. Ryu (✉)
Department of Child and Adolescent Counseling and Psychology, Hanseo University, Seosan-si 31962, South Korea
e-mail: yek0444@hanmail.net

H. Y. Jang
e-mail: besof@hanmail.net

A. K. Bhoi et al. (eds.), *Bio-inspired Neurocomputing*, Studies in Computational Intelligence 903, https://doi.org/10.1007/978-981-15-5495-7_17

1 Introduction

Stress depends not only on age, gender and personality characteristics but also on various factors, such as family, school social relationships and surroundings. Among these, job stress refers to the stresses caused by occupations, such as social relations, workload and organizational atmosphere [1]. Workers were also stressed due to the physical environment conditions, such as the work burden related to the job and dissatisfaction with working conditions. In particular, uncomfortable relations with members of the organization caused workers to be stressed out. Workers' stresses can lead to negative interactions between members and disrupt team cohesion. In addition, the psychological overwhelming of the hierarchical order and command system of the organization leads to a passive work attitude and acts as a barrier to achieving the organization's common goals. These factors hinder the intrinsic motivation and organizational commitment of employees, which can lead to decision to turn over.

Turnover intention means the intention, plan, concern and determination of an employee to leave the work [2]. Employees think about turnover by combining factors related to the individual, overall job satisfaction and need to remain. External factors that may lead to turnover include lost costs, social positions and salary levels resulting from leaving the organization. In addition, internal factors, such as relationships with organizational members, the similar values with the organization, psychological satisfaction and stress, were found to influence turnover intention [3]. The organization must present its psychological stability and vision to its members and share it to draw empathy from its members. This creates an employee's desire to remain in the organization and can lead to positive results in terms of organizational commitment. The importance of organizational commitment in the recent years is emphasized because it shows that organizational commitment is more effective in predicting turnover than job satisfaction [4]. In the long-term view of indicators for predicting organizational stability, organizational commitment of members and the relationships between them are very useful.

Organizational commitment has been studied in many fields, including industrial psychology, sociology and behavioral science. Organizational commitment affects organizational values and goals, organizational culture and atmosphere and the products produced within the organization. In particular, organizational commitment that was affected by the relationship between members of an organization was defined as a willingness for members to invest in time and energy with attachment to the organization [5]. There may be individual differences, such as salary levels and obligations, but it can be said that it is strong in terms of emotional commitment to identify and attach to the organization. Emotional commitment was found to have a positive effect on higher intimacy, rapport and consensus with members of the organization [6]. Good communication between the members of the organization can increase organizational commitment by building consensus and resolving conflicts.

Communication is the most essential element in social life and includes various nonverbal expressions, such as facial expressions, gestures, intonations and postures

[7]. Organizational communication encompasses communication with a variety of members, including peers and subordinates, as well as external customers and suppliers. Organizations should take appropriate communication according to the circumstances, taking into account the circumstances, position and workability. Efficient and clear expression of speech can create reasonable and appropriate interests. In addition, communication between members provides a good understanding of the organization's goals, and prior work consultations have a positive impact on organizational commitment. By strengthening collaboration through mutual communication and empathizing with the achievement of the organization's goals, it is possible to lower the turnover intention of employees.

Therefore, this study examines the mediating effect of organizational commitment on the effect of workers' stress on turnover intention. It also identifies the moderating effects of organizational communication on the relationship between organizational commitment and turnover intention, and stress and turnover intention. Finally, this study was to identify the moderation of organizational communication on the mediating effects of organizational commitment between stress and turnover intention.

This study established the following research questions. First, are the turnover intentions of workers different depending on their general characteristics? Second, what is the correlation between the major variables? Third, does organizational communication of workers moderate the mediating effects of organizational commitment between stress and turnover intention?

2 Methods

2.1 Research Model

In order to find out the moderating effect of organizational communication between stress and turnover intention through organizational commitment research model as shown in Fig. 1 was set up. PROCESS macro model 15 was used to analyze this model.

2.2 Research Subjects

The research subjects were selected by purposive sampling. These were 400 workers who worked in H, N and T cities in Chungcheongnam-do. The data were collected through a survey, and the researcher visited and conducted the survey, which was conducted from September 2019 to October 2019. A total of 374 copies were used for the final analysis, excluding the questionnaires that were unfaithfully answered.

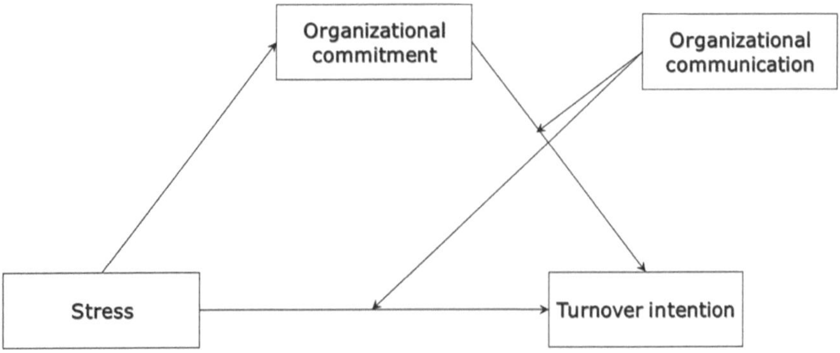

Fig. 1 Research model

The gender distribution of the study subjects was 323 males (86.4%) and 45 females (12.0%). The most common age group had subjects that were in their 20 s with 78 (20.9%), 30 s with 160 (42.8%), 40 s with 84 (22.5%), and 50 s with 31 (8.3%). There were 160 singles (42.8%) and 206 married (55.1%). The education level was 49 high school graduates (13.1%), 45 college graduates (12.0%) and 268 university graduates (71.7%). Depending on their turnover experience, 174 people (46.5%) answered "None", and 195 people (52.1%) answered "Yes". Lastly, their statuses were 183 employees (48.9%), proxy managers were 85 (22.7%), team managers were 46 people (12.3%), and deputy general managers were 50 (13.4%).

2.3 Research Tools

2.3.1 Stress

In order to measure the stress perceived by workers, we used a scale developed by Parker and Decotiis [8] and translated by Lee et al. [9]. The scale consisted of 13 questions. Each item was measured on a five-point Likert scale, ranging from one point of "not at all" to five points of "very so". Higher scores mean higher stress. In this study, Cronbach's α was 0.894.

2.3.2 Organizational Commitment

The organizational commitment was measured on eight questions related to emotional commitment, which was developed by Allen and Meyer [10], and used by Devece et al. [11]. It was a total of five Likert scales, ranging from one point of "not at all" to five points of "very so". Higher scores indicate higher organizational commitment. In this study, the reliability Cronbach's α was 0.899.

2.3.3 Organizational Communication

Organizational communication was measured on the four items related to emotional communication, which was developed by Downs and Hazen [12], and was revised and supplemented by Lee [13]. Each question was a total of five-point Likert scales, ranging from one point of "not at all" to five points of "very so". Higher scores indicate higher organizational communication. In this study, Cronbach's α was 0.868.

2.3.4 Turnover Intention

We used Becker's scale [14] which was modified and supplemented by Lee [15] to measure the turnover intention. The scale consisted of six questions. Each question was a total of five-point Likert scales, ranging from one point of "not at all" to five points of "very so", and the higher the score, the higher the turnover intention. The reliability of this scale, Cronbach's α, was 0.880.

2.4 Data Analysis

The collected data were analyzed using SPSS Win 25.0 and SPSS PROCESS Macro 3.4. The difference in turnover intention according to the general characteristics of the study was analyzed by using t-test and ANOVA, and post-test was performed with Duncan. We also analyzed the correlations among the major variables. Finally, regression analysis was performed using model 15 of the SPSS PROCESS Macro to analyze the moderated mediation effect.

3 Results and Discussion

3.1 Turnover Intention According to General Characteristics

The difference in turnover intention according to the general characteristics of workers is presented in Table 1, and the results were significant in the status items in the workplace. In terms of status, the employee, deputy and section chiefs had higher turnover intention than managers ($F = 2.521$, $p < 0.05$). These findings are consistent with research findings that workers often show frequent turnover rates in order to continue their career development or to access various opportunities [16]. On the other hand, there were no significant results depending on gender, age, marital status and educational background.

Table 1 Differences in turnover intention according to general characteristics

Characteristics		Turnover intention			
		Frequency	Mean(SD)	T-value/F (Duncan)	
Gender	Male	323	2.89(0.80)	−1.656	
	Female	45	3.10(0.71)		
Age	20's	78	2.89(0.96)	1.102	
	30's	160	2.96(0.77)		
	40's	84	2.85(0.69)		
	50's	31	2.69(0.57)		
Marital status	Married	206	2.85(0.73)	−1.659	
	Unmarried	160	2.99(0.84)		
Education	High school	49	3.03(0.91)	0.688	
	College	45	2.90(0.70)		
	University and above	268	2.89(0.78)		
Position	Employee	183	2.98(0.90)	2.521*	A
	Deputy	85	2.96(0.62)		A
	Section chief	46	2.91(0.69)		A
	Manager	50	2.64(0.62)		B

($n = 373$)
*$p < 0.05$

3.2 *Correlation of Main Variables*

The correlation analysis results among the major variables are shown in Table 2. The results show a positive correlation between stress and turnover intention and between organizational commitment and organizational communication. Among them, organizational commitment and turnover intention ($r = -0.538$, $p < 0.01$)

Table 2 Correlation coefficients among major variables

	1	2	3	4
1. Stress	1			
2. Organizational commitment	−0.194**	1		
3. Organizational communication	−0.169**	0.511**	1	
4. Turnover intention	0.472**	−0.538**	−0.357**	1
Mean	2.77	3.30	2.99	2.91
Standard deviation	0.62	0.69	0.76	0.78

**$p < 0.01$

showed the highest correlation, followed by organizational commitment and organizational communication ($r = 0.511$, $p < 0.01$), and stress and turnover intention ($r = 0.472$, $p < 0.01$). The overall correlation coefficient ranged from 0.511 to −0.538, indicating no multicollinearity between major variables.

3.3 *The Moderated Mediating Effects of Organizational Commitment on the Relationship Between Stress and Turnover Intention Through Organizational Communication*

The mediating effect of organizational commitment on the relationship between stress and turnover intention of workers was analyzed by using model 15 of the SPSS PROCESS macro to verify whether the organizational communication exhibits moderated mediation effects. The results were as shown in Table 3.

The analysis shows that stress has a significant negative impact on organizational commitment (−0.2157, $p < 0.001$) and that organizational commitment has a significant negative impact on turnover intention (−0.4498, $p < 0.001$). Stress has a significant effect on organizational commitment, and it has a significant effect on turnover intention. These indicate that organizational commitment plays a mediating role between workers' stress and turnover intention. The higher the stress, the higher the turnover intention, which can be interpreted to be lowered through organizational commitment. These results are consistent with previous studies that organizational commitment has a positive effect on lowering turnover intention of workers [17].

The interaction terms of stress and organizational communication had a significant effect on turnover intention (0.1323, $p < 0.05$). In other words, the stress depends

Table 3 Moderating effect of organizational communication

Variables	Effect	SE	*t*-value	*p*
Mediating variable model (dependent variable: organizational commitment)				
Constant	0.0000	0.0354	0.0000	1.0000
Stress	−0.2157	0.0565	−3.8188	0.0002
Dependent variable model (dependent variable: turnover intention)				
Constant	2.8928	0.0331	89.4866	0.0000
Stress	0.4725	0.0509	9.2803	0.0000
Organizational commitment	−0.4498	0.0526	−8.5568	0.0000
Organizational communication	−0.0788	0.0469	−1.6821	0.0934
Stress × organizational communication	0.1323	0.0625	2.1156	0.0350
Organizational commitment × organizational communication	0.1371	0.0509	2.6925	0.0074

on the organizational communication, which is a moderating variable in a meaningful relationship to increase turnover intention. This means that the higher the organizational communication, the lower the turnover intention.

In addition, the interaction terms of organizational commitment and communication had a significant effect on turnover intention (0.1371, $p < 0.01$). In other words, the organizational commitment depends on the moderating variable, organizational communication, in a meaningful relationship that lowers turnover intention. This means that the higher the organizational communication, the lower the turnover intention.

The conditional effects of stress according to specific values of organizational communication were as shown in Table 4. The results showed that the conditional effects of stress were significant in the organizational communication values between −0.7631 ($M - 1$SD) and 0.7631 ($M + 1$SD).

The conditional effect of stress was significant in areas where organizational communication values were above −1.7125. This means that organizational communication moderates the relationship between stress and turnover intention in the region above −1.7125. 97.1% of the subjects showed significant moderating effect, and 2.9% showed no significant moderating effect.

In order to confirm the form of the moderating effect of organizational communication in the relationship between stress and turnover intention, it is presented in Fig. 2. The value of turnover intention according to stress was higher in order of the low, middle and high groups of organizational communication. In other words, the lower the organizational communication, the higher the turnover intention in the effect of stress on turnover intention.

On the other hand, according to the specific value of organizational communication, the conditional effect of organizational commitment was as shown in Table 5.

Table 4 Conditional effects of stress at values of organizational communication

Organizational communication	Effect	SE	*t*	*p*	LLCI[a]	ULCI[b]
−1SD(−0.7631)	0.3715	0.0746	4.9776	0.0000	0.2248	0.5183
M(0.0000)	0.4725	0.0509	9.2803	0.0000	0.3724	0.5726
+1SD(0.7631)	0.5735	0.0646	8.8818	0.0000	0.4465	0.7005
Conditional effects of stress at values of organizational communication						
Organizational communication	Effect	SE	*t*	*p*	LLCI[a]	ULCI[b]
−1.9904	0.2091	0.1411	1.4818	0.1393	−0.0684	0.4867
		⋮				
−1.7125	0.2459	0.1251	1.9664	0.0500	0.0000	0.4918
		⋮				
2.0096	0.7384	0.1286	5.7410	0.0000	0.4855	0.9913

[a]*LLCI* The lower limit of the indirect effect within the 95% confidence interval
[b]*ULCI* The higher limit of the indirect effect within the 95% confidence interval

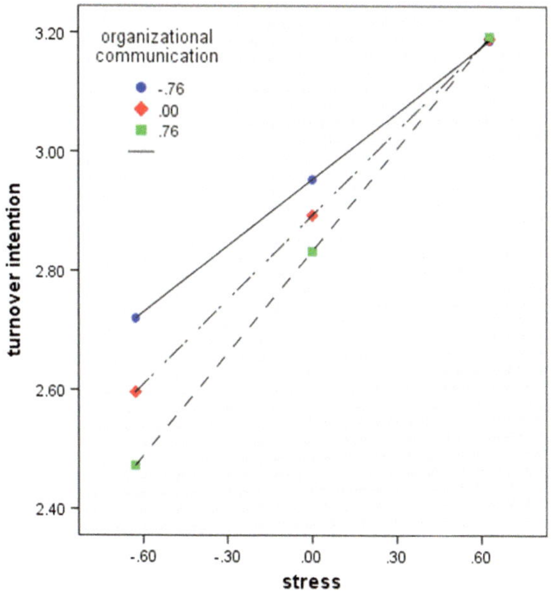

Fig. 2 Moderating effect of organizational communication on the relationship between stress and turnover intention

Table 5 Conditional effects of organizational commitment at values of organizational communication

Organizational communication	Effect	SE	*t*	*p*	LLCI[a]	ULCI[b]
−1SD(−0.7631)	−0.5545	0.0588	−9.4350	0.0000	−0.6700	−0.4389
M(0.0000)	−0.4498	0.0526	−8.5568	0.0000	−0.5532	−0.3465
+1SD(0.7631)	−0.3452	0.0714	−4.8366	0.0000	−0.4855	−0.2048
Conditional effects of organizational commitment at values of organizational communication						
Organizational communication	Effect	SE	*t*	*p*	LLCI[a]	ULCI[b]
−1.9904	−0.7227	0.1044	−6.9231	0.0000	−0.9280	−0.5175
		⋮				
1.7024	−0.2164	0.1100	−1.9664	0.0500	−0.4328	0.0000
		⋮				
2.0096	−0.1743	0.1241	−1.4045	0.1610	−0.4183	0.0697

The results show that the conditional effects of organizational commitment are significant for values ranging from organizational communication −0.7631 ($M - 1$SD) to 0.7631 ($M + 1$SD).

The conditional effect of organizational commitment was significant in areas where organizational communication values were lower than 1.7024. In other

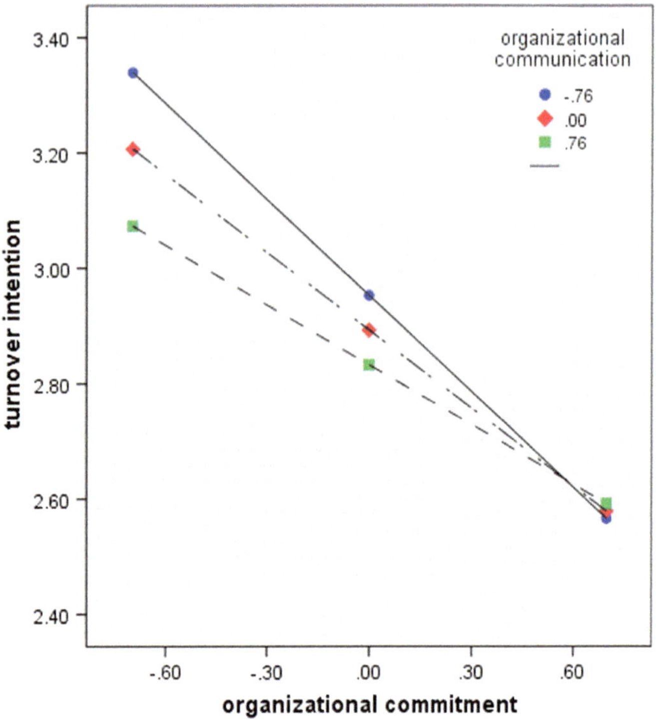

Fig. 3 Moderating effect of organizational communication on the relationship between organizational commitment and turnover intention

words, it means moderating the relationship between organizational commitment and turnover intention in areas where organizational communication was lower than 1.7024. 98.9% of the respondents showed significant, and 1.1% showed no significant.

Figure 3 shows the form of the moderating effect of organizational communication on the relationship between organizational commitment and turnover intention. The value of the turnover intention according to organizational commitment was higher in order of the low, middle and high groups of organizational communication. In other words, the lower the organizational communication, the higher the turnover intention in the effect of organizational commitment on turnover intention.

Finally, the conditional indirect effects of the effect of the independent variable on the dependent variable according to the specific value of the moderating variable were as shown in Table 6. The results of the analysis show that the conditional indirect effects of organizational communication on turnover intention (stress → organizational commitment → turnover intention) are significant at the levels of −0.7631 ($M-1$SD) to 0.7631 ($M+1$SD) of organizational communication.

The moderate mediation effect analysis showed that high stress leads to lower organizational commitment and higher turnover intention. It can also be seen that

Table 6 Moderated mediation effects of organizational communication on the relationship between stress and turnover intention

Conditional indirect effect	Effect	Boot SE	LLCI[a]	ULCI[b]
$M - 1SD(-0.7613)$	0.1196	0.0390	0.0485	0.2008
$M(0.0000)$	0.0970	0.0311	0.0399	0.1619
$M + 1SD(0.7613)$	0.0745	0.0277	0.0260	0.1332

organizational communication exhibits moderated mediation effects by moderating these mediating effects. The most previous studies focused only on the relationship between workers' stress and turnover intention, and few studies have found that the influence of organizational commitment and communication can affect the turnover intention. Based on these findings, there should be a way to increase organizational commitment and lower turnover intention through smooth organizational communication among workers. Through this, it is expected to predict stable operation of the organization and to form a positive organizational culture.

4 Conclusion

The aim of this study was to verify whether organizational communication moderates the effect of reduced organizational commitment through stress on turnover intention. Research has shown that organizational commitment mediates between stress and turnover intention, and in this relationship, organizational communication has a moderated mediation effect. Based on these findings, we discussed to reduce the turnover intention of workers.

Based on the limitations of this study, the suggestions for follow-up studies were as follows: First, this study has a limitation in that it only covers some workers working in Chungnam area. In the future, there will be a need for nationwide sampling and comparisons by industry and position. Second, as social sentiment changes recently, comparative studies by age should be conducted in accordance with the trend that emphasizes individualism, respect for privacy and balance between work and personal life. Such a study could analyze the problems of intergenerational communication and suggest appropriate communication methods. Despite these limitations, this study was the first to discover the moderated mediation effect of organizational communication in relation to turnover intention due to workers' stress. Therefore, the results of this study are meaningful in that they seek organizational commitment to reduce the turnover intention of workers and reveal the importance of communication among members.

References

1. Beehr, T.A., Newman, J.E.: Job stress, employee health, and organizational effectiveness: a facet analysis, model, and literature review 1. Pers. Psychol. **31**(4), 665–699 (1978)
2. Tett, R.P., Meyer, J.P.: Job satisfaction, organizational commitment, turnover intention, and turnover: path analyses based on meta-analytic findings. Pers. Psychol. **46**(2), 259–293 (1993)
3. Porter, L.W., Steers, R.M., Mowday, R.T., Boulian, P.V.: Organizational commitment, job satisfaction, and turnover among psychiatric technicians. J. Appl. Psychol. **59**(5), 603 (1974)
4. Koch, J.L., Steers, R.M.: Job attachment, satisfaction, and turnover among public sector employees. J. Vocat. Behav. **12**(1), 119–128 (1978). https://doi.org/10.1016/0001-8791(78)90013-1
5. Sheldon, M.E.: Investments and involvements as mechanisms producing commitment to the organization. Adm. Sci. Q. **16**(2), 143–150 (1971). https://doi.org/10.2307/2391824
6. Cohen, A.: Organizational commitment and turnover: a met A-analysis. Academy Manage. J. **36**(5) (1993). https://doi.org/10.5465/256650
7. Lee, J.H.: Influence of Communication and Job Satisfaction on Job Performance of SME Employees: Focusing on Mediation Effect of Organization Commitment. In: Dissertation, 112 p. Hoseo University, Seoul. Available from: http://www.riss.kr/link?id=T14666325
8. Parker, D.F., DeCotiis, T.A.: Organizational determinants of job stress. Organizational Behavior and Human Performance **32**, 160–177 (1983). https://doi.org/10.1016/0030-5073(83)90145-9
9. Lee, C.S., Jang, H.Y.: A double mediation of hope and job satisfaction between job stress and turnover intention. Ind. J. Publ. Health Res. Dev. **9**(9), 1246–1251 (2018). https://doi.org/10.5958/0976-5506.2018.01165.8
10. Allen, N.J., Meyer, J.P.: The measurement and antecedents of affective, continuance and normative commitment to the organization. J. Occup. Psychol. **63**(1), 1–18 (1990)
11. Devece, C., Palacios-Marques, D., Alguacil, M.P.: Organizational commitment and its effects on organizational citizenship behavior in a high-unemployment environment. J. Bus. Res. **69**(5), 1857–1861 (2016)
12. Downs, C.W., Hazen, M.: A factor analytic study of communication satisfaction. J. Bus. Commun. **14**(3), 63–73 (1977)
13. Lee, C.S., Choi, K., Jang, H.Y.: The role of organizational communication and hope between authentic leadership and job satisfaction. Int. J. Pure Appl. Math. **120**(6), 5777–5792 (2018)
14. Becker, T.E.: Foci and bases of commitment: are they distinctions worth making? Acad. Manag. J. **35**(1), 232–244 (1992)
15. Lee, C.S., Jang, H.Y.: Mediating effects of hope and organizational effectiveness between authentic leadership and turnover intention. Ind. J. Publ. Health Res. Dev. **9**(8), 461–467 (2018). https://doi.org/10.5958/0976-5506.2018.00777.5
16. Shin, D.G, Shin, K.H, Park, S.Y.: An equilibrium analysis of inter-and intra-sectoral mobility of workers. Employ. Qual. Life. 5–41 Oct 2007
17. Cohen, A., Freund, A.: A longitudinal analysis of the relationship between multiple commitments and withdrawal cognitions. Scand. J. Manage. **21**(3), 329–351 (2005). https://doi.org/10.1016/j.scaman.2004.06.004

Recognition of Activities of Daily Living Based on a Mobile Data Source Framework

Ivan Miguel Pires, Gonçalo Marques, Nuno M. Garcia, Francisco Flórez-Revuelta, Maria Canavarro Teixeira, Eftim Zdravevski, and Susanna Spinsante

Abstract Most mobile devices include motion, magnetic, acoustic, and location sensors. These sensors can be used in the development of a framework for activities of daily living (ADL) and environment recognition. This framework is composed of the acquisition, processing, fusion, and data classification features. This study compares different implementations of artificial neural networks. The obtained results were 85.89% and 100% for the recognition of standard ADL and standing activities with Deep Neural Networks, respectively. Furthermore, the results present 86.50% for identification of the environments using Feedforward Neural Networks. Numerical

I. M. Pires · G. Marques (✉) · N. M. Garcia
Instituto de Telecomunicações, Universidade da Beira Interior, Covilhã 6201-001, Portugal
e-mail: goncalosantosmarques@gmail.com

I. M. Pires
e-mail: impires@it.ubi.pt

N. M. Garcia
e-mail: ngarcia@di.ubi.pt

F. Flórez-Revuelta
Department of Computer Technology, Universidad de Alicante, Alicante, Spain
e-mail: francisco.florez@ua.es

M. C. Teixeira
UTC de Recursos Naturais e Desenvolvimento Sustentável, Polytechnique Institute of Castelo Branco, Castelo Branco, Portugal
e-mail: ccanavarro@ipcb.pt

CERNAS—Research Centre for Natural Resources, Environment and Society, Polytechnique Institute of Castelo Branco, Castelo Branco, Portugal

E. Zdravevski
Faculty of Computer Science and Engineering, University Ss Cyril and Methodius, Skopje, Macedonia
e-mail: eftim.zdravevski@finki.ukim.mk

S. Spinsante
Department of Information Engineering, Marche Polytechnic University, Ancona, Italy
e-mail: s.spinsante@univpm.it

A. K. Bhoi et al. (eds.), *Bio-inspired Neurocomputing*, Studies in Computational Intelligence 903, https://doi.org/10.1007/978-981-15-5495-7_18

results illustrate that the proposed framework can achieve robust performance from the incorporation of data fusion methods using mobile devices.

Keywords Activities of daily living · Artificial intelligence · Data classification · Mobile computing

1 Introduction

Activities of daily living (ADL) recognition using a multiple data source framework [1] is proposed in [2–4]. This framework can be designed using data collected from the mobile sensors such as the magnetometer, the gyroscope, the accelerometer, the global positioning system (GPS) receiver, and the microphone. These sensors allow the capture of several parameters to enable the automatic identification of the activities. The identification of activities and environments is taken into account the sensing of the characteristics of the movement these activities produce and environmental sound characteristics associated with the data acquisition period [5]. Furthermore, this framework can integrate the implementation of a personal digital life coach [6], currently under research.

Based on previous work [7–9] associated with the use of sensors' data for ADL and environments recognition, this study enhances both the support to the GPS receiver, the ability to recognize the driving activity, and a significant number of standing events in comparison with the previous works, including sleeping and watching TV. The data collected using the GPS receiver enables the identification of the geographic information of the user. The proposed method can recognize several ADL, such as running, standing, walking, walking on stairs, driving, and sleeping in distinct environments. These environments include classroom, library, gym, street, kitchen, hall, bedroom, and watching TV. The proposed framework includes numerous modules, such as the data collection performed using a smartphone application and data processing, fusion, and classification methods. The data processing method incorporates data cleaning and feature extraction. The most important achievement of our framework is related to the adoption of multiple mobile devices' sensors for ADL and environment recognition. It may pave the way not only for a ubiquitous personal life coach but also for healthy ageing or disease monitoring and support. This paper presents the performance assessment of several methods with multiple sensors. It allows the adaption of the framework to various mobile devices, as the number of sensors and their capabilities may be different by each smartphone.

The literature review is presented in Sect. 2, focused on the location sensors for ADL and environments recognition. Next, Sect. 3 presents the approach used in this paper. The results, showed in Sect. 4, were obtained with the fusion of the location sensors for standing activities recognition. Section 5 presents the discussion, and Sect. 6 presents the conclusions.

2 Related Work

To date, ADL recognition with some sensors available in smartphones [10–15] and several classification methods have been widely studied. We can conclude that artificial neural networks (ANN) is a reliable solution providing acceptable accuracy levels [16, 17].

Currently, the application of the fusion of the data acquired from multiple sensors available on smartphones, including the gyroscope, the accelerometer, the magnetometer, the GPS receiver, and the microphone for the recognition of ADL and environments is not common in the literature [1]. However, several proposals using subsets of these sensors are available in the literature. This literature review has its main focus on the use of the GPS receiver for ADL and environments identification, where the analysis is available in previous studies [7–9]. The proposed approach uses motion and magnetic sensors for ADL identification and the microphone are used for environmental sounds recognition. The identification of standing activities is made based on the environment information [7–9].

The authors of [18] implemented several methods, such as support vector machine (SVM), Naive Bayes, ANN, i.e. multilayer perceptron (MLP), decision trees, logistic regression, rule-based classifiers, and k-nearest neighbour (k-NN), to identify running, walking, standing, sitting, and going downstairs and upstairs, using the data collected using the accelerometer, the magnetometer, the gyroscope, and GPS. They extracted the mean and standard deviation related to the accelerometer, gyroscope and magnetometer sensors, and the distance, location, and speed from the GPS receiver, reporting accuracies between 69 and 89% [18].

The data collected using the accelerometer, the gyroscope, and the GPS receiver for ADL recognition is presented in [19]. The authors used several features, including mean, energy, standard deviation, the correlation between axis, and entropy extracted from the motion sensors, and distance, location, and speed obtained from the GPS receiver. That study reported accuracies of 99% with MLP, 96% with logistic regression, 94.2% with a J48 decision tree, and 93.3% with SVM.

In [20], with the data collected using the gyroscope, the barometer, the accelerometer, and the GPS, the authors recognized washing dishes, walking on stairs, sitting, standing, cycling, and running with the SVM method. This method was implemented with several features, including mean, standard deviation, and mean squared obtained using the gyroscope and the accelerometer, pressure derived from the barometer, and altitude in metres and speed using GPS data. Finally, the proposed method provides 90% of accuracy.

The authors of [21] used several types of sensors, such as acoustic, location, motion, and medical sensors for the recognition of preparing food, sleeping, standing, jogging, eating, working, and travelling activities. They implemented the C4.5 decision tree, Naive Bayes, RIPPER, random forest, SVM, Bagging, Vote, and AdaBoost [21]. The inputs for these methods were the features extracted from several sensors [21]. The sound features corresponded to the Mel-Frequency Cepstral Coefficients (MFCC), the averages of the spectral centroids, the zero-crossing rates, and the linear predictive coding (LPC) values [21]. The distance between access points is extracted using the Wi-Fi receiver [21]. The features were extracted from the GPS receiver where this location sensor collects the velocity and the category of the nearest place [21]. The acceleration features extracted were the elementary activity and energy expenditure obtained by an algorithm [21]. Finally, the heart-rate and respiration-rate features were minimum, maximum, and average [10]. The reported results obtained by the several methods implemented were 68% for the Naive Bayes, 66% for the C4.5 decision tree, 72% for the RIPPER, 72% for the SVM, 71% for the random forest, 69% for the Bagging, 66% for the AdaBoost, and 77% for the Vote [21].

In [22], with the accelerometer, GPS receiver, camera, and timer used for ADL identification, including standing, lying, sitting, use an elevator, walking, dining, going upstairs and downstairs, moving a kettle, washing dishes, preparing a meal, drying hands, moving plates, washing hands, brushing teeth and combing hair, the authors implemented a decision tree as a classification method. The features used were mean, range of *Y*-axis, standard deviation, the sum of intervals of, signal magnitude area (SMA), the difference of ranges, the interval between *X*- and *Z*-axis, distance, speed, and location extracted using the accelerometer. The proposed approach states from 88.24 to 100% of accuracy [22].

The SVM method was implemented with several features as input, such as the minimum, maximum, mean, standard deviation, correlation between axis and median crossing extracted from the accelerometer data, and the distance, location, and speed obtained from the GPS receiver, to recognize the walking, standing still, and running activities, reporting an accuracy around 97.51% [23].

The accelerometer and the GPS receiver were used to recognize standing, travelling by car, travelling by train, and walking activities with a J48 decision tree, random forest, ZeroR, Logistic, decision table, radial basis function network (RBFN), ANN, Naive Bayes, and Bayesian network [24]. The input features of the methods were the average accuracy, average speed, average rail line closeness, average heading change, average acceleration, magnitudes of the frequency domain, and the signal variance [24]. The average reported accuracies were 85.2% with a J48 decision tree, 85.1% with random forest, 84.8% with ZeroR, 84.7% with logistic, 84.6% with decision table, 84.4% with RBFN, 84.4% with MLP, 84.2% with Naive Bayes, and 84.1% with Bayesian network [24].

The authors of [25] implemented several methods, including J48 decision tree, MLP, and Logistic Regression (LR) for the identification of several ADL. These activities are going downstairs, sitting, jogging, standing, walking, and going upstairs using the accelerometer and the GPS receiver. The input features for the methods implemented were the maximum, minimum, mean, standard deviation, and zero-crossing rate for each axis for the accelerometer, the correlation between the axis of the accelerometer, and the distance, location, and speed acquired from the GPS receiver [25]. The reported accuracies were 92.4% with a J48 decision tree, 91.7% with MLP, and 84.3% with LR [25].

In [26], the accelerometer and the GPS receiver are used for the identification of driving, standing, walking, running, going upstairs and downstairs, use an elevator, and cycling, using Bayesian networks. The input features used are mean, variance, spectral energy, spectral entropy from the accelerometer, and the location retrieved from the GPS receiver, reporting accuracy of 95% [26].

Other authors propose different methods for ADL identification using GPS and other mobile sensors. The authors of [27] tested the use of SVM and HMM based on the information collected using the accelerometer, the gyroscope, and the GPS receiver, recognizing, standing, walking, running, and sitting activities.

In [28], travelling by car or train and cycling activities were recognized based on the International Road Index (IRI) and angle of slope measured by the data acquired from the accelerometer, gyroscope, and GPS receiver.

The authors of [29] combined the accelerometer with the GPS data for the identification of different ADL such as lying, sitting, standing, and falling. They used the following features: SMA, signal magnitude vector (SMV), and tilt angle (TA) using the accelerometer sensor, and the distance, location, and speed retrieved from the GPS.

In [30], the accelerometer, the GPS, the Wi-Fi positioning system (WPS) and the GSM positioning system (GSMPS) were used for the recognition of standing, sitting, lying, walking, jogging, cycling, travelling by bus, train, taxi or car. These authors applied several features, including the numbers of peaks, the number of troughs, the sum of peaks and troughs, the difference from the maximum peak and the maximum trough, the difference from the maximum and the minimum, either peak or trough. The proposed method revealed its effectiveness in terms of energy efficiency (less 53% of battery energy spent than others).

The authors of [31] used only the GPS receiver for the recognition of working, attending lectures, shopping, swimming, training in a gym, visiting friends, eating, playing team sports, going to a pub, to the cinema, to a concert, to the theatre, to a church, and visiting a doctor, based on the density and time-based methods available in the OpenStreetMap (OSM) platform. Based on different levels of threshold, the accuracies reported by the authors were between 72.2 and 95.4% with the density-based method and between 66.1 and 69.6% with a time-based approach [31].

Following the analysis of the methods proposed in the literature, Table 1 shows the ADL recognized with GPS receiver and other sensors, sorted in descending order of their respective number of research works. As shown, the standing, walking, going upstairs and downstairs, sitting, running, and driving/travelling are the most recognized ADL.

Table 2 shows the features extracted using the GPS data in descending order. As observed, the location, speed, and distance are the most common features. On the one hand, the GPS receiver provides the geographic information of the place of activity. On the other hand, speed and distance enable to measure the intensity of the event.

Finally, Table 3 highlights the most popular methods for ADL recognition in descending order of their reported accuracies, such as MLP, logistic regression, and SVM. Therefore, the method that indicates better average accuracy is with the MLP method (93.53%).

Table 1 Distribution of the ADL recognized and the number of studies

ADL	Number of studies
Standing	11
Driving/travelling (i.e. car, train, bus, taxi)	9
Walking	8
Sitting	7
Running; going upstairs; going downstairs	6
Lying; cycling	4
Jogging	3
Washing dishes; preparing food; eating; working; riding an elevator	2
Sleeping; dining; moving a kettle; drying hands; moving plates; washing hands; brushing teeth; combing hair; falling; attending lectures; shopping; swimming; training in a gym; playing team sports; visiting friends; going to a pub; going to the cinema; going to a concert; going to the theatre; visiting a doctor; going to church	1

Table 2 Distribution of the features and the number of studies

Features	Number of studies
Location	9
Speed	8
Distance	6
Altitude difference in metres; velocity; category of the nearest place; International Road Index (IRI); angle of the slope; Points of Interest (POI)	1

Table 3 Distribution of the classification methods, the number of studies, and average accuracy

Methods	Number of studies	Average of reported accuracy (%)
MLP	4	93.53
Logistic regression	3	93.23
SVM	6	90.36
Bayesian network	2	89.55
k-NN	1	89.00
Rule-based classifiers	1	89.00
Decision trees (i.e. J48, C4.5)	6	88.88
ZeroR	1	84.80
Decision table	1	84.60
Radial basis function network (RBFN)	1	84.40
Logistic Regression (LR)	1	84.30
Naïve Bayes	3	83.73
Random forest	2	78.05

(continued)

Table 3 (continued)

Methods	Number of studies	Average of reported accuracy (%)
Vote	1	77.00
RIPPER	1	72.00
Bagging	1	69.00
AdaBoost	1	66.00

3 Methods

The development of a multiple data source framework [2–4] for ADL identification enhances the techniques proposed in previous studies [7–9] with the following modules:

- Data acquisition performed by a mobile application;
- Data processing forked in data cleaning and feature extraction methods;
- Recognition forked in data fusion and classification methods.

3.1 Data Acquisition

The study [8] presented the acquisition of data using the magnetometer, the gyroscope, the accelerometer, the microphone, and the GPS for the identification of several ADL and environments, which is stored in [32] and made publicly available for validation of these conclusions and further research.

Considering that the data acquisition occurs for 5 seconds every 5 min and users are active for 16 hours per day, we estimate that the data collection time is around 16 minutes per day. Thus, the proposed method is feasible either in the most sophisticated or low-cost mobile devices. The ultimate goal of this study is to allow the identification of standing activities, such as driving, sleeping, and watching TV to map the users' lifestyles.

3.2 Data Processing

This study uses the gyroscope, the accelerometer, and the magnetometer data, applying the low pass filter [33] for the reduction of environmental noise effects and invalid data. This study also acquires microphone data, and the data is immediately summarized, therefore not compromising the user's privacy, because it only receives raw data.

Based on both the literature and our previous studies [7–9], the most relevant features extracted in the sensors mentioned above are the features mentioned in [8] plus the distance travelled and the environment recognized.

3.3 Data Fusion Using Mobile Devices' Sensors

The features presented in the previous section for standing activities recognition are merged, creating different datasets as depicted in Fig. 1. Thus, based on these datasets, three different experiments were performed: (1) using just the accelerometer, (2) using the accelerometer combined with the magnetometer, and (3) using the accelerometer, magnetometer, and gyroscope.

3.4 Data Classification

Based on the state of art review, this study focuses on the fusion of the features extracted using location, motion, mechanical, and acoustic sensors available in the off-the-shelf mobile devices. As observed in the literature, the ANN method, one of the most popular classification methods for ADL and environment recognition, reports the best accuracy to recognize standing activities.

The implementations used both normalized and non-normalized data with some frameworks and methods implemented and tested in [8]. In this paper, we tested these methods with different values of maximum training iterations, such as 106, 2 $\times$ 106 and 4 $\times$ 106. Figure 2 presents the results obtained with the proposed framework

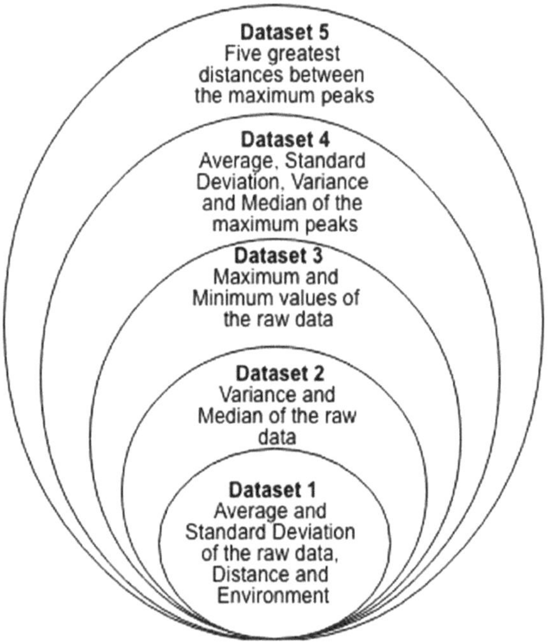

Fig. 1 Datasets created for the analysis of standing activities, using the distance travelled calculated with the data collected using GPS and the previously recognized environment

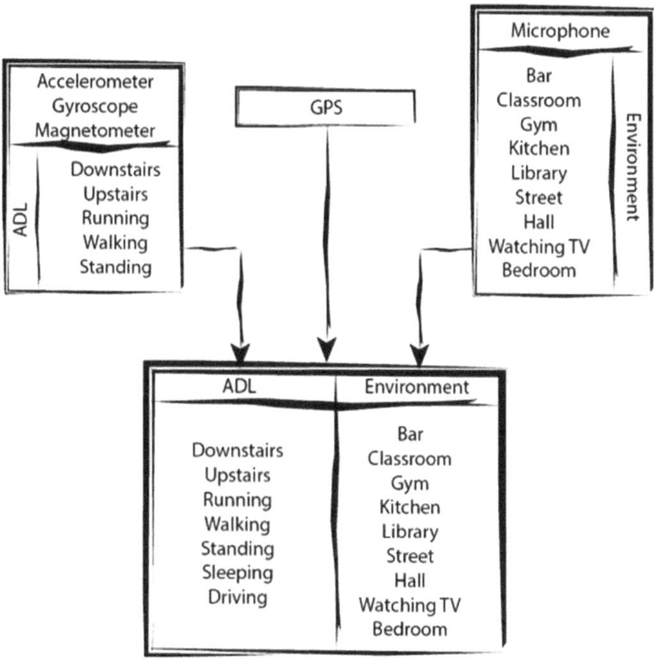

Fig. 2 ADL and environments recognized by the proposed framework

for the identification of ADL and environmental sounds using the motion, magnetic, acoustic, and location sensors.

4 Results

The authors benchmark both MLP and Feedforward Neural Networks (FNN) with Backpropagation and the Deep Neural Networks (DNN) to determine the most accurate method for the identification of standing activities. The mentioned datasets were created based on 6000 records equally distributed for the proposed standing activities, namely watching TV, sleeping, and driving. Based on the analysed combinations of sensors for the identification of standing activities, the following combinations of sensors' data were defined:

1. Environment information extracted using the accelerometer and the GPS sensors;
2. Sensors' data collected in the mix one plus the magnetometer data;
3. Sensors' data collected in combination with two plus the gyroscope data.

The standing activities recognition is based on the features extracted from the combination 1. The experiments revealed that the results obtained with MLP with

Backpropagation are always 100%, except in the dataset 1 (66.67%) with non-normalized data. Also, the results extracted using FNN with Backpropagation are still 100%, except when in both datasets 4 and 5 (96.67%) with non-normalized data. Finally, the DNN method reports always 100% of accuracy, except when it is used the dataset 1 (33.37%) with non-normalized data.

Therefore, for the recognition of standing activities based on the features extracted from the combination 2, the obtained results on the implementation of the MLP with Backpropagation are always around 100%, except when it is used the dataset 1 (66.72%) with non-normalized data. Similarly, the results obtained with the FNN with Backpropagation are always 100%, except when it is used the dataset 1 (94.03%) with non-normalized data. At last, the DNN method results are still 100% with normalized data and 33.42% with non-normalized data.

Finally, for standing activities recognition based on the features extracted from the combination 3, the results revealed an accuracy equals to 100% by MLP method, except when it is used the dataset 1 (55.35%) with non-normalized data. Congruently, the results achieved with the FNN with Backpropagationn are always 100%, except when it is used the dataset 1 (96%) with non-normalized data. Finally, the results obtained with the DNN are still 100% with normalized data and 33.4% with non-normalized data.

Comparing the maximum accuracies reported by the different implementations, Table 4 presents the best results reported with the various combinations of sensors and datasets.

Table 4 Best accuracies obtained in recognition of standing activities

	Framework	Dataset (combination)	Iterations needed for training	Best accuracy achieved (%)
Non-normalized data	Neuroph	2 (1)	106	99.97
		2 (3)	106	100.00
		4 (2)	106	100.00
	tEancog	1 (1)	106	100.00
		3 (2)	106	99.97
		2 (3)	106	99.98
	Deep learning	2 (1)	106	100.00
		2 (2)	106	33.42
		1 (3)	106	33.40
Normalized data	Neuroph	2 (1)	106	100.00
		5 (2)	106	100.00
		1 (3)	106	100.00
	Encog	1 (1, 2 or 3)	106	100.00
	Deep learning	1 (1, 2 or 3)	106	100.00

5 Discussion

This paper complements the research about the design of a framework for ADL and environments recognition [2–4] using all sensors available in mobile derives. The ADL identified is associated with the movement, i.e. standing, walking, running going upstairs and downstairs, several user's environments, i.e. classroom, kitchen, bar, gym, library, street, hall, watching TV, and bedroom, and some standing activities, i.e. watching TV, driving, and sleeping. Table 5 summarizes the literature on the recognition of ADL and environments. Also, the highlighted values in Table 5 represent the ADL/environments recognized by the proposed framework.

As shown in Table 5, the results obtained with either the accelerometer or accelerometer combined with magnetometer and gyroscope, revealed the ability of our framework to recognize 5 of 6 ADL presented in the literature (83%). On the other hand, the proposed framework can identify 4 of 8 events presented in the literature (50%) with the accelerometer, magnetometer, gyroscope, and microphone. Finally, 6 of 7 ADL presented in the literature (86%) are recognized by the proposed framework with the accelerometer, magnetometer, gyroscope, microphone, and GPS receiver.

Table 5 Most recognized ADL and environments based on the state of art review distributed by the sensors used (# represents the number of studies available in the literature that identifies the ADL/environment)

Accelerometer		Accelerometer magnetometer gyroscope		Accelerometer magnetometer gyroscope microphone		Accelerometer magnetometer gyroscope microphone GPS receiver	
ADL	#	ADL	#	ADL environment	#	ADL	#
Walking	63	Walking	21	Emergency vehicles	6	Resting standing	11
Resting Standing	48	Going downstairs	17	Sleeping	5	Driving travelling	9
Going upstairs	45	Going upstairs	17	Walking	5	Walking	8
Going downstairs	44	Resting standing	16	Resting standing	5	Sitting	7
Running	31	Running	13	Street traffic	5	Running	6
Sitting	30	Sitting	11	Ocean	5	Going upstairs	6
				Driving	4	Going downstairs	6
				River	4		

Table 6 Summarization of the accuracy of the final framework for ADL and environments recognition

Stages of the framework	Accelerometer microphone GPS receiver (%)	Accelerometer magnetometer microphone GPS receiver (%)	Accelerometer magnetometer gyroscope microphone GPS receiver (%)	Average accuracy (%)
Recognition of common ADL	85.89	86.49	89.51	87.30
Recognition of environments	86.50	86.50	86.50	86.50
Recognition of standing activities	100.00	100.00	100.00	100.00

As shown in Table 6, the best accuracies of standard ADL identification is 89.51%, in the identification of environments is 86.50%, and in the identification of standing activities is 100%. Thus, we recommend the design suing DNN methods with normalized data and the application of L2 regularization for standard ADL and standing activities recognition and the implementation of the FNN method with non-normalized data for environments identification. Also, the average accuracy of the framework for all devices with different combinations of sensors is 91.27%.

6 Conclusions

The hardware of mobile devices includes numerous sensors that can be used for ADL and environment recognition. It combines the data acquired from an enlarged set of mobile devices' sensors to develop a framework that adapts their functionalities with the number of sensors available in the equipment used.

This paper presents the design of the methods for the different stages of the framework. The framework starts with the data acquisition, data cleaning, and feature extraction methods, and, at the first stage of ADL recognition, the framework uses the DNN for the identification of walking, running, standing, going downstairs, and going upstairs. In the second stage, the framework recognizes some environments with the FNN with Backpropagation, and these are classroom, bar kitchen, gym, street, hall, bedroom, library, and watching TV. Finally, in the third stage, the framework uses DNN for standing activities identification, and these are watching TV, driving, and sleeping.

The ADL and environment recognition are conducted on the features extracted from the different sensors' data (excluding the microphone and the GPS receiver), such as the five most considerable the average, standard deviation, variance and median of the maximum peaks, the standard deviation, average, maximum value, minimum value, variance, distance from the maximum peaks and median of the raw signal. The features extracted from the microphone data are the 26 MFCC coefficients, the standard deviation, average, maximum value, minimum value, variance, and median of the raw signal. Also, the unique feature obtained from the GPS receiver, the distance travelled, also enables to identify the users' location.

For the design of a framework for ADL and environment identification, the authors compared three distinct implementations of ANN, and these are MLP and FNN with Backpropagation and DNN. The proposed study states 87.50% of average accuracy for standard ADL recognition, 86.50% of accuracy for environment identification, and 100% for standing activities recognition. The proposed framework provides 91.27% of average accuracy. Thus, the proposed framework proves its reliability in the identification of ADLs and its environments.

Acknowledgements This work is funded by FCT/MCTES through national funds and when applicable co-funded EU funds under the project UIDB/EEA/50008/2020 (Este trabalho é financiado pela FCT/MCTES através de fundos nacionais e quando aplicável cofinanciado por fundos comunitários no âmbito do projeto UIDB/EEA/50008/2020).

This article/publication is based on work from COST Action IC1303—AAPELE—Architectures, Algorithms and Protocols for Enhanced Living Environments and COST Action CA16226—SHELD-ON—Indoor living space improvement: Smart Habitat for the Elderly, supported by COST (European Cooperation in Science and Technology). More information in www.cost.eu.

References

1. Pedretti, L.W., Early, M.B.: Occupational therapy: Practice skills for physical dysfunction. Mosby St. Louis, MO (2001)
2. Pires, I.M., Garcia, N.M., Pombo, N., Flórez-Revuelta, F.: From data acquisition to data fusion: a comprehensive review and a roadmap for the identification of activities of daily living using mobile devices. Sensors **16**(2) (2016)
3. Pires, I.M., Garcia, N.M., Flórez-Revuelta, F.: Multi-sensor data fusion techniques for the identification of activities of daily living using mobile devices. In Proceedings of the ECMLPKDD (2015)
4. Pires, I.M., Garcia, N.M., Pombo, N., Flórez-Revuelta, F.: Identification of activities of daily living using sensors available in off-the-shelf mobile devices: research and hypothesis. In: Lindgren, H., De Paz, J.F., Novais, P., Fernández-Caballero, A., Yoe, H., Jiménez Ramírez, A., Villarrubia, G. (eds.) Ambient Intelligence-Software and Applications—7th International Symposium on Ambient Intelligence (ISAmI 2016), pp. 121–130. Springer International Publishing, Cham (2016)
5. Salazar, L.H.A., Lacerda, T., Nunes, J.V., von Wangenheim, C.G.: A systematic literature review on usability heuristics for mobile phones. Int. J. Mob. Human-Comput. Interact. IJMHCI **5**(2), 50–61 (2013)
6. Garcia, N.M.: A roadmap to the design of a personal digital life coach. In International Conference on ICT Innovations, pp. 21–27. Springer
7. Pires, I.M., Garcia, N.M., Pombo, N., Flórez-Revuelta, F., Spinsante, S., Teixeira, M.C., Zdravevski, E.: Pattern recognition techniques for the identification of activities of daily living using mobile device accelerometer. Tech. Rep. Peer J. Preprints (2018)
8. Pires, I.M., Garcia, N.M., Pombo, N., Flórez-Revuelta, F., Spinsante, S., Teixeira, M.C.: Identification of activities of daily living through data fusion on motion and magnetic sensors embedded on mobile devices. Pervasive Mob. Comput. **47**, 78–93 (2018)

9. Pires, I.M., Garcia, N.M., Pombo, N., & Flórez-Revuelta, F.: User environment detection with acoustic sensors embedded on mobile devices for the recognition of activities of daily living. arXiv:1711.00124 (2017)
10. Akhoundi, M.A.A., Valavi, E.: Multi-sensor fuzzy data fusion using sensors with different characteristics. arXiv:1010.6096 (2010)
11. Banos, O., Damas, M., Pomares, H., Rojas, I.: On the use of sensor fusion to reduce the impact of rotational and additive noise in human activity recognition. Sensors **12**(6), 8039–8054 (2012)
12. Dernbach, S., Das, B., Krishnan, N.C., Thomas, B.L., Cook, D.J.: Simple and complex activity recognition through smartphones. In: 2012 Eighth International Conference on Intelligent Environments, pp. 214–221, June 2012
13. Hsu, Y., Chen, K., Yang, J., Jaw, F.: Smartphone-based fall detection algorithm using feature extraction. In 2016 9th International Congress on Image and Signal Processing, BioMedical Engineering and Informatics (CISP-BMEI), pp. 1535–1540, Oct 2016
14. Paul, P., George, T.: An effective approach for human activity recognition on smartphone. In: 2015 IEEE International Conference on Engineering and Technology (ICETECH), pp. 1–3, Mar 2015
15. Shen, C., Chen, Y., Yang, G.: On motion-sensor behavior analysis for human-activity recognition via smartphones. In: 2016 IEEE International Conference on Identity, Security and Behavior Analysis (ISBA), pp. 1–6, Feb 2016
16. Doya, K., Wang, D.: Exciting time for neural networks. Neural Networks **61**, xv–xvi (2015)
17. Wang, D.: Pattern recognition: neural networks in perspective. IEEE Expert **8**, 52–60 (1993)
18. Shoaib, M., Scholten, H., Havinga, P.J.M.: Towards physical activity recognition using smartphone sensors. In 2013 IEEE 10th International Conference on Ubiquitous Intelligence and Computing and 2013 IEEE 10th International Conference on Autonomic and Trusted Computing, pp. 80–87, Dec 2013
19. Hung, W., Shen, F., Wu, Y., Hor, M., Tang, C.: Activity recognition with sensors on mobile devices. Int. Conf. Mach. Learn. Cybern. **2**, 449–454 (2014)
20. Altini, M., Vullers, R., Van Hoof, C., van Dort, M., Amft, O.: Self-calibration of walking speed estimations using smartphone sensors. In: 2014 IEEE International Conference on Pervasive Computing and Communication Workshops (PERCOM WORKSHOPS), pp. 10–18, Mar 2014
21. Luštrek, M., Cvetković, B., Mirchevska, V., Kafali, O., Romero, A.E., Stathis, K.: Recognising lifestyle activities of diabetic patients with a smartphone. In: 2015 9th International Conference on Pervasive Computing Technologies for Health-care (PervasiveHealth), pp. 317–324, May 2015
22. Wu, H.H., Lemaire, E.D., Baddour, N.: Change-of-state determination to recognize mobility activities using a blackberry smartphone. In: 2011 Annual International Conference of the IEEE Engineering in Medicine and Biology Society, pp. 5252–5255, Aug 2011
23. Kaghyan, S., Sarukhanyan, H.: Accelerometer and GPS sensor combination based system for human activity recognition. In: Ninth International Conference on Computer Science and Information Technologies Revised Selected Papers, pp. 1–9, Sep 2013
24. Bloch, A., Erdin, R., Meyer, S., Keller, T., Spindler, A.D.: Battery-efficient transportation mode detection on mobile devices. In: 2015 16th IEEE International Conference on Mobile Data Management, vol. 1, pp. 185–190, June 2015
25. Zainudin, M.N.S., Sulaiman, M.N., Mustapha, N., Perumal, T.: Activity recognition based on accelerometer sensor using combinational classifiers. In: 2015 IEEE Conference on Open Systems (ICOS), pp. 68–73, Aug 2015
26. Zou, X., Gonzales, M., Saeedi, S.: A context-aware recommendation system using smartphone sensors. In: 2016 IEEE 7th Annual Information Technology, Electronics and Mobile Communication Conference (IEMCON), pp. 1–6, Oct 2016
27. Kaghyan, S., Sarukhanyan, H.: Multithreaded signal preprocessing approach for inertial sensors of smartphone. In: 2015 Computer Science and Information Technologies (CSIT), pp. 85–89, Sep 2015
28. Seraj, F., Meratnia, N., Havinga, P.J.M.: Rovi: continuous transport infrastructure monitoring framework for preventive maintenance. In: 2017 IEEE International Conference on Pervasive Computing and Communications (Per-Com), pp. 217–226, Mar 2017

29. He, Y., Li, Y., Bao, S.: Fall detection by built-in tri-accelerometer of smartphone. In: Proceedings of 2012 IEEE-EMBS International Conference on Biomedical and Health Informatics, pp. 184–187, Jan 2012
30. Oshin, T.O., Poslad, S.: Lals: a low power accelerometer assisted location sensing technique for smartphones. In: 2013 IEEE Global Communications Conference (GLOBECOM), pp. 127–133, Dec 2013
31. Difrancesco, S., Fraccaro, P., Veer, S.N.V.D., Alshoumr, B., Ainsworth, J., Bellazzi, R., Peek, N.: Out-of-home activity recognition from GPS data in schizophrenic patients. In: 2016 IEEE 29th International Symposium on Computer-Based Medical Systems (CBMS), pp. 324–328, June 2016
32. Github: https://github.com/impires/August_2017-_Multi-sensor_data_fusion_in_mobile_devices_for_the_identification_of_activities_of_dail, 2018. Online. Accessed 20 Mar 2019
33. Graizer, V.: Effect of low-pass filtering and re-sampling on spectral and peak ground acceleration in strong-motion records. In: Proceedings of 15th World Conference of Earthquake Engineering, Lisbon, Portugal, pp. 24–28, 2012

Dynamic Programmable Clock Frequency Using Machine Learning Algorithms to Reduce Power Consumption in Wearables

A. Ajin Roch, S. Karthik, and R. Arthi

Abstract In today's world, wearable devices have become a key device in monitoring all internal activities of human being with the help of sensors. Humans are provided with a wristband device fabricated with various sensors to detect overall body condition in every day-to-day activity. Most of the wearable devices can be used to monitor the beats per minute (BPM) by heart rate sensor (pressure sensor), sleep and body movement activity by three-axis accelerometer sensor, the respiratory rate by biosensors and electrocardiogram (ECG) by sensors (electrodes). It is well known that soon in the near future, all sensors related to human activity and health monitoring play a major role in human beings; especially, for old age people, the wearable devices with well-packed sensor will soon be reaching out the market. The most wearable devices are able to provide almost all sensors in a single package, but fail to work if the power goes off shortly. The proposed model makes an attempt to reduce the power consumption in the wearable device by introducing the programmable clock frequency with machine learning concept. To reduce the power consumption in wearable devices, a programmable clock frequency of 1 GHz has been designed and monitored on a monthly basis. The proposed work uses different frequencies based on different activities of wearable device with generated data set of machine learning algorithms and is analysed using PYNQ boards. The proposed work concludes based on the multilinear-based regression model to check the power consumed for the given parameters observed that step count consumes 96% power during the complete analysis.

Keywords Biosensors · PYNQ · Step count · Heart rate · BPM · Sleep activity

A. Ajin Roch (✉) · R. Arthi
Electronics and Communication Engineering, SRM Institute of Science and Technology, Ramapuram Campus, Chennai 600089, India
e-mail: ajinroca@srmist.edu.in

R. Arthi
e-mail: arthir2@srmist.edu.in

S. Karthik
Electronics and Communication Engineering, SRM Institute of Science and Technology, Vadapalani Campus, Chennai 600026, India
e-mail: karthiks1@srmist.edu.in

A. K. Bhoi et al. (eds.), *Bio-inspired Neurocomputing*, Studies in Computational Intelligence 903, https://doi.org/10.1007/978-981-15-5495-7_19

1 Introduction

Human has no more space for paper works, paper bills and prescriptions. As everything is getting digital form, the E-health application is grabbing the topic in front. E-health field is a multidisciplinary area to get inspired by distinct science and technology. The E-health application is a multipurpose, easily configurable and easy to step up for various needs. Loneliness is a condition in old-age people that can be the reason for dementia, traumatic and situationally induced negative health in a desperate situation. Ubiquitous health care is turning out to be reality where the sensors that meet customized human services with minimal effort and productive patient observing are more popular are called as Next Generation E-health (NGeN) which assist individuals with doing their everyday life exercises paying little heed to any medical issue exclusively with no assistance. E-health services are normally built using the components which have reusable services [1]. In most case, the home environment has been considered for old-age people; obstructiveness, robustness and cost are of much relevance. Smart objects need to be designed and deployed at homes. Wi-Fi technology and practical sensors are age friendly that are effective for practical application. Real-time monitoring systems were used to record, measure and monitor the activities of the patient. Patient empowerment is most important, thereby becoming active participants, and outcomes are better. A skilled practice proctor has been advanced to envisage the fitness condition of patients centred on everyday habitat measurement of palpitate, breathing level, in case of acute illness. Wearable motion sensors that are suitable for practical and effective operation have been used. The sensors are connected to cloud that has access flexibility and make the personalization easy [2].

The expert system is a practical method that helps in developing healthy diary supervising the system. The expert system plays an important role in decreasing the time and cost of interpreting psychological facts generated by observing patients under a regulated clinical setting. The Wi-Fi compliant sensors used for practical operation are high power consuming and have low battery life. The major difficulty in the expert system is the missing of data and execution of time-oriented processing of the data. The other terms, classified as Board sensors and Fuzzy Pattern Tree (FPT), do not require training but have an advantage of getting domain knowledge from clinical partners. The other classifier sensors needed to provide a labelled set of training data, the neural network and support vector machine.

Ubiquitous personalized e-health service architecture consists of sensors, security mechanisms and IoT where the generated data are managed using cloud. In case of real time, there occurs a delay due to transferring of data which becomes improper to the cloud. Therefore, a smart gateway technique called as fog computing would be preferred rather than cloud computing data centre. Fog computing solves the greater latency problem, location awareness, reliability and relocating the data to the finest location [3]. The smart house scheme uses sensors to monitor the activity of the patient based on behavioural patterns and is reported in paper [4].

Mixture models are used to develop the probabilistic model of behavioural patterns using the log back of events. Day-to-day activities can be differentiated by logging to the data collected from an accelerometer, cardio tachometer and smart sensors. The impact of collapses can be recorded and differentiated from regular day-to-day events. Smart sensors borne on the body of an elderly people if, unintentional falls can be noticed. Telehealth is the use of digital information through technologies in the field of communication to access health care. It has provided medical education over a long distance making use of advance telemedicine [5]. It has become possible for physicians to treat patients whenever needed and wherever the patient is, by using a SMART watch by sending mail. To enhance clinical decision support, new research has been carried out to integrate cybercare along with analysis [6]. The sleep problem has become common among older adults that results in various health issues. The wearable activity sensor has been used to analyse the diurnal activity rhythm and sleep patterns in their normal life [7].

During the long-lasting chronicle of the exercises of individuals that decline in degrees of activity, relentlessness and force of the activity cadence, with uplifted breaking down of musicality and daytime lack of involvement. The integrated activity sensor is a wrist-worn device that collects the user's activity data similar to traditional actigraphy [8]. There are a lot of activities to get monitored using sensors, the data from unremarkable, lightweight, and an able, well-organized wearable sensors are used to record the physical activity. An incremental diagnosis method (IDM) generates data under supervised training to determine a medical illness with minimum wearable sensors by using the dynamically adjusted sensor in their natural environment. The progress of less load functional sensors resolves to be comfortable for monitoring the range of activities of inhabitants [9].

The wearable purchaser gadgets like inclination trackers, smart glasses and SMART watches are considered not exclusively to be a glossy item, yet will be in 100 s of million later on. Formal and informal survey predicts that the increase in usage of wearable smart devices in the future will result in the price of these devices as it has a wide application in society. E-health has been aimed towards home use for elderly people by constantly monitoring them in the home environment without disturbing them. At the same time, the system is sensitive in case of security and privacy policy [10].

A loneliness monitoring system for old-age people has been made into experimental values for a period of eight months [11] in which longitudinal linear mixed-effects regression and out-of-sample cross-validation have been studied to prove the accuracy of the system. Embedded-based health monitoring system for old-age people to detect early changes in health has been made possible by continuous monitoring, for which 22 features have been analysed [12] from in-home sensor data with k-nearest neighbour classifier along with fuzzy pattern tree.

In designing architectural model for monitoring patients' health and to pass on information about patient to the doctor, two-layer architecture model has been used in which data and communication layer used with REST Web services [13] to develop home-based telecommunication to effectively pass the information about the chronic diseases of the patients. Data set is the key in machine learning that has been used for blood pressure (BP) with more than 16 million BP measurements; so in day-to-day activity, supervised algorithms [14] shall be implemented over BP sensors with real-time home-based monitoring system, through which cardiovascular risk activities can be reduced. Particle swarm optimization technique for monitoring the patient's [15] state with the past and present measurement of various parameters has been studied. Support vector machine for embedded-based applications [16] shall be effectively utilized by the people with real-time world in the form of smart home.

The major problem faced by smart devices will be the power consumption, if all the sensors are combined in a single package for health monitoring. To improve the power consumption, if all the sensors are consolidated in a solitary bundle for wellbeing observing, diverse clock frequencies for various sensors, with the independent informational collection utilizing AI calculations, have been actualized in the proposed work. In the proposed model, multilinear regression algorithm provides better efficient output under supervised learning.

2 System Design

The proposed system architecture has been designed to perform different programmable-based clockings as shown in Fig. 1. The chosen application based on the wearable model with different clock frequencies has been applied with machine learning algorithms in the ARM processor. The Python library has been used to implement the coding with PYNQ-based field-programmable gate array kit through which the power consumption related-problems for wearables and SMART watches are being rectified.

In the proposed model, different parameters like BPM, step counts, calories burned, temperature and sleep disorder has been considered with real-time values using SMART watches and wearables. The chosen parameter has been monitored using five different devices as shown in Fig. 2.

The nature of our proposed work depicts that the database was taken for a month with three different wearables and two different SMART watches. Different wearables and the SMART watch configuration are shown in Tables 1 and 2, respectively. The tracking data set has been studied for a period of 30 days; though MI band uses

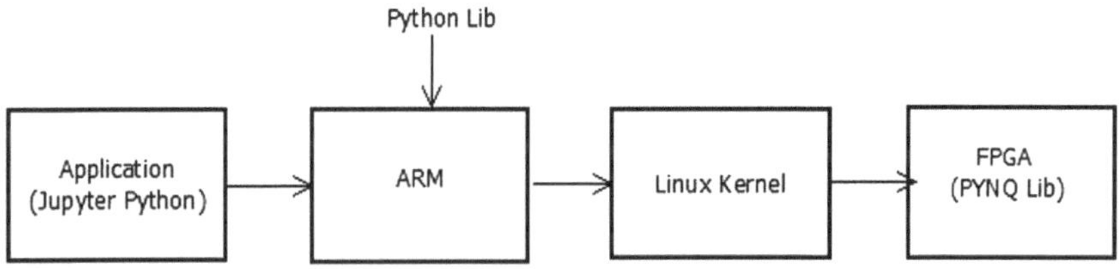

Fig. 1 Architectural view of the model

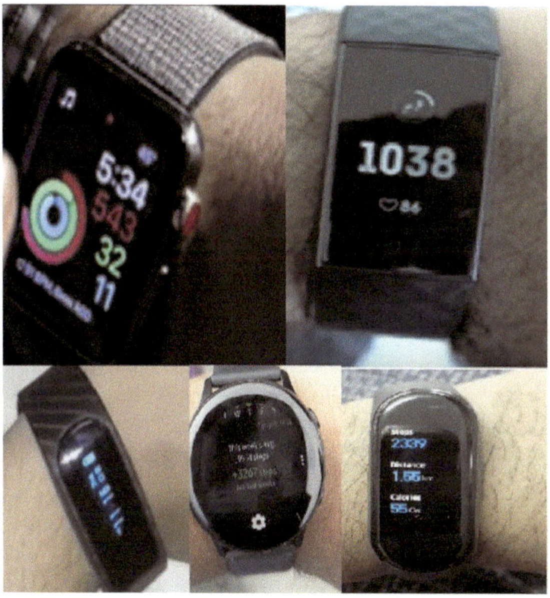

Fig. 2 Devices used for analysis

different colours, it could be able to withstand more than 7 days with full sensors being active for the entire period of testing when compared with other wearables.

It has been noticed here that the wearable technology uses different frequencies ranging from 2400 to 2483 MHz, so a separate study was conducted with these three devices for a period of 30 days and is shown in Table 2.

Two different SMART watches with frequency range of 1 GHz have been compared for a period of 30 days so that the original real-time data shall be provided for the proposed model. While comparing the data among the SMART watch of different brands, power management is literally reduced since these devices have baseband processor, DRAM, Wi-Fi, Bluetooth, RF components, NFC and MCU.

3 Results and Discussion

As proposed by the framework plan, various clocking with various sensors are trained under Supervised Machine Learning Model. The comparison of average power observed for a period of 30 days is as shown in Table 3.

The observed results clearly show that the major factor dealt with the wearable and SMART watches consumes the less amount of power. From the data, it is clear that watch product code series 3 stands for 1 day 13 h and 24 s. In the future, if all the sensors are embedded in smart devices, the power may degrade within a day. In case of old-age people, they will not charge regularly and need continuous monitoring of their overall internal behaviours. So, utilizing the power effectively in this scenario plays a major role. For the same reason, a novel table has been generated for different

Table 1 Wearable configuration

Brand	Code No.	Battery life (days)	Tracked data	Bluetooth version	Voltage (V)	Max input current (mA)	Battery (mAh)
Fitbit	85,913,025	6	Natural action (activity) Kilo calorie (calories burned), distance (aloofness) Sleep (slumber) activity Pace (steps) taken, time, BPM, workouts (physical exertion)	4	3.7	200	90
MI	XMSH07HM	14	Natural action (activity) Kilo calories (calories burned), distance (aloofness) Sleep (slumber) activity Steps (pace) taken, time, BPM, workouts (physical exertion)	5	5.0	250	135
Fast track	Reflex 2.0	10	Natural action (activity) Kilo calorie (calories burned), distance (aloofness) Sleep (slumber) activity Steps (pace) taken, time, BPM, workouts (physical exertion)	4	3.7	200	90

Table 2 SMART watch configuration

Brand	Product code	Battery life	Tracked data	Bluetooth	Frequency band (1 GHz)	Voltage (V)	Input current (mA)	Battery (mAh)	Processor
Apple	Series 3	1 day 13 h	Natural action (activity) Kilo calorie (calories burned), distance (aloofness) Sleep (slumber) activity Steps (pace) taken, time, BPM, workouts (physical exertion)	338S00348	1	5	250	Li-Ion 341	Apple S3-Baseband-QUALCOMM MDM9635M
Samsung	Gear 2	4 days	Natural action (activity) Kilo calorie (calories burned), distance (aloofness) Sleep (slumber) activity Steps (pace) taken, time, BPM, workouts (physical exertion)	4.1	1	5	150	250	1 GHz dual-core Cortex A-7 processor

Table 3 Comparison of average power observed for 30 days

Brand	Steps taken	BPM (Average)	Calories burned (Kcl)	Distance (km)	Power consumed (for 100%)	Workout (mins)	Sleep activity (Hrs)
Fitbit	7024	80	280	5.01	18	30	8
MI	6121	83	249	4.56	9	30	7
Apple	7028	79	276	5.0	72	30	8
Samsung	7027	79	274	5.01	47	30	8
Fast track	7011	78	276	5.02	15	30	7

Table 4 Variation of frequency for various parameters

Clock (MHz)	Normal sleep	Medium sleep	BPM (normal)	BPM (exercise)	Walking (leisure)	Walking (exercise)	Aggressive workout
200	Inactive	Inactive	Proactive	Inactive	Active	Inactive	Inactive
400	Active	Inactive	Inactive	Inactive	Inactive	Inactive	Inactive
600	Proactive	Active	Active	Inactive	Proactive	Inactive	Inactive
800	Inactive	Proactive	Inactive	Proactive	Inactive	Active	Proactive
1000	Inactive	Inactive	Inactive	Active	Inactive	Proactive	Active

parameters of the considered brand with different frequencies selected as given in Table 4.

The tracked data are categorized under active, proactive and inactive for various clock frequencies where the active functionally of device is shown in data set format. The observed values shall be made in high-level performance using the machine learning algorithms. In our approach, multilinear regression for performance monitoring and logistic regression for data-oriented monitoring have been applied. Since both the concept come under supervised model, generated test data shall be used to estimate with good accuracy.

Database gathered for as far back as one month gave an extensive continuous real-time based yield of parameters as shown in Figs. 3 and 4. Utilizing multilinear expectation model, the steps taken and BPM are in persistent checking, 96% of the force has been used by steps taken that was followed with 88% by the BPM has been analysed. The analysed data may change with respect to the climatic conditions; still, it was observed that the overall data gathered provided best outcome with the available wearable and SMART watches. As previously mentioned, multilinear modelling was preferred for the data set to prove the overall power consumption from the gathered information based on various parameters.

Figure 3 shows that percentage of power consumption was higher for steps taken and BPM. The force parameter has been investigated by the shading variety that noticed the rest tracker alongside step taken expends more level of intensity while BPM with exercises devours less power. It is obvious that if it is considered to provide

Fig. 3 Percentage of power consumption

Fig. 4 Order-wise power monitoring of parameters

separate clock frequency for each parameter manually, it can be controlled and the power consumption shall be reduced for the device.

Figure 4 shows the order-wise power monitoring of the parameters and their influence from the generated data set. It has been clearly observed from the picture that steps taken alone consume high power, followed by BPM, workouts and sleep tracker.

4 Conclusion and Future Work

The overall aim of the proposed work was to find the best solution for power-related problems in smart and wearable devices. Even though, nowadays, normal battery resolves the power issues year by year, for medical-related devices, portable batteries could not be able to withstand the high precision-based arithmetic and floating-point-based operations. So, it has become an important key factor for all the embedded circuit designers to work for an efficient device that can balance all the health-care applications in more effective power-based manner. In the proposed work, the database collected for a period of 30 days with the available devices has verified the applications that are responsible for high power consumption. The same can be dealt by giving different clocks for different functions which will tremendously improve the power consumption for which the data set was planned. The future work can be based on providing the clock with different frequencies and to use logistic regression so that the expected data set will be maintained to provide required power consumption by Python-based programming, and it is also clear that PYNQ-based FPGA boards shall be used for implementation.

References

1. Fengou, M.-A., Mantas, G., Komninos, N.: A new framework architecture for next generation e-health services. IEEE J. Biomed. Health Inform. **17**(1) (2013)
2. James, R., Stanley, M., Laiwah, A.: Monitor an expert system that validates and interprets time-dependent partial data based on a cystic fibrosis home monitoring program. IEEE Trans. Bio Med. Eng. **36**(5) (1989)
3. Verma, P., Sood, S.K.: Fog assisted-IOT enabled patient health monitoring in smart homes. IEEE Internet Things J. **5**(3) (2018)
4. Barger, T.S., Brown, D.E., Alwan, M.: Health-status monitoring analysis of behavioural patterns. IEEE Trans. Syst. MAN Cybern. Part A Syst. Humans **35**(1) (2005)
5. Wang, J., Zhang, Z., Lee, S., Simon, R.: An Enhanced fall detection system for elderly person monitoring using consumer home networks. IEEE (2014)
6. Branko, G., Ross, S.: Home telemonitoring of vital signs-technical challenges and future directions. IEEE J. Biomed. Health Inform. **19**(1) (2015)
7. Merilahti, J., Viramo, P., Korhonen, I.: Wearable monitoring of physical functioning and disability changes, circadian rhythms and sleep patterns in nursing home residents. IEEE J. Biomed. Health Inform. **20**(3) (2016)
8. Chandra, S.: Wearable sensors for human activity monitoring: a review. IEEE Sens. J. **15**(3) (2015)
9. Junnila, S., Kailanto, H., Merilahti, J., Vainio, A.M., Vehkaoja, A., Zakrzewski, M. and Hyttinen, J.: Wireless, multipurpose in home health monitoring platform: two case trials. IEEE Trans. Inform. Technol. Biomed. **14**(2) (2010)
10. Yan, H., Huo, H., Gidlund, M: Wireless sensor network based e-health system- implementation and experiment results. IEEE Trans. Consum. Electron. **56**(4) (2010)
11. Austin, J., Dodge, H.H., Riley, T.: A smart-home system to unobtrusively and continuously assess loneliness in older adults. Point Care Technol. **4** (2016)
12. Skubic, M., Guevara, R.D., Rantz, M.: Automated health alerts using in-home sensor data for embedded health assessment. Wearable Sens. Health Monitor. Syst. **4** (2015)

13. Lasierra, N., Alesanco, A., Garcia, J.: Designing an architecture for monitoring patients at home: ontologies and web services for clinical and technical management integration. IEEE J. Biomed. Health Inform. **8**(3) (2014)
14. Quer, G., Nikzad, N., Chieh, Normand, A.A.: Home monitoring of blood pressure: short-term changes during serial measurements for 56398 subjects. IEEE J. Biomed. Health Inform. **22**(5) (2018)
15. Arpaia, P., Manna, C., Montenero, G., Addio, G.D.: In-time prognosis based on swarm intelligence for home-care monitoring: a case study on pulmonary disease. IEEE Sens. J. **12**(3) (2012)
16. Fleury, A., Vacher, M., Noury, N.: SVM-based multimodal classification activities of daily living in health smart homes: sensors, algorithms, and first experimental results. IEEE Trans. Inform. Technol. Biomed. **14**(2) (2010)

Solar Cell Parameter Extraction by Using Harris Hawks Optimization Algorithm

Ashutosh Sharma, Akash Saxena, Shalini Shekhawat, Rajesh Kumar, and Akhilesh Mathur

Abstract Solar energy is growing faster in this modern era. Many researchers have been attracted towards the research on solar energy because it is a clean source of energy. Mostly two problems are occurred to generate energy from this source: (a) having a beneficial model to characterize solar cells and (b) very less available information about PV cells. Due to these issues, PV module performance affected. In order to extract the parameters of the PV cells and modules, numerous algorithms have been suggested. Many of them often fail to find the best solutions. In this chapter, an application of Harris hawks optimization (HHO) algorithm is reported to extract solar cell parameters. The wide applicability of this algorithm has already been examined over different conventional benchmark functions and on some real problem. This fact motivated authors to implement this algorithm on parameter extraction problem. The main motivation behind the implementation of HHO on solar cell parameter extraction is the efficacy of this algorithm to deal with complex optimization problems. Results of HHO are compared with other well-known algorithm results which shows that HHO produces better results.

Keywords Solar cell parameter extraction · Single-diode model · Double-diode model · Harris hawks optimization

A. Sharma · A. Saxena (✉)
Department of Electrical Engineering, Swami Keshvanand Institute of Technology, Management & Gramothan, Jaipur, India
e-mail: aakash.saxena@hotmail.com

A. Sharma
e-mail: ashusharma199566@gmail.com

S. Shekhawat
Department of Mathematics, Swami Keshvanand Institute of Technology, Management & Gramothan, Jaipur, India

R. Kumar · A. Mathur
Department of Electrical Engineering, Malaviya National Institute of Technology, Jaipur, India

A. K. Bhoi et al. (eds.), *Bio-inspired Neurocomputing*, Studies in Computational Intelligence 903, https://doi.org/10.1007/978-981-15-5495-7_20

1 Introduction

Renewable energy sources are very essential sources in the modern age to fulfil all the energy demands. Due to depletion of conventional energy sources, it is not possible to supply whole demand by only conventional sources. For this, researchers are looking for renewable energy sources, which produces larger amount of energy in a very short period of time and in less needed space. Among all the renewable energy sources, solar PV energy is a leading option. Solar PV energy is growing faster in the area of electric power generation [1]. Solar PV energy is very productive and can contribute on high level to supply the energy demand. Solar energy has some advantages like its infinite availability, no harmful waste and also they do not disturb the ecological balance [2]. Among all renewable energy sources, solar energy has highest power density with the global mean of 170 W/m^2 [3]. There are some important points about the replacement of solar PVs with conventional sources as follows:

1. Solar energy is free and in limitless amount, while fossil fuels are depleted because of that price is also increasing [4, 5].
2. Solar energy sources are contributing to minimize the issue of global warming, while conventional energy sources which use fossil fuels increase the global warming problem.
3. Solar PVs do not pollute the environment during the period of operation, while fossil fuels create pollution on high level [6, 7].
4. Solar PVs have less amount of operational cost as well as maintenance cost, while conventional energy sources charge high cost during the operational and maintenance period [8].
5. Solar PVs produce high amount of energy with very less covered area, while for conventional energy sources large area is required [2].

With these above-mentioned advantages mostly, every country tries to implement solar PVs and tries to complete the demand with this renewable energy source [9].

In many countries to boost up this energy, government allows financial inducement [10–14].

A photovoltaic cell is a p-n junction diode which is typically represented by an analogous electrical circuit [15, 16]. In the process to construct PV module, many PV cells are grouped together. It is evident to operate photovoltaic device at maximum power point (MPP) for efficient utilization. In this way, for the comprehensive interpretation, simulation and investigation the designer must know the proper model of PV cell. A model can be assumed with one, two or more than two diodes by the designer. Due to fast convergence rate, simplicity and acceptable accuracy, single-diode model has been known as the most popular diode model [17]. But for temperature variations, this model shows some limitations. As one more diode is added parallel to another diode, double-diode model becomes more accurate as irradiance gets reduced. But by increasing one more diode, the unknown parameters also increase and the system becomes complex. With the addition of one more diode, the loss in depletion region is also occurred by the recombination of carriers [18].

In this, single-diode model contains five parameters, in which three parameters are extracted and the remaining two are calculated. Similarly, in double-diode model there are seven parameters in which five are extracted and the remaining two are calculated. To optimize the value of parameters of PV cells, many optimization techniques have been applied in recent years. Usually an error function should be minimized to provide optimal cell parameters by several optimization algorithms between the set of experimental and simulated I–V data of the PV system. Earlier, many metaheuristic algorithms have been applied to optimize the parameter values of PV cells. To provide optimal cell parameters, usually an error function is minimized by applying some optimization algorithm including the set of simulated data and experimental I–V data of the PV system. Various optimization algorithms provide excellent results in solving real engineering problems. Some examples are [19–22]. Appelbaum et al. [23] presented a comparative analysis of three methods on single and multi-junction solar cell. The first method is a variant of Newton–Raphson method, whereas the second method is a variant of Levenberg–Marquardt algorithm. Both first and second approaches follow the gradient descent method, whereas the third method is based on genetic algorithm. As the fundamental data are not easily available at open- and short-circuit points, the authors used a piece-wise I–V curve fitting method with four-parameter PV model and then five parameters are extracted using explicit extraction method [24]. Because of the complexity to overcome local minima and premature convergence, most optimization techniques give sub-optimal solution. To overcome these problems, the use of artificial bee colony (ABC) algorithm was proposed by Oliva et al. [1]. To extract optimal parameters, flower pollination algorithm (FPA) [25] was proposed. By using three numerous sources of data, this method is tested. Literature reviews on solar cell modelling and parameter estimation methods are presented in [26, 27]. The objective function given in [28] is used to minimize by a modified version of ABC algorithm which is highly effective version. The authors also compared the algorithm with many other algorithms proving their algorithm much faster and accurate.

In this work, Harris hawks optimization algorithm is applied which is proposed by Ali Asghar Heidari et al. This algorithm is recently proposed algorithm and provides better efficiency. This algorithm is working in two phases like exploration phase and exploitation phase. A comparison of HHO and other well-known optimization algorithms is presented in this chapter. Fast convergence rate and optimality of the results are the main motivation behind the selection of HHO.

2 Solar Cell Modelling

Understanding a scientific model that precisely portrays the electrical conduct of sun-oriented cell is very essential. To elucidate the solar cell and its I-V characteristics, different equivalent circuit models have been developed and proposed. Most commonly, three diode models are used in practice, i.e., ideal, single and double-diode models.

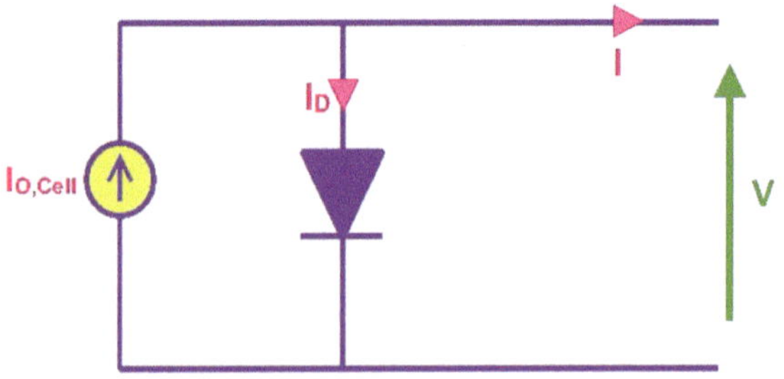

Fig. 1 An ideal PV cells' equivalent circuit

2.1 *Ideal Diode Model*

In Fig. 1, ideal diode model is represented. The mathematical representation of output current (I) of this cell is as follows:

$$I = I_{O,\text{Cell}} - I_D \tag{1}$$

where diode current is ID and can be expressed as Shockley diode equation, and $I_{O,\text{Cell}}$ is the photovoltaic current which is generated due to incident light. Now, Eq. (1) can be written as [25]:

$$I = I_{O,\text{Cell}} - I_{S,\text{Cell}}\left[\exp\left(\frac{qV}{\xi kT}\right) - 1\right] \tag{2}$$

where $I_{S,\text{Cell}}$ is diode's reverse saturation current, an electron charge is q which is 1.602176×10^{-19} C, k is Boltzmann constant, i.e., 1.380650×10^{-23} J/K, absolute temperature of diode junctions is T (in Kelvin) and ideality factor is denoted by ξ.

To configure a module, a few PV cells are associated either in series or in parallel as a general rule. NC is considered as number of cells which are connected in series because more prominent output voltage is given by series connection. In this manner, the more feasible comparable circuit models of PV modules are examined.

2.2 *Single-diode Model*

It is considered due to its simplicity and better accuracy point of view. As shown in Fig. 2, single-diode model is comprised of five parameters, namely current source, single-diode, one ideality factor and two resistors.

$$I = I_{\text{PV}} - I_O\left[\exp\left(\frac{e(V + R_S I)}{\alpha_1 k N_C T}\right) - 1\right] - \frac{V + R_S I}{R_P} \tag{3}$$

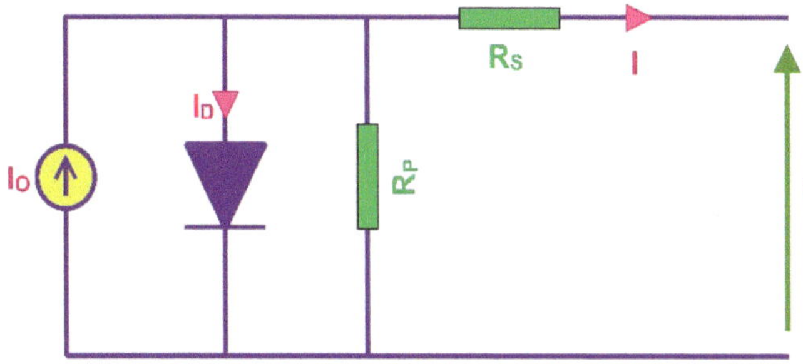

Fig. 2 Single-diode model's equivalent circuit of PV module

Case 1: for monocrystalline cell

$$I = I_{\text{SD}}^{\text{Mono}} + I_O\left[\exp\left(\frac{e(V + R_S I)}{\alpha k N_C^{\text{Mono}} T}\right) - 1\right] - \frac{V + R_S I}{R_P} \tag{4}$$

Considering open-circuit condition where $I = 0$ and $V = V_{\text{OC, SD}}^{\text{Mono}}$

$$I_{\text{SD}}^{\text{Mono}} = \frac{V_{\text{OC, SD}}^{\text{Mono}}}{R_P} + I_O\left[\exp\left(\frac{e\left(V_{\text{OC, SD}}^{\text{Mono}}\right)}{\alpha_1 k N_C^{\text{Mono}} T}\right) - 1\right] \tag{5}$$

For any iteration t

$$\text{SD}_{\text{error OC}}^{\text{Mono}}(t) = \frac{V_{\text{OC, SD}}^{\text{Mono}}}{R_P(t)} + I_O\left[\exp\left(\frac{e\left(V_{\text{OC, SD}}^{\text{Mono}}\right)}{\alpha(t) k N_C^{\text{Mono}} T}\right) - 1\right] - I_{\text{SD}}^{\text{Mono}} \tag{6}$$

Considering short-circuit condition where $I = I_{\text{SC,SD}}^{\text{Mono}}$ and $V = 0$

$$I_{\text{SD}}^{\text{Mono}} = \frac{I_{\text{SC,SD}}^{\text{Mono}} R_S}{R_P} + I_O\left[\exp\left(\frac{e\left(I_{\text{SC, SD}}^{\text{Mono}} R_S\right)}{\alpha k N_C^{\text{Mono}} T}\right) - 1\right] \tag{7}$$

For any iteration t

$$\text{DD}_{\text{error SC}}^{\text{Mono}}(t) = \frac{I_{\text{SC,SD}}^{\text{Mono}} R_S(t)}{R_P(t)} + I_O\left[\exp\left(\frac{e\left(I_{\text{SC, SD}}^{\text{Mono}} R_S(t)\right)}{\alpha(t) k N_C^{\text{Mono}} T}\right) - 1\right] - I_{\text{SD}}^{\text{Mono}} \tag{8}$$

Considering maximum power condition where $I = I_{\text{MPP, SD}}^{\text{Mono}}$ and $V = V_{\text{MPP, SD}}^{\text{Mono}}$

$$I_{\text{SD}}^{\text{Mono}} = \frac{V_{\text{MPP, SD}}^{\text{Mono}} + I_{\text{MPP, SD}}^{\text{Mono}} R_S}{R_P} + I_O\left[\exp\left(\frac{e\left(V_{\text{MPP, SD}}^{\text{Mono}} + I_{\text{MPP, SD}}^{\text{Mono}} R_S\right)}{\alpha k N_C^{\text{Mono}} T}\right) - 1\right] \tag{9}$$

For any iteration t

$$\mathrm{SD}^{\mathrm{Mono}}_{\mathrm{error\,MPP}}(t) = \frac{V^{\mathrm{Mono}}_{\mathrm{MPP,\,SD}} + I^{\mathrm{Mono}}_{\mathrm{MPP,\,SD}} R_S(t)}{R_P(t)} + I_O\left[\exp\left(\frac{e\left(V^{\mathrm{Mono}}_{\mathrm{MPP,\,SD}} + I^{\mathrm{Mono}}_{\mathrm{MPP,\,SD}} R_S(t)\right)}{\alpha(t) k N_C^{\mathrm{Mono}} T}\right) - 1\right] - I^{\mathrm{Mono}}_{\mathrm{SD}} \tag{10}$$

$$J^{\mathrm{SD}}_{\mathrm{Mono}} = \frac{\left(\mathrm{SD}^{\mathrm{Mono}}_{\mathrm{error\,OC}}\right)^2 + \left(\mathrm{SD}^{\mathrm{Mono}}_{\mathrm{error\,SC}}\right)^2 + \left(\mathrm{SD}^{\mathrm{Mono}}_{\mathrm{error\,MPP}}\right)^2}{3} \tag{11}$$

Case 2: for polycrystalline cell

$$I = I^{\mathrm{Poly}}_{\mathrm{SD}} + I_O\left[\exp\left(\frac{e(V + R_S I)}{\alpha k N_C^{\mathrm{Poly}} T}\right) - 1\right] - \frac{V + R_S I}{R_P} \tag{12}$$

Considering open-circuit condition where $I = 0$ and $V = V^{\mathrm{Poly}}_{\mathrm{OC,\,SD}}$

$$I^{\mathrm{Poly}}_{\mathrm{SD}} = \frac{V^{\mathrm{Poly}}_{\mathrm{OC,\,SD}}}{R_P} + I_O\left[\exp\left(\frac{e\left(V^{\mathrm{Poly}}_{\mathrm{OC,\,SD}}\right)}{\alpha_1 k N_C^{\mathrm{Poly}} T}\right) - 1\right] \tag{13}$$

For any iteration t

$$\mathrm{SD}^{\mathrm{Poly}}_{\mathrm{error\,OC}}(t) = \frac{V^{\mathrm{Poly}}_{\mathrm{OC,\,SD}}}{R_P(t)} + I_O\left[\exp\left(\frac{e\left(V^{\mathrm{Poly}}_{\mathrm{OC,\,SD}}\right)}{\alpha(t) k N_C^{\mathrm{Poly}} T}\right) - 1\right] - I^{\mathrm{Poly}}_{\mathrm{SD}} \tag{14}$$

Considering short-circuit condition where $I = I^{\mathrm{Poly}}_{\mathrm{SC,SD}}$ and $V = 0$

$$I^{\mathrm{Poly}}_{\mathrm{SD}} = \frac{I^{\mathrm{Poly}}_{\mathrm{SC,SD}} R_S}{R_P} + I_O\left[\exp\left(\frac{e\left(I^{\mathrm{Poly}}_{\mathrm{SC,\,SD}} R_S\right)}{\alpha k N_C^{\mathrm{Poly}} T}\right) - 1\right] \tag{15}$$

For any iteration t

$$\mathrm{DD}^{\mathrm{Poly}}_{\mathrm{error\,SC}}(t) = \frac{I^{\mathrm{Poly}}_{\mathrm{SC,SD}} R_S(t)}{R_P(t)} + I_O\left[\exp\left(\frac{e\left(I^{\mathrm{Poly}}_{\mathrm{SC,\,SD}} R_S(t)\right)}{\alpha(t) k N_C^{\mathrm{Poly}} T}\right) - 1\right] - I^{\mathrm{Poly}}_{\mathrm{SD}} \tag{16}$$

Considering maximum power condition where $I = I^{\mathrm{Poly}}_{\mathrm{MPP,\,SD}}$ and $V = V^{\mathrm{Poly}}_{\mathrm{MPP,\,SD}}$

$$I_{\text{SD}}^{\text{Poly}} = \frac{V_{\text{MPP, SD}}^{\text{Poly}} + I_{\text{MPP, SD}}^{\text{Poly}} R_S}{R_P} + I_O \left[\exp\left(\frac{e\left(V_{\text{MPP, SD}}^{\text{Poly}} + I_{\text{MPP, SD}}^{\text{Poly}} R_S\right)}{\alpha k N_C^{\text{Poly}} T} \right) - 1 \right] \tag{17}$$

For any iteration t

$$\text{SD}_{\text{error MPP}}^{\text{Poly}}(t) = \frac{V_{\text{MPP, SD}}^{\text{Poly}} + I_{\text{MPP, SD}}^{\text{Poly}} R_S(t)}{R_P(t)} + I_O \left[\exp\left(\frac{e\left(V_{\text{MPP, SD}}^{\text{Poly}} + I_{\text{MPP, SD}}^{\text{Poly}} R_S(t)\right)}{\alpha(t) k N_C^{\text{Poly}} T} \right) - 1 \right] - I_{\text{SD}}^{\text{Poly}} \tag{18}$$

$$J_{\text{Poly}}^{\text{SD}} = \frac{\left(\text{SD}_{\text{error OC}}^{\text{Poly}}\right)^2 + \left(\text{SD}_{\text{error SC}}^{\text{Poly}}\right)^2 + \left(\text{SD}_{\text{error MPP}}^{\text{Poly}}\right)^2}{3} \tag{19}$$

Case 3: for thin film

$$I = I_{\text{SD}}^{\text{Thin}} + I_O \left[\exp\left(\frac{e(V + R_S I)}{\alpha k N_C^{\text{Thin}} T} \right) - 1 \right] - \frac{V + R_S I}{R_P} \tag{20}$$

Considering open-circuit condition where $I = 0$ and $V = V_{\text{OC, SD}}^{\text{Thin}}$

$$I_{\text{SD}}^{\text{Thin}} = \frac{V_{\text{OC, SD}}^{\text{Thin}}}{R_P} + I_O \left[\exp\left(\frac{e\left(V_{\text{OC, SD}}^{\text{Thin}}\right)}{\alpha_1 k N_C^{\text{Thin}} T} \right) - 1 \right] \tag{21}$$

For any iteration t

$$\text{SD}_{\text{error OC}}^{\text{Thin}}(t) = \frac{V_{\text{OC, SD}}^{\text{Thin}}}{R_P(t)} + I_O \left[\exp\left(\frac{e\left(V_{\text{OC, SD}}^{\text{Thin}}\right)}{\alpha(t) k N_C^{\text{Thin}} T} \right) - 1 \right] - I_{\text{SD}}^{\text{Thin}} \tag{22}$$

Considering short-circuit condition where $I = I_{\text{SC,SD}}^{\text{Thin}}$ and $V = 0$

$$I_{\text{SD}}^{\text{Thin}} = \frac{I_{\text{SC,SD}}^{\text{Thin}} R_S}{R_P} + I_O \left[\exp\left(\frac{e\left(I_{\text{SC, SD}}^{\text{Thin}} R_S\right)}{\alpha k N_C^{\text{Thin}} T} \right) - 1 \right] \tag{23}$$

For any iteration t

$$\text{DD}_{\text{error SC}}^{\text{Thin}}(t) = \frac{I_{\text{SC,SD}}^{\text{Thin}} R_S(t)}{R_P(t)} + I_O \left[\exp\left(\frac{e\left(I_{\text{SC, SD}}^{\text{Thin}} R_S(t)\right)}{\alpha(t) k N_C^{\text{Thin}} T} \right) - 1 \right] - I_{\text{SD}}^{\text{Thin}} \tag{24}$$

Considering maximum power condition where $I = I_{\text{MPP, SD}}^{\text{Thin}}$ and $V = V_{\text{MPP, SD}}^{\text{Thin}}$

$$I_{\mathrm{SD}}^{\mathrm{Thin}} = \frac{V_{\mathrm{MPP,\,SD}}^{\mathrm{Thin}} + I_{\mathrm{MPP,\,SD}}^{\mathrm{Thin}} R_S}{R_P} + I_O\left[\exp\left(\frac{e\left(V_{\mathrm{MPP,\,SD}}^{\mathrm{Thin}} + I_{\mathrm{MPP,\,SD}}^{\mathrm{Thin}} R_S\right)}{\alpha k N_C^{\mathrm{Thin}} T}\right) - 1\right] \tag{25}$$

For any iteration t

$$\mathrm{SD}_{\mathrm{error\,MPP}}^{\mathrm{Thin}}(t) = \frac{V_{\mathrm{MPP,\,SD}}^{\mathrm{Thin}} + I_{\mathrm{MPP,\,SD}}^{\mathrm{Thin}} R_S(t)}{R_P(t)} + I_O\left[\exp\left(\frac{e\left(V_{\mathrm{MPP,\,SD}}^{\mathrm{Thin}} + I_{\mathrm{MPP,\,SD}}^{\mathrm{Thin}} R_S(t)\right)}{\alpha(t) k N_C^{\mathrm{Thin}} T}\right) - 1\right] - I_{\mathrm{SD}}^{\mathrm{Thin}} \tag{26}$$

$$J_{\mathrm{Thin}}^{\mathrm{SD}} = \frac{\left(\mathrm{SD}_{\mathrm{error\,OC}}^{\mathrm{Thin}}\right)^2 + \left(\mathrm{SD}_{\mathrm{error\,SC}}^{\mathrm{Thin}}\right)^2 + \left(\mathrm{SD}_{\mathrm{error\,MPP}}^{\mathrm{Thin}}\right)^2}{3} \tag{27}$$

2.3 *Double-diode Model*

Another approach to depict solar cell electrical conduct is to model it by adding another diode parallel to a modifying diode. Furthermore, a resistor represents the solar cell metal contacts and semiconductor material mass opposition associated in arrangement with the cell shunt components [26]. Figure 3 shows the equivalent circuit adopting double-diode model for the PV module, and mathematical equations of the double-diode models are derived as:

$$I = I_{\mathrm{PV}} - I_{O_1}\left[\exp\left(\frac{e(V + R_S I)}{\alpha_1 k N_C T}\right) - 1\right] - I_{O_2}\left[\exp\left(\frac{e(V + R_S I)}{\alpha_2 k N_C T}\right) - 1\right] - \frac{V + R_S I}{R_P} \tag{28}$$

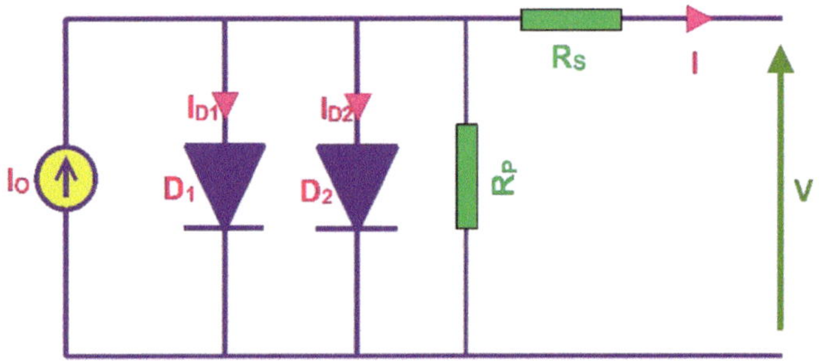

Fig. 3 Single-diode model's equivalent circuit of PV module

Case 4: for monocrystalline cell

$$I = I_{\text{DD}}^{\text{Mono}} + I_{O_1}\left[\exp\left(\frac{e(V + R_S I)}{\alpha_1 k N_C^{\text{Mono}} T}\right) - 1\right] - I_{O_2}\left[\exp\left(\frac{e(V + R_S I)}{\alpha_2 k N_C^{\text{Mono}} T}\right) - 1\right] - \frac{V + R_S I}{R_P} \tag{29}$$

Considering open-circuit condition where $I = 0$ and $V = V_{\text{OC, DD}}^{\text{Mono}}$

$$I_{\text{DD}}^{\text{Mono}} = \frac{V_{\text{OC, DD}}^{\text{Mono}}}{R_P} + I_{O_1}\left[\exp\left(\frac{e\left(V_{\text{OC, DD}}^{\text{Mono}}\right)}{\alpha_1 k N_C^{\text{Mono}} T}\right) - 1\right] + I_{O_2}\left[\exp\left(\frac{e\left(V_{\text{OC, DD}}^{\text{Mono}}\right)}{\alpha_2 k N_C^{\text{Mono}} T}\right) - 1\right] \tag{30}$$

For any iteration t

$$\text{DD}_{\text{error OC}}^{\text{Mono}}(t) = \frac{V_{\text{OC, DD}}^{\text{Mono}}}{R_P(t)} + I_{O_1}\left[\exp\left(\frac{e\left(V_{\text{OC, DD}}^{\text{Mono}}\right)}{\alpha_1(t) k N_C^{\text{Mono}} T}\right) - 1\right] + I_{O_2}\left[\exp\left(\frac{e\left(V_{\text{OC, DD}}^{\text{Mono}}\right)}{\alpha_2(t) k N_C^{\text{Mono}} T}\right) - 1\right] - I_{\text{DD}}^{\text{Mono}} \tag{31}$$

Considering short-circuit condition where $I = I_{\text{SC,DD}}^{\text{Mono}}$ and $V = 0$

$$I_{\text{DD}}^{\text{Mono}} = \frac{I_{\text{SC,DD}}^{\text{Mono}} R_S}{R_P} + I_{O_1}\left[\exp\left(\frac{e\left(I_{\text{SC, DD}}^{\text{Mono}} R_S\right)}{\alpha_1 k N_C^{\text{Mono}} T}\right) - 1\right] + I_{O_2}\left[\exp\left(\frac{e\left(I_{\text{SC, DD}}^{\text{Mono}} R_S\right)}{\alpha_2 k N_C^{\text{Mono}} T}\right) - 1\right] \tag{32}$$

For any iteration t

$$\text{DD}_{\text{error SC}}^{\text{Mono}}(t) = \frac{I_{\text{SC,DD}}^{\text{Mono}} R_S(t)}{R_P(t)} + I_{O_1}\left[\exp\left(\frac{e\left(I_{\text{SC, DD}}^{\text{Mono}} R_S(t)\right)}{\alpha_1(t) k N_C^{\text{Mono}} T}\right) - 1\right] + I_{O_2}\left[\exp\left(\frac{e\left(I_{\text{SC, DD}}^{\text{Mono}} R_S(t)\right)}{\alpha_2(t) k N_C^{\text{Mono}} T}\right) - 1\right] - I_{\text{DD}}^{\text{Mono}} \tag{33}$$

Considering maximum power condition where $I = I_{\text{MPP, DD}}^{\text{Mono}}$ and $V = V_{\text{MPP, DD}}^{\text{Mono}}$

$$I_{\mathrm{DD}}^{\mathrm{Mono}} = \frac{V_{\mathrm{MPP,\,DD}}^{\mathrm{Mono}} + I_{\mathrm{MPP,\,DD}}^{\mathrm{Mono}} R_S}{R_P} + I_{O_1}\left[\exp\left(\frac{e\left(V_{\mathrm{MPP,\,DD}}^{\mathrm{Mono}} + I_{\mathrm{MPP,\,DD}}^{\mathrm{Mono}} R_S\right)}{\alpha_1 k N_C^{\mathrm{Mono}} T}\right) - 1\right] + I_{O_2}\left[\exp\left(\frac{e\left(\mathrm{V}_{\mathrm{MPP,\,DD}}^{\mathrm{Mono}} + \mathrm{I}_{\mathrm{MPP,\,DD}}^{\mathrm{Mono}} R_S\right)}{\alpha_2 k N_C^{\mathrm{Mono}} T}\right) - 1\right] \tag{34}$$

For any iteration t

$$\mathrm{DD}_{\mathrm{error\,MPP}}^{\mathrm{Mono}}(t) = \frac{V_{\mathrm{MPP,\,DD}}^{\mathrm{Mono}} + I_{\mathrm{MPP,\,DD}}^{\mathrm{Mono}} R_S(t)}{R_P(t)} + I_{O_1}\left[\exp\left(\frac{e\left(V_{\mathrm{MPP,\,DD}}^{\mathrm{Mono}} + I_{\mathrm{MPP,\,DD}}^{\mathrm{Mono}} R_S(t)\right)}{\alpha_1(t) k N_C^{\mathrm{Mono}} T}\right) - 1\right] + I_{O_2}\left[\exp\left(\frac{e\left(V_{\mathrm{MPP,\,DD}}^{\mathrm{Mono}} + I_{\mathrm{MPP,\,DD}}^{\mathrm{Mono}} R_S(t)\right)}{\alpha_2(t) k N_C^{\mathrm{Mono}} T}\right) - 1\right] - I_{\mathrm{DD}}^{\mathrm{Mono}} \tag{35}$$

$$J_{\mathrm{Mono}}^{\mathrm{DD}} = \frac{\left(\mathrm{DD}_{\mathrm{error\,OC}}^{\mathrm{Mono}}\right)^2 + \left(\mathrm{DD}_{\mathrm{error\,SC}}^{\mathrm{Mono}}\right)^2 + \left(\mathrm{DD}_{\mathrm{error\,MPP}}^{\mathrm{Mono}}\right)^2}{3} \tag{36}$$

Case 5: for polycrystalline cell

$$I = I_{\mathrm{DD}}^{\mathrm{Poly}} + I_{O_1}\left[\exp\left(\frac{e(V + R_S I)}{\alpha_1 k N_C^{\mathrm{Poly}} T}\right) - 1\right] - I_{O_2}\left[\exp\left(\frac{e(V + R_S I)}{\alpha_2 k N_C^{\mathrm{Poly}} T}\right) - 1\right] - \frac{V + R_S I}{R_P} \tag{37}$$

Considering open-circuit condition where $I = 0$ and $V = V = V_{\mathrm{OC,\,DD}}^{\mathrm{Poly}}$

$$I_{\mathrm{DD}}^{\mathrm{Poly}} = \frac{V_{\mathrm{OC,\,DD}}^{\mathrm{Poly}}}{R_P} + I_{O_1}\left[\exp\left(\frac{e\left(V_{\mathrm{OC,\,DD}}^{\mathrm{Poly}}\right)}{\alpha_1 k N_C^{\mathrm{Poly}} T}\right) - 1\right] + I_{O_2}\left[\exp\left(\frac{e\left(V_{\mathrm{OC,\,DD}}^{\mathrm{Poly}}\right)}{\alpha_2 k N_C^{\mathrm{Poly}} T}\right) - 1\right] \tag{38}$$

For any iteration t

$$\mathrm{DD}_{\mathrm{error\,OC}}^{\mathrm{Poly}}(t) = \frac{V_{\mathrm{OC,\,DD}}^{\mathrm{Poly}}}{R_P(t)} + I_{O_1}\left[\exp\left(\frac{e\left(V_{\mathrm{OC,\,DD}}^{\mathrm{Poly}}\right)}{\alpha_1(t) k N_C^{\mathrm{Poly}} T}\right) - 1\right]$$

$$+ I_{O_2}\left[\exp\left(\frac{e\left(V_{\text{OC, DD}}^{\text{Poly}}\right)}{\alpha_2(t)kN_C^{\text{Poly}}T}\right)-1\right]-I_{\text{DD}}^{\text{Mono}} \quad (39)$$

Considering short-circuit condition where $I = I_{\text{SC,DD}}^{\text{Mono}}$ and $V = 0$

$$I_{DD}^{\text{Poly}} = \frac{I_{\text{SC,DD}}^{\text{Poly}}R_S}{R_P} + I_{O_1}\left[\exp\left(\frac{e\left(I_{\text{SC, DD}}^{\text{Poly}}R_S\right)}{\alpha_1 kN_C^{\text{Poly}}T}\right)-1\right] + I_{O_2}\left[\exp\left(\frac{e\left(I_{\text{SC, DD}}^{\text{Poly}}R_S\right)}{\alpha_2 kN_C^{\text{Poly}}T}\right)-1\right] \quad (40)$$

For any iteration t

$$\text{DD}_{\text{error SC}}^{\text{Poly}}(t) = \frac{I_{\text{SC,DD}}^{\text{Poly}}R_S(t)}{R_P(t)} + I_{O_1}\left[\exp\left(\frac{e\left(I_{\text{SC, DD}}^{\text{Poly}}R_S(t)\right)}{\alpha_1(t)kN_C^{\text{Poly}}T}\right)-1\right] + I_{O_2}\left[\exp\left(\frac{e\left(I_{\text{SC, DD}}^{\text{Poly}}R_S(t)\right)}{\alpha_2(t)kN_C^{\text{Poly}}T}\right)-1\right]-I_{\text{DD}}^{\text{Poly}} \quad (41)$$

Considering maximum power condition where $I = I_{\text{MPP, DD}}^{\text{Mono}}$ and $V = V_{\text{MPP, DD}}^{\text{Mono}}$

$$I_{\text{DD}}^{\text{Poly}} = \frac{V_{\text{MPP, DD}}^{\text{Poly}} + I_{\text{MPP, DD}}^{\text{Poly}}R_S}{R_P} + I_{O_1}\left[\exp\left(\frac{e\left(V_{\text{MPP, DD}}^{\text{Poly}} + I_{\text{MPP, DD}}^{\text{Poly}}R_S\right)}{\alpha_1 kN_C^{\text{Poly}}T}\right)-1\right] + I_{O_2}\left[\exp\left(\frac{e\left(V_{\text{MPP, DD}}^{\text{Poly}} + I_{\text{MPP, DD}}^{Poly}R_S\right)}{\alpha_2 kN_C^{\text{Poly}}T}\right)-1\right] \quad (42)$$

For any iteration t

$$\text{DD}_{\text{error MPP}}^{\text{Poly}}(t) = \frac{V_{\text{MPP, DD}}^{\text{Poly}} + I_{\text{MPP, DD}}^{\text{Poly}}R_S(t)}{R_P(t)} + I_{O_1}\left[\exp\left(\frac{e\left(V_{\text{MPP, DD}}^{\text{Poly}} + I_{\text{MPP, DD}}^{\text{Poly}}R_S(t)\right)}{\alpha_1(t)kN_C^{\text{Poly}}T}\right)-1\right] + I_{O_2}\left[\exp\left(\frac{e\left(V_{\text{MPP, DD}}^{\text{Poly}} + I_{\text{MPP, DD}}^{\text{Poly}}R_S(t)\right)}{\alpha_2(t)kN_C^{\text{Poly}}T}\right)-1\right]-I_{\text{DD}}^{\text{Poly}} \quad (43)$$

$$J_{\text{Mono}}^{\text{DD}} = \frac{\left(\text{DD}_{\text{error OC}}^{\text{Poly}}\right)^2 + \left(\text{DD}_{\text{error SC}}^{\text{Poly}}\right)^2 + \left(\text{DD}_{\text{error MPP}}^{\text{Poly}}\right)^2}{3} \tag{44}$$

Case 5: for thin film

$$I = I_{\text{DD}}^{\text{Thin}} + I_{O_1}\left[\exp\left(\frac{e(V + R_S I)}{\alpha_1 k N_C^{\text{Thin}} T}\right) - 1\right] - I_{O_2}\left[\exp\left(\frac{e(V + R_S I)}{\alpha_2 k N_C^{\text{Thin}} T}\right) - 1\right] - \frac{V + R_S I}{R_P} \tag{45}$$

Considering open-circuit condition where $I = 0$ and $V = V = V_{\text{OC, DD}}^{\text{Mono}}$

$$I_{\text{DD}}^{\text{Thin}} = \frac{V_{\text{OC, DD}}^{\text{Thin}}}{R_P} + I_{O_1}\left[\exp\left(\frac{e\left(V_{\text{OC, DD}}^{\text{Thin}}\right)}{\alpha_1 k N_C^{\text{Thin}} T}\right) - 1\right] + I_{O_2}\left[\exp\left(\frac{e\left(V_{\text{OC, DD}}^{\text{Thin}}\right)}{\alpha_2 k N_C^{\text{Thin}} T}\right) - 1\right] \tag{46}$$

For any iteration t

$$\text{DD}_{\text{error OC}}^{\text{Thin}}(t) = \frac{V_{\text{OC, DD}}^{\text{Thin}}}{R_P} + I_{O_1}\left[\exp\left(\frac{e\left(V_{\text{OC, DD}}^{\text{Thin}}\right)}{\alpha_1(t) k N_C^{\text{Thin}} T}\right) - 1\right] + I_{O_2}\left[\exp\left(\frac{e\left(V_{\text{OC, DD}}^{\text{Thin}}\right)}{\alpha_2(t) k N_C^{\text{Thin}} T}\right) - 1\right] - I_{\text{DD}}^{\text{Thin}} \tag{47}$$

Considering short-circuit condition where $I = I_{\text{SC,DD}}^{\text{Mono}}$ and $V = 0$

$$I_{\text{DD}}^{\text{Thin}} = \frac{I_{\text{SC,DD}}^{\text{Thin}} R_S}{R_P} + I_{O_1}\left[\exp\left(\frac{e\left(I_{\text{SC, DD}}^{\text{Thin}} R_S\right)}{\alpha_1 k N_C^{\text{Thin}} T}\right) - 1\right] + I_{O_2}\left[\exp\left(\frac{e\left(I_{\text{SC, DD}}^{\text{Thin}} R_S\right)}{\alpha_2 k N_C^{\text{Thin}} T}\right) - 1\right] \tag{48}$$

For any iteration t

$$\text{DD}_{\text{error SC}}^{\text{Thin}}(t) = \frac{I_{\text{SC,DD}}^{\text{Thin}} R_S(t)}{R_P(t)} + I_{O_1}\left[\exp\left(\frac{e\left(I_{\text{SC, DD}}^{\text{Thin}} R_S(t)\right)}{\alpha_1(t) k N_C^{\text{Thin}} T}\right) - 1\right] + I_{O_2}\left[\exp\left(\frac{e\left(I_{\text{SC, DD}}^{\text{Thin}} R_S(t)\right)}{\alpha_2(t) k N_C^{\text{Thin}} T}\right) - 1\right] - I_{\text{DD}}^{\text{Thin}} \tag{49}$$

Considering maximum power condition where $I = I_{\text{MPP, DD}}^{\text{Mono}}$ and $V = V_{\text{MPP, DD}}^{\text{Mono}}$

$$I_{\text{DD}}^{\text{Thin}} = \frac{V_{\text{MPP, DD}}^{\text{Thin}} + I_{\text{MPP, DD}}^{\text{Thin}} R_S}{R_P} + I_{O_1}\left[\exp\left(\frac{e\left(V_{\text{MPP, DD}}^{\text{Poly}} + I_{\text{MPP, DD}}^{\text{Poly}} R_S\right)}{\alpha_1 k N_C^{\text{Poly}} T}\right) - 1\right] + I_{O_2}\left[\exp\left(\frac{e\left(V_{\text{MPP, DD}}^{\text{Thin}} + I_{\text{MPP, DD}}^{\text{Thin}} R_S\right)}{\alpha_2 k N_C^{\text{Thin}} T}\right) - 1\right] \tag{50}$$

For any iteration t

$$\text{DD}_{\text{error MPP}}^{\text{Thin}}(t) = \frac{V_{\text{MPP, DD}}^{\text{Thin}} + I_{\text{MPP, DD}}^{\text{Thin}} R_S(t)}{R_P(t)} + I_{O_1}\left[\exp\left(\frac{e\left(V_{\text{MPP, DD}}^{\text{Thin}} + I_{\text{MPP, DD}}^{\text{Thin}} R_S(t)\right)}{\alpha_1(t) k N_C^{\text{Thin}} T}\right) - 1\right] + I_{O_2}\left[\exp\left(\frac{e\left(V_{\text{MPP, DD}}^{\text{Thin}} + I_{\text{MPP, DD}}^{\text{Thin}} R_S(t)\right)}{\alpha_2(t) k N_C^{\text{Thin}} T}\right) - 1\right] - I_{\text{DD}}^{\text{Thin}} \tag{51}$$

$$J_{\text{Thin}}^{\text{DD}} = \frac{\left(\text{DD}_{\text{error OC}}^{\text{Thin}}\right)^2 + \left(\text{DD}_{\text{error SC}}^{\text{Thin}}\right)^2 + \left(\text{DD}_{\text{error MPP}}^{\text{Thin}}\right)^2}{3} \tag{52}$$

3 Numerical Data

In this dissertation, three different types of solar cell modules have been contemplated which are polycrystalline, monocrystalline and thin film. The typical data which are given by the manufacturer are given in Table 1. It is obvious to describe that the data of current and voltage are given by the designer at the standard test conditions, i.e., 25 °C temperature and 1000 W/m^2 irradiance, at three important points (short-circuit point, open-circuit point and maximum power point).

4 Objective Function

The main aim of this optimization is to extract PV cell parameters precisely so that the data given in Table 1 for all three major points (open-circuit, short-circuit and maximum power points) are satisfied. Mostly, maximum power point validates in the past studies but all three major points are considered in the problem formulation. The details of optimized parameters and calculated parameters by using single and double-diode models are given in Table 2.

Table 1 At standard test condition (STC) electrical parameters for PV cells

Type	Thin film [27]	Monocrystalline [27]	Polycrystalline [28]
Number of cells in series, S_C	36	36	54
Temperature coefficient of I_{SC}, $K_{I,SC}$ (amp/C)	0.35×10^3	0.8×10^3	3.18×10^3
Temperature coefficient of V_{OC}, $K_{V,OC}$ (volt/C)	−0.1	−0.0725	−0.123
Voltage at maximum power, V_{MPP} (volt)	16.60	17.20	26.3
Current at maximum power, I_{MPP} (amp)	2.41	4.95	7.61
Open-circuit voltage, V_{OC} (volt)	23.30	22.20	32.9
Short-circuit current, I_{SC} (amp)	2.68	5.45	8.21

Table 2 Calculated parameters and extracted variables for both diode models of solar cells

Model	Extracted parameters	Calculated parameters
Single-diode model	ψ, R_{SE}, R_{SH}	I_{rsc} [Eq. (6)] I_{dmc} [Eq. (7)]
Double-diode model	ψ_1, ψ_2, R_{SE}, R_{SH}, I_{rsc1}	I_{rsc2} [Eq. (13)] I_{dmc} [Eq. (14)]

Single-diode Model:
From Eq. (4), open-circuit point error is:

$$\text{err}_{OC} = I_{rsc}\left[\exp\left(\frac{qV_{OC}}{\psi k S_C T}\right) - 1\right] + \frac{V_{OC}}{R_{SH}} - I_{dmc} \tag{53}$$

From Eq. (5), short-circuit point error is:

$$\text{err}_{SC} = I_{SC} + I_{rsc}\left[\exp\left(\frac{qI_{SC}R_{SE}}{\psi k S_C T}\right) - 1\right] + \frac{I_{SC}R_{SE}}{R_{SH}} - I_{dmc} \tag{54}$$

From Eq. (8), maximum power point error is:

$$\text{err}_{MPP} = I_{dmc} - I_{rsc}\left[\exp\left(\frac{q(V_{MPP} + I_{MPP}R_{SE})}{\psi k S_C T}\right) - 1\right] - \frac{(V_{MPP} + I_{MPP}R_{SE})}{R_{SH}} - I_{MPP} \tag{55}$$

Double-diode Model:
From Eq. (11), open-circuit error is:

$$\text{err}_{\text{OC}} = I_{\text{rsc}_1}\left[\exp\left(\frac{qV_{\text{OC}}}{\psi_1 k S_C T}\right) - 1\right] + I_{\text{rsc}_2}\left[\exp\left(\frac{qV_{\text{OC}}}{\psi_2 k S_C T}\right) - 1\right] + \frac{V_{\text{OC}}}{R_{\text{SH}}} - I_{\text{dmc}} \tag{56}$$

From Eq. (12), short-circuit error is:

$$\text{err}_{\text{OC}} = I_{\text{SC}} + I_{\text{rsc}_1}\left[\exp\left(\frac{qI_{\text{SC}}R_{\text{SE}}}{\psi_1 k S_C T}\right) - 1\right] + I_{\text{rsc}_2}\left[\exp\left(\frac{qI_{\text{SC}}R_{\text{SE}}}{\psi_2 k S_C T}\right) - 1\right] + \frac{I_{\text{SC}}R_{\text{SE}}}{R_{\text{SH}}} - I_{\text{dmc}} \tag{57}$$

Maximum power point error is given by using Eq. (15):

$$\begin{aligned}\text{err}_{\text{OC}} = I_{\text{dmc}} - I_{\text{rsc}_1}\left[\exp\left(\frac{q(V_{\text{MPP}} + I_{\text{MPP}}R_{\text{SE}})}{\psi_1 k S_C T}\right) - 1\right] \\ - I_{\text{rsc}_2}\left[\exp\left(\frac{q(V_{\text{MPP}} + I_{\text{MPP}}R_{\text{SE}})}{\psi_2 k S_C T}\right) - 1\right] - \frac{V_{\text{MPP}} + I_{\text{MPP}}R_{\text{SE}}}{R_{\text{SH}}} - I_{\text{MPP}}\end{aligned} \tag{58}$$

Now, sum of square errors is taken as the main aim of optimization method. So that, the aggregate error has to be minimized as:

$$\text{ERR} = \text{err}_{\text{OC}}^2 + \text{err}_{\text{SC}}^2 + \text{err}_{\text{MPP}}^2 \tag{59}$$

5 Harris Hawks Optimization (HHO)

Harris hawks optimization algorithm is a nature-inspired algorithm which is recently proposed by Heidari et al. [29] in 2018. Inspiration of this algorithm comes from cooperative behaviour, surprise pounce and chasing style of Harris hawks. In this algorithm, two phases are involved, namely exploratory and exploitative phases which can be shown in Fig. 4.

5.1 Exploration Phase

In this subsection, exploration mechanism is explained. Harris hawks normally track and then detect the prey by their high-powered eyes, but prey cannot seen easily so hawks wait, observe the prey, after then monitor and at last attack the prey. Here in HHO candidate solutions are Harris hawks and the best solution in each step is considered as near to optimum position. Now, mathematical formulation is shown as:

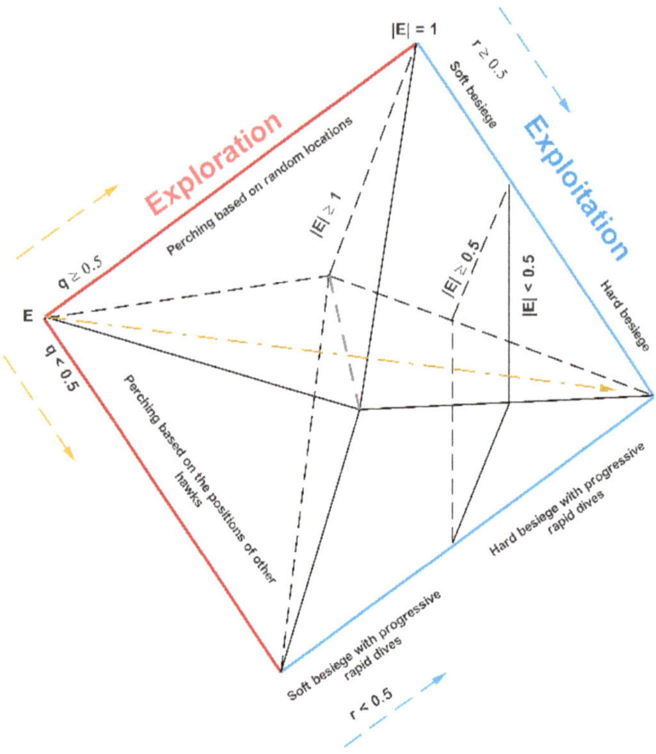

Fig. 4 Different phases of Harris hawks optimization

$$Y(t+1) = \begin{cases} Y_{\text{rand}}(t) - r_1 |Y_{\text{rand}}(t) - 2r_2 Y(t)| & q \geq 0.5 \\ (Y_{\text{rabbit}}(t) - Y_m(t)) - r_3(\text{LB} + r_4(\text{UB} - \text{LB})) & q < 0.5 \end{cases} \quad (60)$$

where in the next iteration t the position of hawks is $Y(t + 1)$, position of rabbit is Y_{rabbit}, Y_t is the current position of hawks and r_1; r_2; r_3; r_4 and q are the random numbers in between the range of 0 and 1. The average position of hawks is described as:

$$Y_m(t) = \frac{1}{N} \sum_{t=1}^{N} Y_i(t) \quad (61)$$

where the location of each hawk is indicated by $Yi(t)$ in iteration t and the total number of hawks is denoted by N.

5.2 Exploration to Exploitation Transition

The exploratory behaviour can be transferred into exploitative behaviour based on escaping energy of prey which can change between numerous exploitative behaviours. Significantly during the escaping behaviour, the energy of a prey diminishes. To show this fact, prey's energy is modelled as:

$$E = 2E_O\left(1 - \frac{t}{T}\right) \tag{62}$$

where E denoted the escaping energy of the prey, initial state of energy is indicated by E_O and the maximum iteration is shown as T.

5.3 Exploitation Phase

In this subsection, surprise pounce is performed by Harris hawks by attacking the prey which was observed in previous phase and on other side prey tries to escape from unsafe places. Hence, in this part several chasing techniques occur in real situations. In these, four main strategies are defined which are as follows:

1. Soft besiege
2. Hard besiege
3. Soft besiege with progressive rapid dives
4. Hard besiege with progressive rapid dives.

6 Results

6.1 Results for Single-diode Model

In this part, for numerous solar cells simulation results are analysed and summarized with the application of HHO algorithm. Comparison of HHO is conducted with sine cosine algorithm (SCA) [30], grey wolf optimizer (GWO) [31], ant lion optimizer (ALO) [32], whale optimization algorithm (WOA) [33], salp swarm algorithm (SSA) [34] and crow search algorithm (CSA) [35]. The values of search agents and maximum iterations are kept same for fair comparison of all optimization techniques. The convergence curves of HHO for single-diode model and double-diode models are shown in Fig. 11 with other metaheuristic algorithms. Convergence curve shows the behaviour of the algorithm, and with this it can be concluded that HHO algorithm converges faster than other competitor algorithms. The extracted parameter values of single-diode model for monocrystalline cell, polycrystalline cell and thin film have been shown in Table 3. Also, the error analysis exhibits in Table 5.

Table 3 Parameters of solar cell for single-diode model

Module	Parameters	SCA [30]	GWO [31]	ALO [32]	WOA [33]	SSA [34]	CSA [35]	HHO
Thin film	ψ	1.11E+00	1.43E+00	1.82E+00	1.63E+00	1.61E+00	1.57E+00	1.24E+00
	R_{SE}	2.01E−01	3.82E−01	7.97E−01	6.58E−01	6.97E−01	5.45E−01	5.88E−01
	R_{SH}	7.08E+01	9.24E+01	1.75E+02	1.10E+02	1.18E+02	8.65E+01	7.97E+01
	I_{rac}	1.07E−06	2.73E−06	5.47E−06	3.16E−06	2.70E−06	1.08E−06	8.62E−07
	I_{dmc}	2.69E+00	2.69E+00	2.69E+00	2.70E+00	2.70E+00	2.70E+00	2.70E+00
Polycrystalline cell	ψ	1.24E+00	1.09E+00	1.10E+00	1.06E+00	9.96E−01	1.57E+00	1.10E+00
	R_{SE}	5.56E−02	8.44E−02	1.45E−01	1.51E−01	2.90E−01	5.45E−01	1.88E−01
	R_{SH}	8.87E+01	6.40E+01	8.13E+01	8.53E+01	1.15E+02	8.65E+01	1.04E+02
	I_{rac}	9.66E−07	2.54E−07	5.96E−07	1.41E−07	3.42E−07	1.08E−06	1.25E−07
	I_{dmc}	8.22E+00	8.22E+00	8.23E+00	8.22E+00	8.24E+00	2.70E+00	8.23E+00
Monocrystalline cell	Ψ	1.70E+00	1.61E+00	1.58E+00	1.51E+00	1.38E+00	1.36E+00	1.51E+00
	R_{SE}	1.11E−01	1.73E−01	1.42E−01	2.16E−01	2.99E−01	3.33E−01	2.21E−01
	R_{SH}	1.13E+02	1.23E02	9.79E+01	1.23E+02	1.15E+02	1.32E+02	1.14E+02
	I_{rac}	1.72E−05	1.70E−05	1.04E−05	7.59E−06	7.31E−06	6.02E−07	7.17E−06
	I_{dmc}	5.46E+00	5.46E+00	5.46E+00	5.46E+00	5.47E+00	5.46E+00	5.46E+00

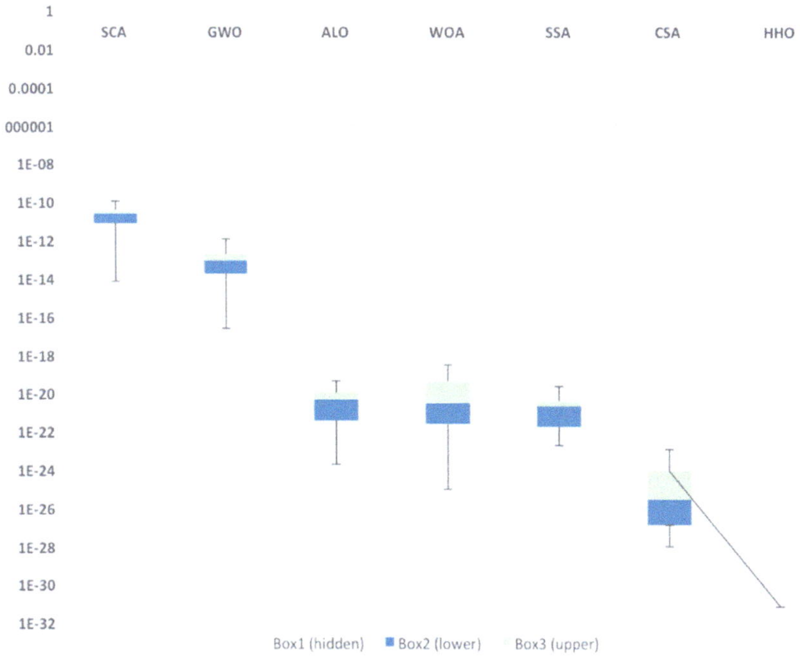

Fig. 5 Box plot of single-diode model thin film

Error analysis is presented in terms of mean, standard deviation (SD), min and max, obtained from the optimization process. Error's mean value of HHO algorithm for single-diode model reaches 10^{-32} which is the best obtained results among all other compared algorithms as per in Table 5. Standard deviation values are also very optimal which is a good indicator of better solution quality. Table 7 determines the *t*-test results. *T*-test finds out the significant difference in the mean values of two groups. Table 8 shows the *t*-test2 results which is a hypothesis test. Box plots of thin-film, polycrystalline and monocrystalline solar cell of single-diode model are presented in Figs. 5, 6 and 7, respectively. Data with box plots drawn are shown in Table 6 for both single-diode model and double-diode model.

6.2 Results for Double-diode Model

Similarly, in double-diode model five parameters are extracted and remaining two are calculated. Table 4 presents the extracted parameter values as well as calculated parameter values obtained for three PV cells, i.e., monocrystalline cell, polycrystalline cell and thin film. Comparison of error analysis is shown in Table 5 in which also the mean and standard deviation values are very optimal than other algorithms. Box plot diagrams of double-diode model are shown in Figs. 8, 9 and 10 for

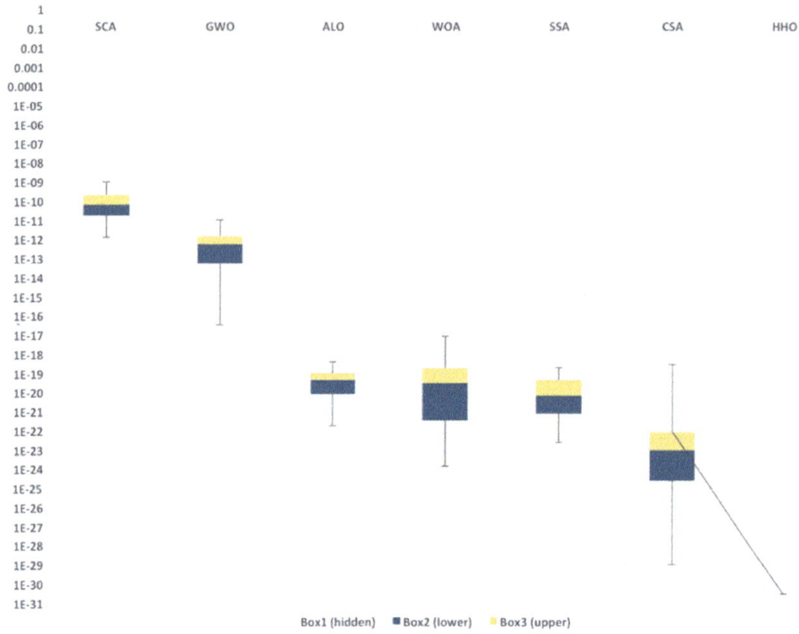

Fig. 6 Box plot of single-diode model polycrystalline film

monocrystalline solar cell, polycrystalline solar cell and thin film, respectively, and the box plot is shown in Table 6. For double-diode model, *t*-test and *t*-test2 results are shown in Table 7 and in Table 8, respectively (Fig. 11).

7 Results

This chapter is an effort to report the application of a recently developed Harris hawks optimization algorithm for carrying out the optimal solution of solar cell parameters. Parameter extraction has been done using data-sheet information at three different crucial points of I–V characteristic of solar cells. Detailed implementation of HHO has been explained, and obtained results are compared with different well-known algorithms. After carrying out the comparison of results obtained using HHO, it is revealed that the algorithm proves to be an optimal solution provider than other compared algorithms in terms of convergence characteristics and solution quality. The effectiveness of HHO has been shown by the convergence curve of error values. Box plots have also been shown for all three different cells for single-diode model and double-diode models.

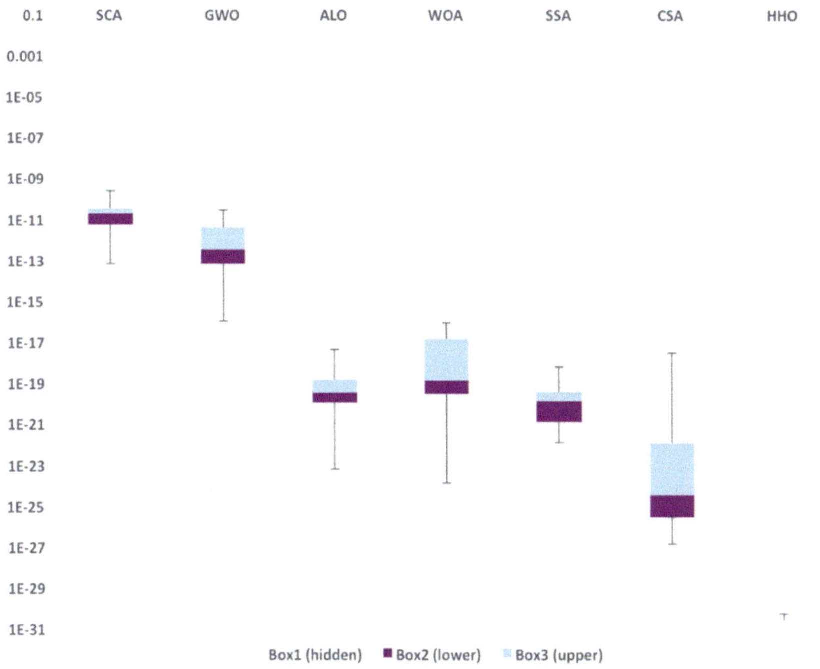

Fig. 7 Box plot of single-diode model monocrystalline film

Table 4 Double-diode model of solar cell parameters

Module	Parameters	SCA [30]	GWO [31]	ALO [32]	WOA [33]	SSA [34]	CSA [35]	HHO
Thin film	I_{rsc1}	2.81E−08	1.80E−07	8.29E−07	6.61E−07	5.46E−07	4.38E−07	4.63E−07
	Ψ_1	1.61E+00	1.83E+00	1.93E+00	1.88E+00	1.81E+00	1.77E+00	1.84E+00
	Ψ_2	1.02E+00	1.43E+00	1.84E+00	1.65E+00	1.70E+00	1.47E+00	1.55E+00
	R_{SE}	1.36E−01	4.60E−01	8.84E−01	7.37E−01	6.52E−01	6.09E−01	6.26E−10
	R_{SH}	6.59E+01	9.52E+01	1.80E+02	1.32E+02	1.27E+02	8.96E+01	1.06E+02
	I_{rsc2}	1.16E−06	2.23E−06	3.67E−06	2.89E−06	1.66E−06	4.40E−07	1.78E−06
	I_{dmc}	2.69E+00	2.69E+00	2.69E+00	2.70E+00	2.69E+00	2.70E+00	2.70E+00
Polycrystalline cell	I_{rsc1}	7.76E−08	3.47E−07	4.70E−07	5.22E−07	5.12E−07	5.04E−07	3.87E−07
	Ψ_1	1.65E+00	1.79E+00	1.77E+00	1.60E+00	1.76E+00	1.68E+00	1.64E+00
	Ψ_2	1.29E+00	9.34E−01	9.64E−01	1.03E+00	1.02E+00	1.08E+00	9.42E−01
	R_{SE}	9.84E−03	1.72E−01	2.62E−01	1.66E−01	2.18E−01	2.30E−01	2.13E−01
	R_{SH}	7.88E+01	7.63E+01	1.22E+02	1.22E+02	1.14E+02	1.26E+02	1.10E+02
	I_{rsc2}	5.62E−07	1.01E−07	3.07E−07	3.13E−07	4.50E−07	1.62E−08	2.58E−07
	I_{dmc}	8.21E+00	8.23E+00	8.23E+00	8.22E+00	8.23E+00	8.23E+00	8.23E+00
Monocrystalline cell	I_{rsc1}	6.49E−08	2.83E−07	3.66E−07	4.12E−07	4.59E−07	4.50E−07	4.73E−07
	Ψ_1	1.71E+00	1.79E+00	1.72E+00	1.75E+00	1.69E+00	1.68E+00	1.64E+00
	Ψ_2	1.91E+00	1.66E+00	1.45E+00	1.19E+00	1.26E+00	1.28E+00	1.22E+00
	R_{SE}	1.09E−02	1.29E−01	2.14E−01	3.34E−01	3.00E−01	3.30E−01	2.89E−01
	R_{SH}	1.01E+02	1.25E+02	1.18E+02	1.24E+02	1.10E+02	1.24E+02	1.10E+02
	I_{rsc2}	2.26E−05	1.53E−05	1.02E−05	2.05E−06	2.26E−06	1.03E−06	1.91E−06
	I_{dmc}	5.45E+00	5.46E+00	5.46E+00	5.47E+00	5.47E+00	5.47E+00	5.47E+00

Table 5 Comparative error analysis of single-diode and double-diode models

Single-diode model		SCA [30]	GWO [31]	ALO [32]	WOA [33]	SSA [34]	CSA [35]	HHO
Thin film	Min	6.39E−15	2.14E−17	1.76E−24	8.88E−26	1.66E−23	9.01E−29	0.00E+00
	Max	1.04E−10	1.11E−12	4.10E−20	2.73E−19	2.16E−20	1.09E−23	6.57E−32
	Mean	2.86E−11	1.80E−13	9.33E−21	3.80E−20	3.62E−21	1.25E−24	9.86E−33
	SD	2.86E−11	2.75E−13	1.25E−20	7.42E−20	5.59E−21	2.88E−24	2.41E−32
Polycrystalline cell	Min	1.16E−12	3.01E−17	1.65E−22	1.28E−24	2.24E−23	9.47E−30	0.00E+00
	Max	9.41E−10	9.37E−12	3.59E−19	7.65E−18	1.89E−19	2.81E−19	2.63E−31
	Mean	1.46E−10	1.33E−12	6.80E−20	6.37E−19	3.12E−20	1.94E−20	3.94E−32
	SD	2.20E−10	2.38E−12	8.62E−20	1.77E−18	5.17E−20	6.60E−20	9.63E−32
Monocrystalline cell	Min	5.90E−14	9.17E−17	5.79E−24	1.20E−24	1.12E−22	1.25E−27	0.00E+00
	Max	2.21E−10	2.51E−11	3.71E−18	8.22E−17	5.98E−19	2.81E−18	5.26E−31
	Mean	3.16E−11	3.09E−12	2.72E−19	1.04E−17	5.11E−20	1.46E−19	1.31E−31
	SD	5.00E−11	6.06E−12	8.19E−19	2.16E−17	1.33E−19	6.27E−19	1.60E−31
Double-diode model		SCA	GWO	ALO	WOA	SSA	CSA	HHO
Thin film	Min	4.84E−14	8.01E−17	2.97E−23	5.30E−26	5.99E−24	2.37E−30	0.00E+00
	Max	3.80E−10	6.28E−13	7.01E−19	1.64E−17	2.09E−17	6.77E−20	0.00E+00
	Mean	8.86E−11	1.26E−13	4.81E−20	1.41E−18	1.11E−18	3.64E−21	0.00E+00
	SD	1.09E−10	1.73E−13	1.55E−19	3.89E−18	4.66E−18	1.51E−20	0.00E+00
Polycrystalline cell	Min	6.16E−14	4.22E−20	1.16E−22	3.47E−21	1.44E−24	2.53E−27	0.00E+00
	Max	3.97E−09	8.37E−12	8.17E−19	1.76E−16	1.05E−18	3.00E−19	2.63E−31

(continued)

Table 5 (continued)

Double-diode model

		SCA	GWO	ALO	WOA	SSA	CSA	HHO
	Mean	6.60E−10	1.45E−12	1.19E−19	1.18E−17	1.19E−19	1.88E−20	6.57E−32
	SD	1.04E−09	2.20E−12	1.95E−19	3.92E−17	2.39E−19	6.69E−20	1.17E−31
Monocrystalline cell	Min	5.47E−14	2.28E−16	2.27E−23	1.05E−22	1.50E−23	2.53E−28	0.00E+00
	Max	1.18E−09	1.87E−11	1.32E−18	1.28E−16	3.29E−19	7.98E−21	2.63E−31
	Mean	1.97E−10	3.18E−12	1.73E−19	1.39E−17	3.74E−20	5.95E−22	5.26E−32
	SD	2.93E−10	5.07E−12	3.41E−19	3.46E−17	8.05E−20	1.81E−21	1.08E−31

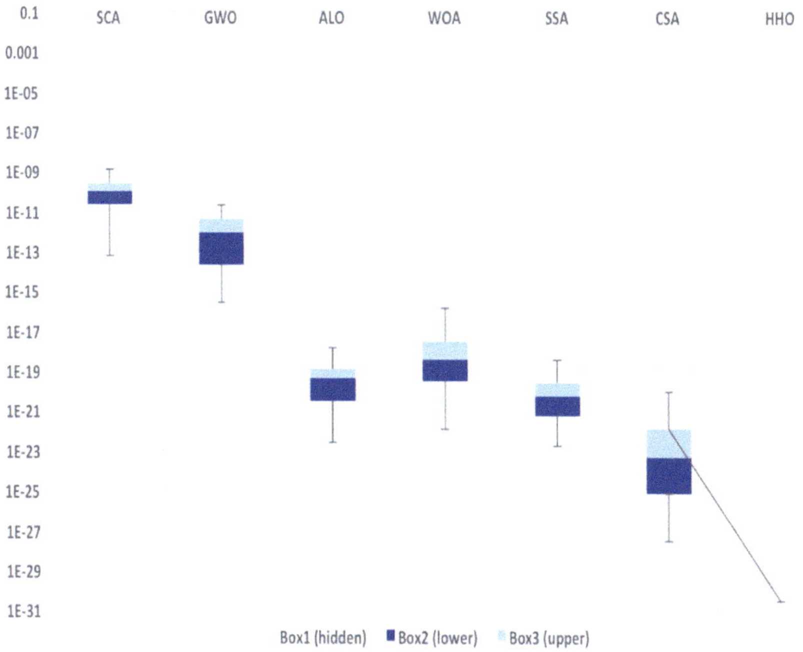

Fig. 8 Box plot of double-diode model monocrystalline film

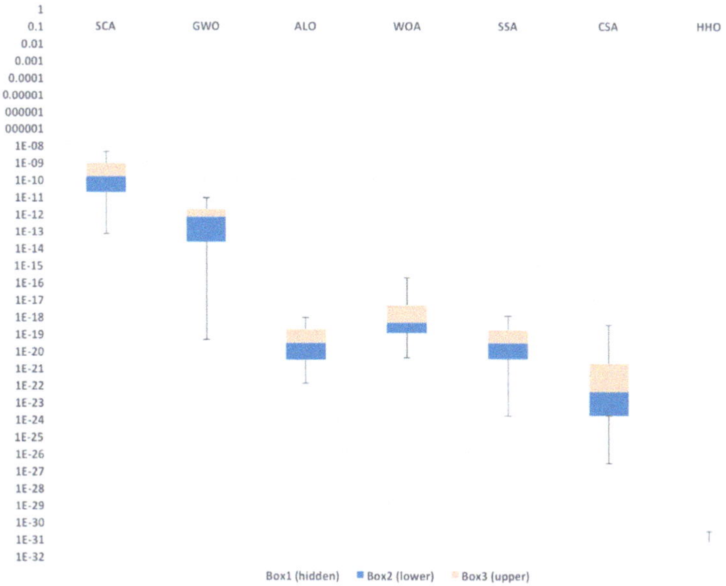

Fig. 9 Box plot of double-diode model polycrystalline film

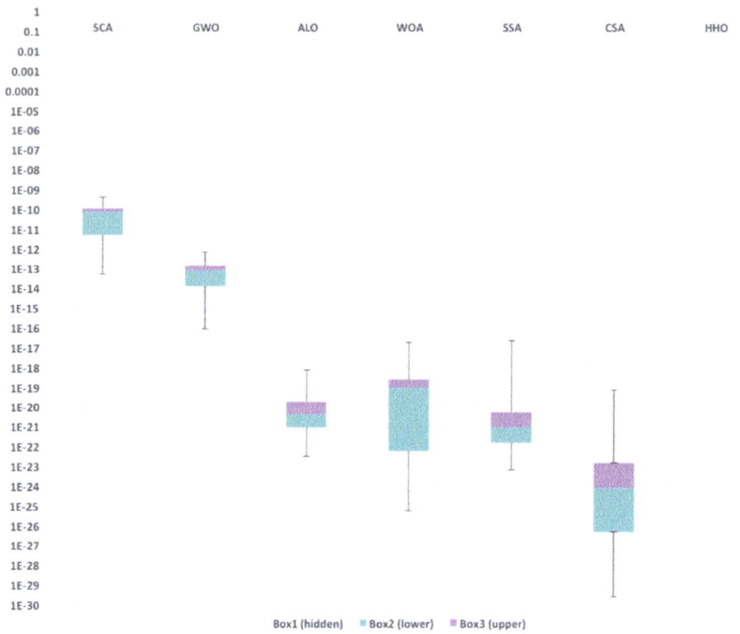

Fig. 10 Box plot of double-diode model thin film

Table 6 Single-diode model and double-diode model box plot

Single-diode model						Single-diode model					
Thin film						Thin film					
	Min	Q1	Median	Q3	Max		Min	Q1	Median	Q3	Max
SCA	6.39E−15	6.94E−12	2.24E−11	3.50E−11	1.04E−10	SCA	4.84E−14	4.66E−12	6.48E−11	1.04E−10	3.80E−10
GWO	2.14E−17	1.61E−14	8.32E−14	1.70E−13	1.11E−12	GWO	8.01E−17	1.12E−14	7.66E−14	1.33E−13	6.28E−13
ALO	1.76E−24	3.25E−22	4.34E−21	1.03E−20	4.10E−20	ALO	2.97E−23	8.26E−22	3.81E−21	1.73E−20	7.01E−19
WOA	8.88E−26	2.39E−22	2.76E−21	3.84E−20	2.73E−19	WOA	5.30E−26	5.50E−23	8.24E−20	2.30E−19	1.64E−17
SSA	1.66E−23	1.57E−22	2.00E−21	3.51E−21	2.16E−20	SSA	5.99E−24	1.40E−22	7.98E−22	5.02E−21	2.09E−17
CSA	9.01E−29	1.22E−27	2.79E−26	8.04E−25	1.09E−23	CSA	2.37E−30	4.41E−27	7.46E−25	1.38E−23	6.77E−20
HHO	0.00E+00	0.00E+00	0.00E+00	0.00E+00	6.57E−32	HHO	0.00E+00	0.00E+00	0.00E+00	0.00E+00	0.00E+00
Polycrystalline cell						Polycrystalline cell					
	Min	Q1	Median	Q3	Max		Min	Q1	Median	Q3	Max
SCA	1.16E−12	1.61E−11	5.86E−11	1.95E−10	9.41E−10	SCA	6.16E−14	1.77E−11	1.53E−10	8.07E−10	3.97E−09
GWO	3.01E−17	4.91E−14	4.99E−13	1.32E−12	9.37E−12	GWO	4.22E−20	2.12E−14	6.55E−13	1.72E−12	8.37E−12
ALO	1.65E−22	7.38E−21	4.24E−20	9.39E−20	3.59E−19	ALO	1.16E−22	2.68E−21	2.84E−20	1.71E−19	8.17E−19
WOA	1.28E−24	2.97E−22	2.86E−20	1.77E−19	7.65E−18	WOA	3.47E−21	1.02E−19	4.14E−19	4.30E−18	1.76E−16
SSA	2.24E−23	7.58E−22	6.63E−21	4.17E−20	1.89E−19	SSA	1.44E−24	2.92E−21	2.73E−20	1.60E−19	1.05E−18
CSA	9.47E−30	2.30E−25	9.09E−24	7.67E−23	2.81E−19	CSA	2.53E−27	1.53E−24	3.86E−23	1.65E−21	3.00E−19
HHO	0.00E+00	0.00E+00	0.00E+00	0.00E+00	2.63E−31	HHO	0.00E+00	0.00E+00	0.00E+00	6.57E−32	2.63E−31
Monocrystalline cell						Monocrystalline cell					
	Min	Q1	Median	Q3	Max		Min	Q1	Median	Q3	Max
SCA	5.90E−14	4.85E−12	1.70E−11	2.93E−11	2.21E−10	SCA	5.47E−14	1.98E−11	8.91E−11	2.19E−10	1.18E−09

(continued)

Table 6 (continued)

Monocrystalline cell						Monocrystalline cell					
	Min	Q1	Median	Q3	Max		Min	Q1	Median	Q3	Max
GWO	9.17E−17	5.78E−14	3.05E−13	3.55E−12	2.51E−11	GWO	2.28E−16	1.78E−14	7.35E−13	3.52E−12	1.87E−11
ALO	5.79E−24	9.77E−21	3.02E−20	1.32E−19	3.71E−18	ALO	2.27E−23	2.79E−21	3.85E−20	1.10E−19	1.32E−18
WOA	1.20E−24	2.52E−20	1.22E−19	1.25E−17	8.22E−17	WOA	1.05E−22	2.76E−20	3.17E−19	2.69E−18	1.28E−16
SSA	1.12E−22	1.09E−21	1.12E−20	3.21E−20	5.98E−19	SSA	1.50E−23	4.72E−22	4.68E−21	2.30E−20	3.29E−19
CSA	1.25E−27	2.68E−26	3.34E−25	1.13E−22	2.81E−18	CSA	2.53E−28	6.31E−26	4.22E−24	1.11E−22	7.98E−21
HHO	0.00E+00	0.00E+00	0.00E+00	2.63E−31	5.26E−31	HHO	0.00E+00	0.00E+00	0.00E+00	0.00E+00	2.63−31

Table 7 *t*-test results of single-diode model and double-diode model

Single-diode model

	Thin film		Polycrystalline cell		Monocrystalline cell	
	h-value	*p*-value	*h*-value	*p*-value	*h*-value	*p*-value
SCA	1	2.61E−04	1	8.02E−03	1	1.09E−02
GWO	1	8.53E−03	1	2.23E−02	1	3.41E−02
ALO	1	3.37E−03	1	2.25E−03	0	1.55E−01
WOA	1	3.35E−02	0	1.24E−01	1	4.46E−02
SSA	1	9.24E−03	1	1.41E−02	0	1.03E−01
CSA	0	6.64E−02	0	2.03E−01	0	3.10E−01

Double-diode model

	Thin film		Polycrystalline cell		Monocrystalline cell	
	h-value	*p*-value	*h*-value	*p*-value	*h*-value	*p*-value
SCA	1	1.72E−03	1	1.04E−02	1	7.22E−03
GWO	1	4.11E−03	1	8.37E−03	1	1.12E−02
ALO	0	1.81E−01	1	1.37E−02	1	3.55E−02
WOA	0	1.22E−01	0	1.93E−01	0	8.78E−02
SSA	0	3.00E−01	1	3.86E−02	0	5.15E−02
CSA	0	2.95E−01	0	2.24E−01	0	1.58E−01

Table 8 *t*-test2 results of single-diode model and double-diode model

Single-diode model

	Thin film		Polycrystalline cell		Monocrystalline cell	
	h-value	*p*-value	*h*-value	*p*-value	*h*-value	*p*-value
SCA	1	6.78E−05	1	0.005257	1	0.007582
GWO	1	0.00566	1	0.017365	1	0.028117
ALO	1	0.001839	1	0.001114	0	0.146668
WOA	1	0.027532	0	0.116285	1	0.037952
SSA	1	0.00622	1	0.010259	0	0.094731
CSA	0	0.058872	0	0.195333	0	0.303516

Double-diode model

	Thin film		Polycrystalline cell		Monocrystalline cell	
	h-value	*p*-value	*h*-value	*p*-value	*h*-value	*p*-value
SCA	1	0.000798	1	0.007124	1	0.004638
GWO	1	0.002344	1	0.005532	1	0.007807
ALO	0	0.173476	1	0.00986	1	0.02944
WOA	0	0.11358	0	0.184668	0	0.079886
SSA	0	0.293309	1	0.032322	1	0.044471
CSA	0	0.288318	0	0.216755	0	0.149697

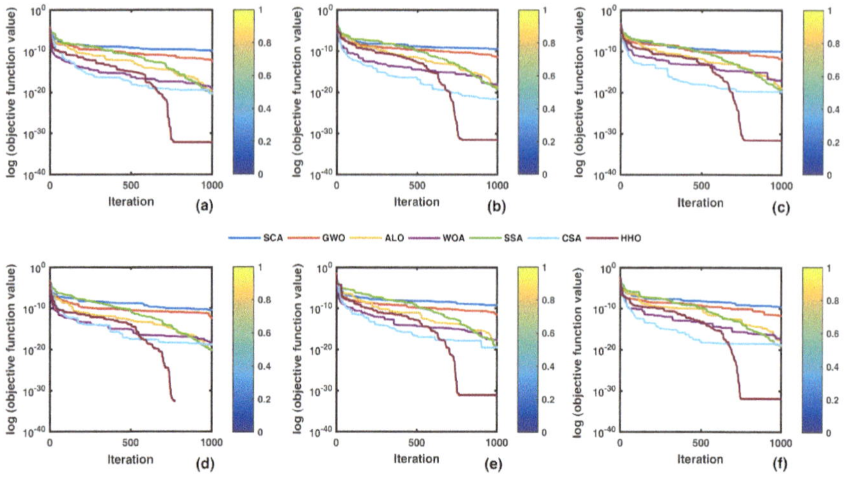

Fig. 11 Convergence curve of HHO and other compared algorithms

References

1. Oliva, D., Cuevas, E., Pajares, G.: Parameter identification of solar cells using artificial bee colony optimization. Energy **72**, 93–102 (2014)
2. Chauhan, A., Saini, R.P.: A review on integrated renewable energy system based power generation for stand-alone applications: configurations, storage options, sizing methodologies and control. Renew. Sustain. Energy Rev. **38**, 99–120 (2014)
3. Energy at the crossroads (2015). http://home.cc.umanitoba.ca/vsmil/pdf_pubs/oecd.pdf
4. Shafiee, S., Topal, E.: When will fossil fuel reserves be diminished? Energy Policy **37**(1), 181–189 (2009)
5. Apergis, N., Payne, J.E.: Renewable energy, output, CO_2 emissions, and fossil fuel prices in Central America: Evidence from a nonlinear panel smooth transition vector error correction model. Energy Econ. **42**, 226–232 (2014)
6. Shivalkar, R.S., Jadhav, H.T., Deo, P.: Feasibility study for the net metering implementation in rooftop solar PV installations across reliance energy consumers. In: 2015 International Conference on Circuits, Power and Computing Technologies [ICCPCT-2015], pp. 1–6. IEEE, New York (2015)
7. Wang, Y., Zhou, S., Huo, H.: Cost and CO_2 reductions of solar photovoltaic power generation in China: perspectives for 2020. Renew. Sustain. Energy Rev. **39**, 370–380 (2014)
8. Sundareswaran, K., Peddapati Sankar, P., Srinivasa Rao Nayak, Simon, S.P., Palani, S.: Enhanced energy output from a PV system under partial shaded conditions through artificial bee colony. IEEE Trans. Sustain. Energy **6**(1), 198–209 (2014)
9. The Solar Singularity is Nigh (2015). http://www.greentechmedia.com/articles
10. Dusonchet, L., Telaretti, E.: Economic analysis of different supporting policies for the production of electrical energy by solar photovoltaics in western European Union countries. Energy Policy **38**(7), 3297–3308 (2010)
11. Solangi, K.H., Islam, M.R., Saidur, R., Rahim, N.A., Fayaz, H.: A review on global solar energy policy. Renew. Sustain. Energy Rev. **15**(4), 2149–2163 (2011)
12. Branker, K., Pearce, J.M.: Financial return for government support of largescale thin-film solar photovoltaic manufacturing in Canada. Energy Policy **38**(8), 4291–4303 (2010)

13. Campoccia, A., Dusonchet, L., Telaretti, E., Zizzo, G.: Comparative analysis of different supporting measures for the production of electrical energy by solar PV and Wind systems: four representative European cases. Sol. Energy **83**(3), 287–297 (2009)
14. Timilsina, G.R., Kurdgelashvili, L., Narbel, P.A.: Solar energy: markets, economics and policies. Renew. Sustain. Energy Rev. **16**(1), 449–465 (2012)
15. Mohamed, M.A., Eltamaly, A.M.: Modeling and Simulation of Smart Grid Integrated with Hybrid Renewable Energy Systems. Springer, Berlin (2018)
16. Deshmukh, M.K., Deshmukh, S.S.: Modeling of hybrid renewable energy systems. Renew. Sustain. Energy Rev. **12**(1), 235–249 (2008)
17. Shannan, N.M.A.A., Yahaya, N.Z., Singh, B.: Singlediode model and two-diode model of PV modules: A comparison. In: 2013 IEEE International Conference on Control System, Computing and Engineering, pp. 210–214. IEEE, New York (2013)
18. Ishaque, K., Salam, Z., Taheri, H.: Simple, fast and accurate two-diode model for photovoltaic modules. Sol. Energy Mater. Sol. Cells **95**(2), 586–594 (2011)
19. Shekhawat, S., Saxena, A.: Development and applications of an intelligent crow search algorithm based on opposition based learning. ISA Trans. (2019)
20. Saxena, A.: A comprehensive study of chaos embedded bridging mechanisms and crossover operators for grasshopper optimisation algorithm. Expert Syst. Appl. **132**, 166–188 (2019)
21. Saxena, A., Shekhawat, S., Kumar, R.: Application and development of enhanced chaotic grasshopper optimization algorithms. Modell. Simul. Eng. 14 p (2018)
22. Saxena, A., Kumar, R., Mirjalili, S.: A harmonic estimator design with evolutionary operators equipped grey wolf optimizer. Expert Syst. Appl. **145**, 113125 (2020)
23. Appelbaum, J., Peled, A.: Parameters extraction of solar cells–A comparative examination of three methods. Sol. Energy Mater. Sol. Cells **122**, 164–173 (2014)
24. Bai, J., Liu, S., Hao, Y., Zhang, Z., Jiang, M., Zhang, Y.: Development of a new compound method to extract the five parameters of PV modules. Energy Convers. Manag. **79**, 294–303 (2014)
25. Alam, D.F., Yousri, D.A., Eteiba, M.B.: Flower pollination algorithm based solar PV parameter estimation. Energy Convers. Manag. **101**, 410–422 (2015)
26. Chin, V.J., Salam, Z., Ishaque, K.: Cell modelling and model parameters estimation techniques for photovoltaic simulator application: A review. Appl. Energy **154**, 500–519 (2015)
27. Jordehi, A.R.: Parameter estimation of solar photovoltaic (PV) cells: A review. Renew. Sustain. Energy Rev. **61**, 354–371 (2016)
28. Jamadi, M., Merrikh-Bayat, F., Bigdeli, M.: Very accurate parameter estimation of single-and double-diode solar cell models using a modified artificial bee colony algorithm. Int. J. Energy Environ. Eng. **7**(1), 13–25 (2016)
29. Heidari, A.A., Mirjalili, S., Faris, H., Aljarah, I., Mafarja, M., Chen, H.: Harris hawks optimization: algorithm and applications. Future Gener. Comput. Syst. **97**, 849–872 (2019)
30. Mirjalili, S.: SCA: a sine cosine algorithm for solving optimization problems. Knowl.-Based Syst. **96**, 120–133 (2016)
31. Mirjalili, S., Mirjalili, S.M., Lewis, A.: Grey wolf optimizer. Adv. Eng. Software **69**, 46–61 (2014)
32. Mirjalili, S.: The ant lion optimizer. Adv. Eng. Softw. **83**, 80–98 (2015)
33. Mirjalili, S., Lewis, A.: The whale optimization algorithm. Adv. Eng. Softw. **95**, 51–67 (2016)
34. Mirjalili, S., Gandomi, A.H., Mirjalili, S.Z., Saremi, S., Faris, H., Mirjalili, S.M.: Salp swarm algorithm: a bio-inspired optimizer for engineering design problems. Adv. Eng. Software **114**, 163–191 (2017)
35. Askarzadeh, A.: A novel metaheuristic method for solving constrained engineering optimization problems crow search algorithm. Comput. Struct. **169**, 1–12 (2016)

Classification of Microsoft Office Vulnerabilities: A Step Ahead for Secure Software Development

Supriya Raheja and Geetika Munjal

Abstract Classification of software vulnerability no doubt facilitates the understanding of security-related information and accelerates vulnerability analysis. In the absence of proper classification, it hinders its understanding and also renders the strategy of developing mitigation mechanism for clustered vulnerabilities. Now, software developers and researchers have agreed on the fact that incorporating security in software engineering phases including requirement and design may yield maximum benefits. In this paper, we have attempted to classify software vulnerabilities of Microsoft Office, so that this can help in building secure software. Vulnerabilities are firstly classified on well-established security properties like authentication and authorization. Vulnerability data is collected from various authentic sources, including Common Weakness Enumeration (CWE) and Common Vulnerabilities and Exposures (CVE). From these databases, only those vulnerabilities are included whose mitigation is possible at the design phase. Then, this vulnerability data is preprocessed to handle missing values and noise removal; further, the data is classified using various supervised machine learning techniques. Classification models are compared for three security metrics: integrity, confidentiality and availability. All the classifiers achieved highest accuracy for integrity.

Keywords Vulnerabilities · Software flaws · Classification · Microsoft Office vulnerabilities · Analysis

1 Introduction

Software development is a complex process. Due to the high customer demand, design and program complexity of a software increase which creates difficulties for a software developer. Program complexity subsequently increases the security flaws.

S. Raheja (✉) · G. Munjal
The Amity University, Noida, India
e-mail: supriya.raheja@gmail.com

G. Munjal
e-mail: munjal.geetika@gmail.com

A. K. Bhoi et al. (eds.), *Bio-inspired Neurocomputing*, Studies in Computational Intelligence 903, https://doi.org/10.1007/978-981-15-5495-7_21

The use of common and familiar code in the software development also increases the probability of exploitation by the attackers as they have the expertise and tools to exploit the bugs present in the code. Sometimes, the developer leaves a software bug which may allow an attacker to exploit an application. Even, software vulnerabilities have adverse concerns that can aggravate the common software system misadventure. The exploitation of these software vulnerabilities may affect the survival of people which can have possibly terrible effects. In addition, the execution of vulnerable software can violate the security policy of an organization either implicitly or explicitly. So, in the current era, security flaws should be the major concern for the developers. But, most of the software developers compromise with the software vulnerabilities to achieve the organization's goal.

1.1 Vulnerability

Vulnerability is a flaw in the security of system. This fragility exists because of procedures followed in designing and implementation of security system or internal controls which permit an attacker to intentionally exploit the system. It compromises the security of the system or violates the system's security policy. A famous example of vulnerability is the Java vulnerability which occurred due to a coding error. In this, the software developer decided the package membership with the first component of the package name, and developer has restricted the package name with the first period in the full name, rather than the last period in the full name [1, 2].

Vulnerability consists of three basic elements:

i. Defect in the design of the system
ii. Attacker's access to the flaw
iii. Attacker's capability to exploit the flaw.

If an attacker wants to exploit vulnerability, then they must have at least one technique that can connect to weakness of the system [3].

1.1.1 Classification of Vulnerabilities

Vulnerabilities are classified into different classes based on the asset class; they are belongings such as personnel, physical site, organizational, hardware, network and software.

- Personnel class: includes vulnerabilities due to improper recruitment process and inadequate security knowledge.
- Physical site class: includes vulnerabilities based on the area subject to natural disasters and unpredictable power outages.
- Organizational class: includes vulnerabilities due to lack of improper audits and the absence of continuous planning about security.

- Hardware class: includes vulnerabilities which can be prone to humidity, easily prone to dust, soiling and malicious storage.
- Network class: contains vulnerabilities due to unprotected communication lines and insecure network architecture.
- Software class: includes vulnerabilities due to insufficient testing, lack of audit trail and many causes due to which these vulnerabilities occur like complexity, familiarity, connectivity, password management flaws, software bugs, unchecked user input, OS design flaws and many more [4].

In this work, the focus is on software-based vulnerabilities. The main sources of these vulnerabilities are software bugs, insufficient testing, unchecked user inputs, lack of audit trail, etc. [2]. Finding of vulnerabilities becomes an everlasting issue in all kind of software products. In 2018 only, a total of 16,555 vulnerabilities are in the record of CVE [5], in which more than 10,000 were found due to software bugs only. Even, in the last few years, there is an increase in data of undisclosed or zero-day [5] vulnerabilities. Software developers cannot ignore the publicly published vulnerabilities as they are still creating the highest threat for software companies. As stated by the SANS report, most of the security breaches were exploited because of known vulnerabilities [3].

The security of user data majorly depends on the security of product which they are using. Attackers may compromise the user data by exploiting the vulnerabilities. There are various software products which are mainly preferred by the users as well as by developers. The Microsoft software products are one of them. So, the exploitation of these software products may affect the mass users. Like, if a user works with a susceptible version of Microsoft Office, malicious program present in the file may exploit the vulnerability which further infects the user's document. Microsoft security development [6] team highlighted the following security-related issues: (i) choice of granularity, which is actually involved in data collection and making prediction based on it. It also focuses on multi-level prediction as compared to binary-level because binary-level prediction model adds little knowledge thus by adding more security defects. (ii) Statistical learner choice, it helps in choosing learning method which is even more important than 'data.' (iii) Classification performance, it suggests that performance in terms of precision and recall with threshold of '0.7' is fair for prediction models. This chapter addresses the software vulnerabilities of the most usable product of Microsoft, i.e., Microsoft Office.

The most preferable base for vulnerability database is the Common Vulnerabilities and Exposures (CVE) by researchers. This database is maintained by the National Institute of Standards and Technology (NIST) [7]. The usage of such databases like CVE makes the software developers, security experts, security vendors and research-based organizations to use the vulnerability data proficiently. It helps the software developers to recognize the infected software products in terms of particular vulnerability from the CVE database. But it is a tremendous task for the developer to identify and search the vulnerability from the list.

A software developer must give emphasis on the secure code which can reduce the chances of exploitation. Analysis of these vulnerabilities can provide a base to the software developers for enhancing the security of software product. In this work, we are emphasizing the Microsoft Office. Through the analysis and classification, developers can prioritize the vulnerabilities. The severity of any vulnerability can be acknowledged with the help of a vulnerability score defined by common vulnerability severity score (CVSS). These vulnerability scores are majorly influenced by three impact metrics, including integrity, confidentiality and availability. Being directly proportional to given metrics, reducing these metrics, one can reduce the vulnerability score. Therefore, knowledge of which vulnerability increases these factors may provide the scope of improvement to developers and also to prioritize their responses.

The prime objective of this work is to categorize different vulnerabilities to define their accuracy in terms of these three metrics. Authors have been considering the last twelve years database of Microsoft Office vulnerabilities (January 2006–June 2018). The other objective of this work is to provide the comparative evaluation of various classification techniques in analyzing the vulnerable data so that appropriate classification techniques can be used for the required task.

The rest of the chapter is organized as follows: In Sect. 2, we discuss the background and related work of software vulnerabilities. Section 3 discusses the Microsoft Office vulnerabilities and database used in detail. Section 4 discusses different types of classification techniques. The results of different classifiers are demonstrated in Sect. 5, and finally, conclusions and future scope are discussed in Sect. 6.

2 Background and Related Work

While developing a software program, change may introduce security vulnerabilities. Software development teams have made use of various strategies to verify if code contains security vulnerabilities; for example, the Microsoft has also introduced security development life cycle to enhance its product's security; however, for large software products, a single approach to fill security feature is not sufficient. Thus, the vulnerability detection and removal techniques using more effective techniques must be ordered to the most skeptical areas of the product. Generally, vulnerabilities have been classified into broad categories such as buffer overflows, format string vulnerabilities, code execution and memory corruption. [3]. There are two major issues associated with these broad categories: First, it is difficult for a developer to assign a vulnerability to a single category; second, the dissimilarities are too common to be useful for any analysis. Correct classification of vulnerabilities can lead to improved relationship between the incidents and their exploitations. This can help developers to determine the effectiveness of countermeasures; however, such analysis has not been often applied.

In one of the studies, authors [7] have proposed an ontology-based approach to retrieve vulnerability-related data and establish a correlation between them; the data is taken from the National Vulnerability Database [8]. Authors also provided machine recognizable CVE vulnerability knowledge and reusable security vulnerabilities interoperability. Authors proposed a classification framework which transforms the vulnerability dictionary into CVE. Authors have used clustering algorithm for analysis purpose [4]. Li et al. have explored other possibilities of classifying vulnerabilities based on the CVE using self-organizing maps [9].

A system is introduced to observe vulnerabilities of software used in organization [10]. This paper also explains the CPE dictionary which is used to identify the CPE identifier. These identifiers are consigned to software products based on the collected information. The CVE database is also used to recognize vulnerabilities in software based on the allotted CPE identifiers. The projected approach was not human interactive. Searching for CVEs by a software developer may lead to a poor efficacy.

Several researchers have worked on the identification and classification of software faults. Carlstead et al. conducted a research on operating system. Authors have worked on the protection errors in operating system [11]. Later, Landwher et al. [9] in 1993 have suggested how to enhance the security in the existing operating systems. Authors have published a list of security flaws in different operating systems and classified each flaw based on its genesis or on other factors. Marick [12] have presented a survey on different studies on software faults. Aslam et al. [13] in 1995 and then in 1996 have developed a classification scheme to understand the software faults that can challenge the security mechanisms used during development of software products. The proposed classification scheme has divided the software faults mainly into two broad categories: coding faults and emerging faults. Coding faults are due to syntactical errors, design errors or missing requirements, whereas emergent faults mainly occur because of improper installation of software. In this case, faults may present even when the software functions properly according to specifications. Bishop [14] and Bailey have shown that this classification technique does not provide the accurate results when a fault exists in more than one classification category. Leveson [15] has suggested that by understanding the nature of vulnerabilities, the system design can be improved which can further reduce the risk of running critical or sensitive systems.

Authors have discussed the characteristic-based classification for bugs [14], but this bug classification method was not effective to classify software vulnerabilities. In [11], authors have classified the vulnerabilities when they are introduced on the basis on software development life cycle. But the scheme was not effective; when the vulnerability belongs to more than one phase, then it was difficult to determine the specific class to which that belongs. In [16], authors have classified the vulnerabilities according to genesis to categorize them, if they are done for purpose or inadvertent. They further classified the intentional vulnerabilities into malicious or non-malicious. In [17], authors have classified the vulnerabilities on the basis of their immediate impact on software. Akyol presented the same work but specific for the network protocols.

In [16], authors have classified vulnerabilities based on their disclosure procedure. In authors have classified vulnerabilities in terms of impact, type of attack and source of vulnerability. Authors have classified vulnerabilities of mobile operating system based on elevation of privilege notions. Authors have applied machine learning models to classify vulnerabilities. In [18], authors have applied the Naïve Bayes algorithm to classify vulnerabilities in terms of text info. Further, in [19], LDA and SVM algorithms have been applied to classify vulnerabilities from National Vulnerability Database (NVD). Authors have proposed an approach for classification of software vulnerabilities based on different vulnerability characteristics including resources consumption, strict timing requirement, etc. They classify and give statistical analysis of the vulnerabilities. Massaccis specifically reported by Google Project Zero.

Various studies [20–22] compared the effectiveness of distinct modeling methods for fault prediction, and they have demonstrated that different modeling methods result in different prediction accuracies where accuracy results are dependent on the dataset. To find the impact of different modeling methods on Microsoft Office, we have applied several of these modeling techniques in our experiments. It will also help to evaluate if an algorithm choice makes a difference in terms of class prediction or not.

From the last some years, the software engineering community has increased the utilization of machine learning technology in software analysis [23, 24]. This will benefit the researchers in taking advantage from structure present in the data beyond the linear combinations modeled by regression. This also includes discovering new vulnerabilities by auditing; however, still lots of vulnerabilities are unidentified. For example, in [25], author has tried to related commits to CVEs. In [26], the author has worked on relatedness of commit message and issue reports where machine learning approach is used for natural language processing. The approach uses machine learning techniques for natural languages. Sabetta and Bezzi [27] identified usage of source code change along with commit message. In all the above studies [25, 26], the results have demonstrated high precision and low recall related to the commit message, thus highlighting the commit parameter in software vulnerability analysis.

Alhazmi and Malaiya [5] have built a model for future vulnerabilities where they targeted operating systems. The study is based on the logistic model that is tested on the existing data; the results are validated on the parameters including average error and average bias.

In one of the studies, several existing databases of vulnerability are compared on the basis of vulnerability features in each of them. In various databases, many important features are missing. For example, 'discovery date' is hard to find [28] in database.

There are quite a number of researches which are focusing on the classification of vulnerabilities to help out the software developers and even no one has explored particularly about the classification of Microsoft products. There are many more vulnerability classification approaches existing in the literature, but they are very general with respect to software vulnerabilities. Supriya et al. [27] have mentioned the classification of CVE-based vulnerabilities of Linux kernel. As per our literature survey, no one has ever classified the vulnerabilities of Microsoft Office which is majorly used by the users for their day-to-day task. Therefore, in this research work, authors are focusing on the classification of Microsoft Office (MSO) vulnerabilities using machine learning technique. For this work, authors have precisely considered the CVE database of Microsoft Office vulnerabilities. Accuracy of classification is defined in terms of three metrics: integrity, confidentiality and availability. In the next section, a detailed description of vulnerabilities and different classification techniques are discussed.

3 MS Office Vulnerabilities Dataset and Their Attributes

Database is the backbone of any system. Similarly, vulnerability databases are an important part of any software security knowledgebase. There are numerous vulnerability databases available which exchange software vulnerability information. However, the impact of categorization depends on the used vulnerability database. Extracting valuable and important information from these available databases is a tremendous task for a developer [10, 29–32]. Classifying these vulnerability databases assists developers to detect a mitigation method and also to prioritize and remediate vulnerabilities in real time. This research work classifies the Microsoft Office vulnerability based on three major security metrics using machine learning techniques; these metrics directly impact the vulnerability score. Severity of any software vulnerability depends on the vulnerability score, and these metrics are directly proportional to the vulnerability score. The vulnerability score can help the developers to identify the risk level of the vulnerability in the early stage of software development life cycle (SDLC).

3.1 Vulnerabilities in MS Office

MS Office documents are exposed to different types of vulnerabilities which can be exploited by attackers and intruders. From the database, authors have identified the main vulnerabilities of MS Office which are exploited in the last twelve years. Different exploited vulnerabilities are illustrated in Fig. 1. This section describes the each vulnerability as follows:

1. *Code Execution*: It is used to define the ability of an attacker to execute the instructions on a target machine. The attacker executes the commands of their own choice. It is a kind of software bug which provides a way to execute an arbitrary code. In most of the cases of exploitations, attackers inject and execute the shellcode to run arbitrary commands manually.
2. *Buffer overflow*: It is also known as buffer overrun. It is an anomaly program that allows a program to overrun the buffer's boundary in parallel to writing data in a buffer, and it also overwrites the adjacent memory locations. As a special case,

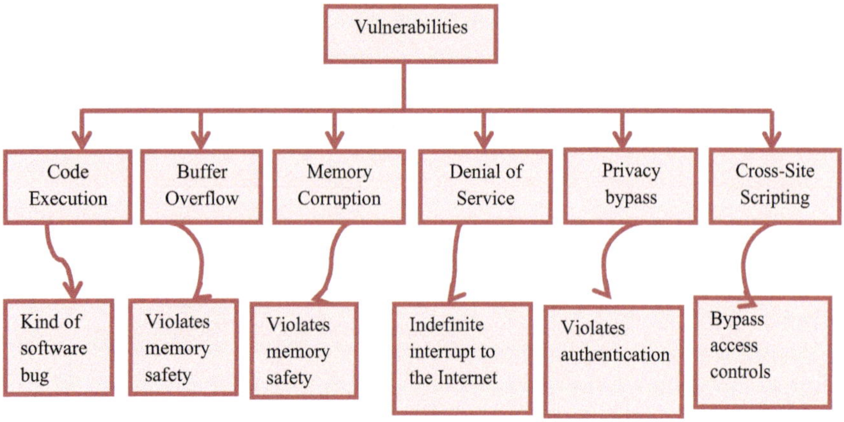

Fig. 1 Important vulnerabilities in MS Office since the last 12 years

buffer overflow is violation of memory safety. This vulnerability is exploited by inputs that are designed to execute code or to change normal execution of the program. This may result in incorrect results, memory access errors, a crash or a breach of system security. Therefore, this software vulnerability is majorly found in software products and can easily be maliciously exploited by an attacker [33].

3. *Memory Corruption*: The contents of memory locations are unintentionally modified due to the programming errors which lead the corruption of the memory in a computer program. Such type of errors violates memory safety. I n fiture, if the infected memory is used in the software, it may cause either to software crash or can reflect some kind of abnormal action. Around 10% of the software application on the Windows system fails due to memory corruption [34].
4. *Privacy Bypass*: Most of the applications require authentication to gain access on confidential data or private information. Based on studies, it can be concluded that there is not a single authentication procedure exists which provides adequate security. There are different ways by which authentication process can be easily bypassed by ignoring or neglecting the security threats. It can be bypassed by ignoring the login page or can directly call a page which allows to be accessed only after the valid authentication [3].
5. *Denial of Service*: It is an attempt made by attacker that makes a machine or network resource unavailable to the intended users such as to interrupt or suspend services of a host who is connected to the Internet. It is accomplished by flooding the targeted machine or resource to overload systems and to prevent all legitimate requests from being fulfilled [8].
6. *XXS*: Cross-site scripting vulnerability is used by attackers to bypass access controls. This scripting is carried out on Web sites. Their effect may range to significant security risk, depending on the sensitivity of the data which is handled by the vulnerable site and the nature of any security mitigation implemented by site's owner [35].

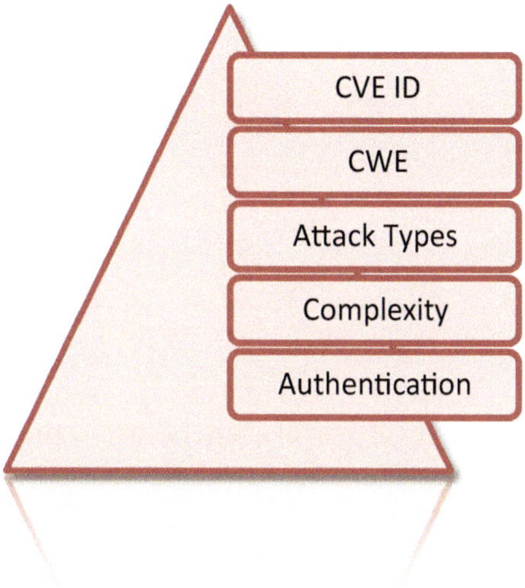

Fig. 2 Attributes used in dataset

All these types of vulnerabilities are exploited by the attackers and may result into risk for software. In the dataset, authors have represented these MS Office vulnerabilities under one attribute, i.e., attack type. This paper considers all important attributes which affect the impact of integrity, confidentiality and availability metrics as given in Fig. 2.

3.2 Dataset and Attributes Used

CVE is a process which is used to allocate unique numbers to widely well-known vulnerabilities present in software applications and to offer the vulnerability statistics in detail. It also provides information about the affected products. Nowadays, CVE is the standard among software companies, antivirus sellers, researchers and security experts to share the information about the known vulnerabilities. Assignment of unique identifiers, i.e., 'CVE ID,' is managed by the MITRE [19, 36, 37]. Figure 3 shows a CVE record acquired from the CVE database (CVE-2018-0950) for Microsoft Office 2018. NIST provides the CVE documents in NVD xml format which contains additional information like severity metrics, vulnerability type and affected software list. Figure 4 shows the severity and metrics of CVE feed (CVE-2018-0950). CVSS Base score is provided in NVD feeds to the vulnerability [33–35]. From Fig. 4, we can see that CVSS Base score is provided in NVD feeds to the vulnerability.

In this work, authors have taken the records of Microsoft Office vulnerabilities from the NVD feeds from the period 2006–2018 to analyze and classify the Microsoft Office vulnerabilities. In this work, authors have explored the vulnerability details of the vulnerabilities found in MS Office document. A sample snapshot of dataset is shown in Fig. 5. These attributes are briefly described in this section:

CVE-2018-0950 Detail

Current Description

An information disclosure vulnerability exists when Office renders Rich Text Format (RTF) email messages containing OLE objects when a message is opened or previewed, aka "Microsoft Office Information Disclosure Vulnerability." This affects Microsoft Word, Microsoft Office. This CVE ID is unique from CVE-2018-1007.

Source: MITRE

Fig. 3 CVE feed for the Microsoft Office 2018 [35]

Impact

CVSS v3.0 Severity and Metrics:

Base Score: 6.5 MEDIUM

Vector: AV:N/AC:L/PR:N/UI:R/S:U/C:H/I:N/A:N (V3.0 legend)

Impact Score: 3.6

Exploitability Score: 2.8

Attack Vector (AV): Network

Attack Complexity (AC): Low

Privileges Required (PR): None

User Interaction (UI): Required

Scope (S): Unchanged

Confidentiality (C): High

Integrity (I): None

Availability (A): None

Fig. 4 Severity and metrics for CVE feed (CVE-2018-0950) [35]

1	CVE ID	CWE ID	Attack Type(s)	Published Date	Update Date	Score	Gained Access	Access	Complexity	Authentication	Conf.	Integ.
2	CVE-2016-0145	119	Exec Code Overflow Mem. Corr.	4/12/2016	4/14/2016	9.3	None	Remote	Medium	Not required	Complete	Complete
3	CVE-2016-0134	119	Exec Code Overflow Mem. Corr.	3/9/2016	3/11/2016	9.3	None	Remote	Medium	Not required	Complete	Complete
4	CVE-2016-0127	119	Exec Code Overflow Mem. Corr.	4/12/2016	4/14/2016	9.3	None	Remote	Medium	Not required	Complete	Complete
5	CVE-2016-0057	264	Priv Bypass	3/9/2016	3/11/2016	7.2	None	Local	Low	Not required	Complete	Complete
6	CVE-2016-0056	119	Exec Code Overflow Mem. Corr.	2/10/2016	2/16/2016	9.3	None	Remote	Medium	Not required	Complete	Complete
7	CVE-2016-0055	119	Exec Code Overflow Mem. Corr.	2/10/2016	2/16/2016	9.3	None	Remote	Medium	Not required	Complete	Complete
8	CVE-2016-0053	119	Exec Code Overflow Mem. Corr.	2/10/2016	2/12/2016	9.3	None	Remote	Medium	Not required	Complete	Complete
9	CVE-2016-0052	119	Exec Code Overflow Mem. Corr.	2/10/2016	2/17/2016	9.3	None	Remote	Medium	Not required	Complete	Complete
10	CVE-2016-0012	200	Bypass +Info	1/13/2016	1/14/2016	4.3	None	Remote	Medium	Not required	Partial	None
11	CVE-2016-0010	119	Exec Code Overflow Mem. Corr.	1/13/2016	1/14/2016	9.3	None	Remote	Medium	Not required	Complete	Complete
12	CVE-2015-6172	20	Exec Code	12/9/2015	12/9/2015	9.3	None	Remote	Medium	Not required	Complete	Complete
13	CVE-2015-6124	119	Exec Code Overflow Mem. Corr.	12/9/2015	12/9/2015	9.3	None	Remote	Medium	Not required	Complete	Complete
14	CVE-2015-6118	119	Exec Code Overflow Mem. Corr.	12/9/2015	12/9/2015	9.3	None	Remote	Medium	Not required	Complete	Complete
15	CVE-2015-6108	119	Exec Code Overflow Mem. Corr.	12/9/2015	12/9/2015	9.3	None	Remote	Medium	Not required	Complete	Complete
16	CVE-2015-6107	119	Exec Code Overflow Mem. Corr.	12/9/2015	12/9/2015	9.3	None	Remote	Medium	Not required	Complete	Complete
17	CVE-2015-6106	119	Exec Code Overflow Mem. Corr.	12/9/2015	12/9/2015	9.3	None	Remote	Medium	Not required	Complete	Complete
18	CVE-2015-6093	119	Exec Code Overflow Mem. Corr.	11/11/2015	11/12/2015	9.3	None	Remote	Medium	Not required	Complete	Complete
19	CVE-2015-6092	119	Exec Code Overflow Mem. Corr.	11/11/2015	11/12/2015	9.3	None	Remote	Medium	Not required	Complete	Complete
20	CVE-2015-6091	119	Exec Code Overflow Mem. Corr.	11/11/2015	11/12/2015	9.3	None	Remote	Medium	Not required	Complete	Complete
21	CVE-2015-2545	20	Exec Code	9/8/2015	9/9/2015	9.3	None	Remote	Medium	Not required	Complete	Complete
22	CVE-2015-2510	119	Exec Code Overflow	9/8/2015	9/9/2015	9.3	None	Remote	Medium	Not required	Complete	Complete
23	CVE-2015-2477	119	Exec Code Overflow Mem. Corr.	8/14/2015	8/18/2015	9.3	None	Remote	Medium	Not required	Complete	Complete
24	CVE-2015-2470	189	Exec Code	8/14/2015	8/18/2015	9.3	None	Remote	Medium	Not required	Complete	Complete

Fig. 5 A snapshot of few entries of MSO vulnerability dataset

Table 1 Different CVSS scores of vulnerabilities with their state condition

Score range	State
1–3	Low
4–6	Medium
7–9	High
9–10	Critical

i. CVE ID: Common Vulnerability and Exposure. This is a unique identifier for public-known cybersecurity vulnerabilities.
ii. CWE ID: Common Weakness Enumeration. This includes the set of software weaknesses that can lead to software vulnerability and errors. This enumeration aims to stop the vulnerabilities at the source by educating software acquirers, designers, architects, etc.
iii. CVSS Score: Common Vulnerability Scoring System (CVSS) provides standardized vulnerability scores, i.e., when an algorithm is used for scoring the vulnerability as mentioned in Table 1. CVSS leverages a single vulnerability allowable time to validate and remediate a given vulnerability. It is also used to prioritize risk [19].
iv. Attack complexity: This shows that how much the conditions are beyond the attacker's control to exploit the vulnerability. This complexity has two possible values:

    (a) Low: This says that no such conditions exist. Successfully, vulnerability can be exploited.
    (b) High: Successful attack cannot be done. It might be target specific, and reconnaissance is required.

v. Confidentiality impact: This gives the measure of confidentiality information managed by software component due to successfully exploited vulnerability. The following are the values showing the impact of confidentiality:

    (a) Complete: total loss of confidentiality.
    (b) Partial: some loss of confidentiality but do not have full control.
    (c) None: There is no loss of confidentiality.

vi. Integrity Impact: This gives the measure of trustworthiness or veracity of information.

    (a) Complete: total loss of protection. Attacker can modify all files protected by component.
    (b) Partial: Modification is possible, but attacker does not have control over the consequences of modification. Data modification does not have serious impact on component.
    (c) None: There is no loss of integrity.

vii. Availability impact: Accessibility of information resources measures the impact of availability of the component.

(a) Complete: Total loss of an availability resulting in attacks is now completely able to deny access to resources.
(b) Partial: reduced interruption in resource availability
(c) None: There is no loss of availability.

4 Classification of Vulnerabilities

Categorizing vulnerabilities will help to explain different possible common vulnerability vectors and the features responsible for them. It can also help in detection and remediation of vulnerabilities. This section briefly discusses various classifiers used in the literature for various domains.

Classification is the task of obtaining the correct class label for an unknown sample [8, 38]. The set of labels is defined in advance, and each sample is considered in separation from the other samples. To match the test sample from the pre-stored labels from the samples in the dataset, many similarity measures are used such as the distance measure, information gain and entropy.

4.1 Bayes Classifiers

Some classifiers follow Bayes theory which is based on statistical classification. Using class belongingness probabilities, they can help in predicting class of data sample, which will further help in the classification process. This classifier follows Bayes' theorem. Naive Bayesian classifiers believe in contribution of class information about remaining in sustainment of participating attribute [38]. This property is known as class conditional independence. The computation involved in the process simplifies the task, in this sense, is considered 'naive.' Various Bayes classifiers are Bayes net and Naive Bayes [18, 21]. The probabilistic strategies which are counted under linear classifier are logistic regression and Naïve Bayes classifier. Naïve Bayes classifier acts as linear classifier, i f the values in likelihood factor $p\ (X_i|c)$ are exponents, where p is the probability of a single attribute 'i' from a set X attributes. $X_m = [x_1, x_2, x_3, \ldots, x_m]$ with 'm' representing count of attributes, c is one of the class or target labels from the set of 'k' class labels $\{c_1, c_2, c_3, \ldots, c_k\}$.

Naïve Bayes classifier assumes class conditional independence. The value of a feature about a class is independent of the values of the other features. It is based upon the principle of Bayes theorem. Bayes classifier uses the probabilistic approach to assign the class/target label to the new sample. Posterior conditional probability $p(c_i|X_m)$ in Bayesian classifier is computed using Bayes rule.

$$p(c_i|X_m) = \frac{p(X_m|c_i)p(c_i)}{p(X_m)} \tag{1}$$

Posterior probabilities are calculated in Eq. (1) for $i = 1, 2, 3, \ldots, m$ using the training data. Naïve Bayes classifier is easy and fast to predict class on test data even when the data is multi-class. Due to its simplicity, this method may outperform on complex models. However, it is quite sensitive to the presence of redundant and irrelevant predicted attributes.

4.2 K-Nearest Neighbor

Another type of classifiers widely used and tested in literature is the nearest neighbor classifier which employs normalized Euclidean distance to search the nearest sample (instance) to the sample in a test data from training samples. It predicts the same class as that given in a training sample. If there are many instances in the training data having the same distance, then the instance found first is used to predict the class. The classifiers which fall into this category are KNN and IB1. They both are instance-based learners.

K-nearest neighbor (KNN) is also said to be ***lazy learner*** which is sensitive to the local structure of the data and does not build model explicitly. Here, the value of k needs to be specified in advance. This approach becomes infeasible in case of high-dimensional data, and unknown samples become expensive [12].

Another nearest neighbor classifier is instance-based (IB1) classifier; they use specific instances during the prediction time of the classifier. To match the samples, they use similarity functions. Instance-based learning algorithms are based on the assumption that the similar samples (instances) have similar classifications. It leads to their local bias for classifying novel samples according to their most similar neighbor's classification. IB1 classifier can be used for two-class or multi-class datasets. It fails to classy suitably and gives less accurate results if the features are logically inadequate for describing the target class. IB1 classifier is useful for classification especially in feature selection problems, as it gives a better accuracy of the reduced data where features are adequate in describing the target class than the accuracy of complete data.

The performance of the nearest neighbor classifier is comparable of decision tree classifiers. It can interpret, generate and classify new instance from the model rapidly. In lazy classifiers, generalization of training data is delayed until a query is made to the system. The advantage of lazy classifier is that it performs approximation for target function locally, as the system needs to solve multiple problems simultaneously. After that, the system deals with the changes made in the problem domain. This classifier is successful for large datasets. IBK, Kstar and LWL are some algorithms of lazy classifier.

Tree-based classifiers are decision tree and random forest; both have proven to be robust and successful tools for solving many problems in the field of machine learning. Random forest can be used for two classes as well multi-class problems where the number of classes is more than two [22]. In this situation, there is little need to fine-tune the algorithmic parameters involved in random forest. However, random forest algorithm does not work well when there is extreme data imbalance in data. ***Decision tree*** generates a tree with a set of rules to represent different classes from the given dataset. Classifiers utilizing decision tree to classify an unknown sample generate the tree in two phases. One is the tree construction phase and other is tree pruning. In the construction phase, the tree keeps all the training examples at the root and then partitions them recursively based upon some criterion. Different approaches to construct a decision tree are Iterative Dichotomiser 3 (ID3), C4.5 a successor of ID3 presented by Quinlan and classification and regression trees (CART). All of them follow a top to bottom greedy approach to construct a tree. Training set attributes are recursively partitioned into smaller subsets as the tree keeps on building. The goal of the decision tree is to create a model that predicts the value of a target variable by learning simple decision rules inferred from the available attributes. This method is able to handle both numerical and categorical data and easily able to handle multi-output problems.

Meta-classifier uses automatic learning algorithms on meta-data. This meta-data is further used to know how the automatic learning can introduce flexibility in solving different learning problems which can further improve the performance of the existing learning algorithms [39]. Meta-classifiers has are good for on datasets with different levels of class imbalance. A meta-classifier 'bagging' is developed using several training datasets; these datasets are built from the original training dataset. It finds various models and averages them to produce a final ensemble design. Another meta-classifier is under-sampled stratified bagging where the total training set size is double the number of the minority class. Some meta-classifiers are cost sensitive.

Some meta-classifier algorithms in WEKA [40] are on the name Adaboost, bagging, stacking, vote, logic boost, etc. Another important classifier discussed in the literature is rule based. It follows technique of assigning each element of population set to one of the defined classes. If every element in the population is assigned to a class it really belongs, it will be considered as a perfect test. However, in case of errors, statistical techniques will be applied to validate the classification. Various rule-based algorithms are decision table, decision trees and random forest [12]. In this work, we have applied various types of classifier on MSO vulnerability dataset to analyze an impact of various metrics.

4.3 Evaluating Classifiers

The classification models are evaluated based on its accuracy, speed, robustness, scalability and interpretability [41].

The output of a classifier is taken in the form of a confusion matrix shown in Table 2 depicting the predicted class with true and false numbers.

False positives are the samples that are erroneously predicted as positive, but actually belong to a negative class. Similarly, false negatives are the samples that are wrongly predicted as negative, but truly belong to a positive class. True positives are the samples that are accurately predicted as positive and actually belong to the

Table 2 Confusion matrix

	Predicted class	
Actual class	Positive	Negative
Positive	True positive	False negative
Negative	False positive	True negative

positive class. Likewise, true negatives are the samples which are accurately predicted as negative and truly belong to the negative class. All these measures of the confusion matrix are further helpful in evaluating the accuracy, specificity and sensitivity of a classifier. Higher values of true positive and true negative show the effectiveness of the algorithm. All these parameters are validated based on a k cross-validation where the data is divided into k folds, where $k - 1$ folds are selected as the training data and the reserved fold is used to calculate accuracy, specificity and sensitivity.

Accuracy is the most common used performance measure. It evaluates the effectiveness by taking the ratio of correct prediction divided by total number of predictions made as in an Eq. (2).

$$\text{Acc} = \frac{\text{TN} + \text{TP}}{\text{TN} + \text{TP} + \text{FN} + \text{FP}} \tag{2}$$

Other performance measure is **error rate** which is the complement of accuracy rate. It evaluates the effectiveness by taking the ratio of incorrect prediction divided by total number of predictions made. **Recall** is also called as the sensitivity or true positive rate. It is defined as the ratio of the samples belonging to positive class which was predicted correctly as positive. **Specificity is** defined as the ratio of the samples belonging to negative class which was predicted correctly as negative. **Precision** is defined as the proportion of the samples belonging to positive class which was predicted correctly as positive.

Cross-validation technique is applied to validate the results; this technique splits the original data into one or more subsets [42, 43]. During training of the classifier, one of the subsets is used as test set to test the accuracy of the classifier and other subsets are used for parameter estimation. It keeps on rotating the test set based upon an applied cross-validation technique. Several types of cross-validation techniques can be used such as k-fold cross-validation and leave-one-out cross-validation.

All the experiments are done in Waikato Environment for Knowledge Analysis (WEKA) [21] which is a landmark in data mining including classification. It provides a comprehensive collection of machine learning algorithms. It allows users to explore various machine learning methods over naive datasets.

5 Performance Evaluations of Classifiers on MSO Vulnerabilities

Various classifiers are applied on MSO vulnerabilities dataset. Some classifiers possess similar accuracy in all of three security services and some differentiate. However, the highest accuracy is held by integrity. Here, accuracy representation for various classifiers individually depicts the accuracy for integrity, confidentiality and availability.

5.1 Using Bayes Classifiers

The Bayes classifier minimizes the probability of misclassification in statistical classification. This classifier possesses high scalability, which requires a number of parameters linear in the number of variables in a learning problem. Various versions of Bayes classifier are used from WEKA [21] including BayesNet, NaveBayes and Nave Bayes Updatable classifier. These classifiers depict the accuracy in terms of integrity, confidentiality and availability. Bayes net gave a highest accuracy for all three parameters as compared to nave Bayes and Nave Bayes updatable version as illustrated in Fig. 6.

5.2 Using Lazy Classifiers

The lazy classifiers have depicted accuracy as shown in Fig. 7. IBK algorithm of lazy classifier has shown 99% of accuracy in integrity, 98% for availability and

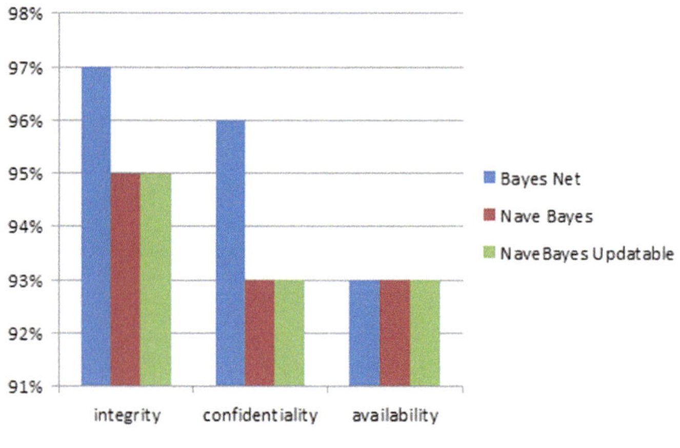

Fig. 6 Accuracy by Bayes classifier

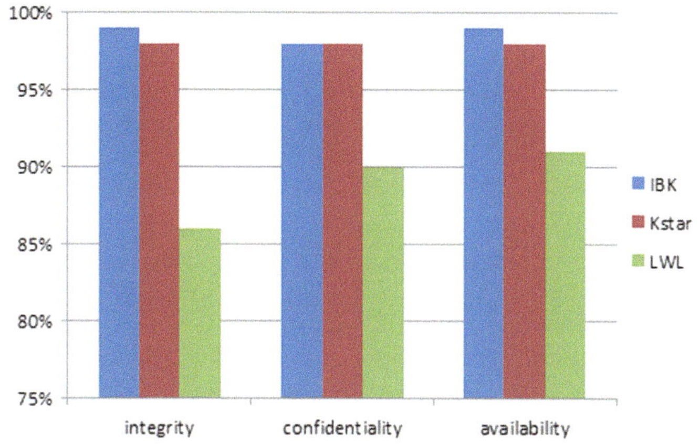

Fig. 7 Accuracy by lazy classifier

98% for confidentiality. KStar, performance is same for all three parameters, i.e., for integrity, confidentiality and availability. Locally weighted learning (LWL) is giving 91% accuracy for availability and 90% for confidentiality and 86% for integrity. These three algorithms are applied under cross-validation (CV) testing option to get validated results.

5.3 *Using Meta-Classifiers*

Various kinds of application are applied for meta-classifier as illustrated in Fig. 8. The algorithms applied are Adaboost M1, attribute selected, bagging, classification

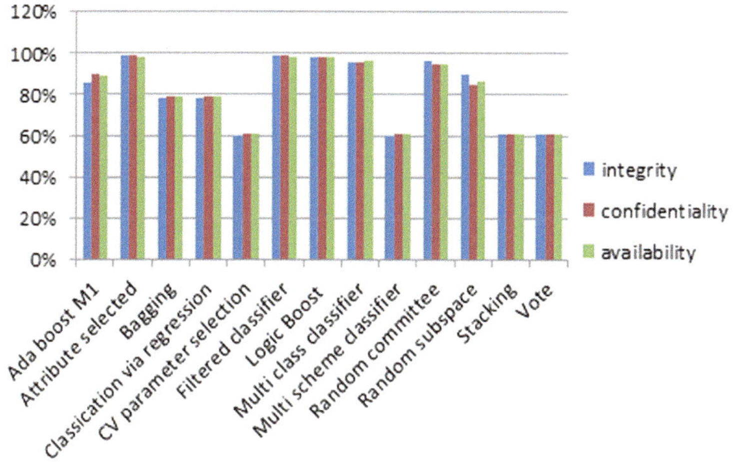

Fig. 8 Accuracy by meta-classifier

via regression, CV parameter selection, filtered classifier, logic boost, multi-class classifier, multi-scheme classifier, random committee, random subspace, stacking and vote. All these classifier algorithms are applied in this vulnerability. Results of these algorithms are represented in terms of accuracy for integrity, confidentiality and availability impact.

Filtered classifier and attribute selected possess 99% of accuracy in integrity and confidentiality. The logic boost depicts same amount of accuracy, i.e., 98% for integrity, confidentiality and availability. The bagging, classification via regression and CV parameter selection have 79% accuracy for confidentiality and availability and 79% for confidentiality and availability and 61% accuracy for confidentiality and availability. The random committee possesses 99% accuracy for integrity. Stacking and vote have 61% for all integrity, confidentiality and availability impact. This shows that there are various results for the classifier for meta, and majority is still with integrity.

5.4 *Using Rules Classifiers*

The classifier algorithms applied in rules are decision tables, JRip, OneR, PART and ZeroR as illustrated in Fig. 9. These depict the amount of accuracy for integrity, confidentiality and availability. In the rules classifier, decision table and PART have 99% of accuracy for integrity and JRip shows 99% of accuracy for confidentiality. ZeroR has 61% of accuracy for both confidentiality and availability. Still, the majority is with integrity.

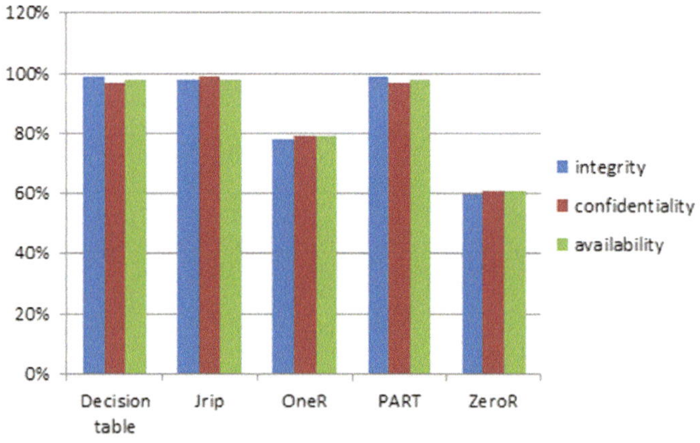

Fig. 9 Accuracy by rules classifier

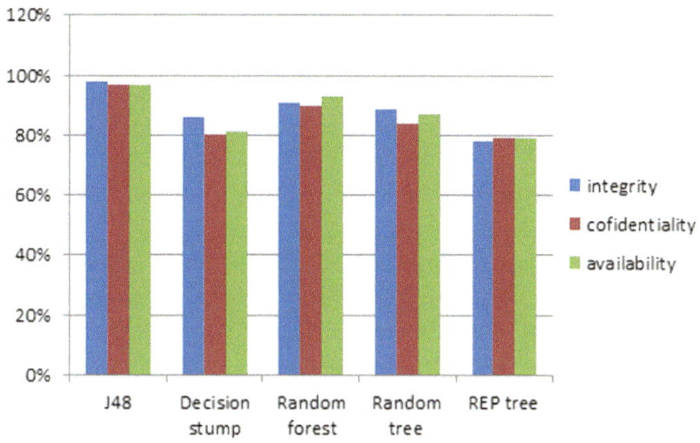

Fig. 10 Accuracy by trees classifiers

5.5 Using Tress-Based Classifiers

The tree classifiers used are J48, decision stump, random forest, random tree and REP tree to give the accuracy for integrity, confidentiality and availability. J48 have the highest accuracy which is about 98% for integrity. Decision stump gives 86% accuracy in integrity. Random forest gives 93% for availability. Random tree has 89% for integrity, and REP tree shows 79% for confidentiality and availability. These results are shown in Fig. 10. These classifiers give the accuracy results by all algorithms in them. Each classifier gives accuracy in terms of integrity, confidentiality and availability.

6 Conclusion and Future Work

In this work, authors studied the Common Vulnerability and Exposures list for different vulnerabilities from the National Vulnerability Database. They further explored the CVE database for Microsoft Office-related vulnerabilities. This work classified the Microsoft Office vulnerabilities using various supervised machine learning techniques. Classification results are evaluated on the basis of three metrics: integrity impact, availability impact and confidentiality impact which enable the software developers for the secure development of a software product. This classification results help the software developers to identify the particular impact of vulnerability on software which in advance can decide assessment and mitigation techniques for that particular vulnerability of Microsoft Office.

Our results depend on the dataset we use and parameter chosen. This suggests that performance can be enhanced with further experimentation and also combining metrics or type of attack which may be: integrity, confidentiality or availability and learning methods. In some of the models, we discover only marginal improvements in all of the assessment parameters. These improvements will further reduce manual work in identifying vulnerability in software and help software developers in taking intelligent decisions while building software. Such studies help in understanding trends and patterns existing in software vulnerabilities and also managing the security of computer systems. In our study, we have also observed that in many databases, most of the entries are blank which may give misguiding results.

As a future work, code-specific parameters can be explored in improving software security. Other technique which may also help in enhancing software security is data resampling techniques and ensemble models of learning. Our current study is limited to classify licensed software vulnerabilities; in future, it can also be extended to classify vulnerabilities in open-source software. The work can also be extended by conducting more experiments using other machine learning models including unsupervised techniques to find similar patterns in vulnerabilities and attacks

References

1. Krsul, I.V: Software vulnerability analysis. Ph.D. dissertation, Purdue University (1998)
2. Krsul, I., Spafford, E.: A Classification of Software Vulnerabilities That Result From Incorrect Environmental Assumptions, Report Purdue University (2015)
3. Alqahtani, S.S., Eghan, E.E., Rilling, J.: Tracing known security vulnerabilities in software repositories–A semantic web enabled modeling approach. Sci. Comp. Prog. pp. 153–175 (2016)
4. Howard, M., LeBlanc, D., Viega, J.: 19 Deadly Sins of Software Security. McGrawHill/Osborne, Emeryville, CA (2005)
5. Alhazmi, O.H., Malaiya, Y.K.: Prediction capabilities of vulnerability discovery models. In: Proceedings of Annual Reliability and Maintainability Symposium (RAMS), pp. 1–10 (2006)
6. Howard, M., Lipner, S.: The Security Development Lifecycle. Microsoft Press (2006)
7. Guo, M., Wang, J.A.: An ontology-based approach to model common vulnerabilities and exposures in information security. In: Proceedings of ASEE 2009 Southeast Section Conference, Marietta, GA, USA, pp. 5–7 (2009)
8. Munjal, G., Kaur, S.: Comparative study of ANN for pattern classification. WSEAS Trans. Comput. **6**, 236–241 (2007)
9. Li, W., Yi, P., Wu, Y., Pan, L., Li, J.: A new intrusion detection system based on KNN classification algorithm in wireless sensor network. J. Electr. Comput. Eng. (2014). https://doi.org/10.1155/2014/240217
10. Syed, R., Zhong, H.: Cybersecurity vulnerability management: An ontology-based conceptual model. In: Twenty-Fourth Americas Conference on Information Systems, New Orleans, LA, USA, pp. 16–18 (2018)
11. Carlstead, J., Bibsey, II, R., Popek, G.: Pattern-Directed Protection Evaluation, Tech. Report., Information Sciences Institute, University of Southern California (1975)
12. Marick, B.: A survey of software fault surveys. Tech. Rep. UIUCDCS-R-90-1651, University of Illinois at Urbana-Champaign (December 1990)
13. Aslam, T., Krsul, I., Spafford, E.: Use of A Taxonomy of Security Faults, Tech. Report Number: 96–051, Department of Computer Science Engineering, Purdue University (1996)
14. Bishop, M., Bailey, D.: A Critical Analysis of Vulnerability Taxonomies. Tech. Rep. CSE-96–11, Department of Computer Science at the University of California at Davis (1996)
15. Leveson, N.: High-pressure steam engines and computer software. In: Computer 27, 10 (October), Keynote Address IEEE/ACM International Conference in Software Engineering Melbourne Australia (1992)

16. Christey, S., Wysopal, C.: Responsible Vulnerability Disclosure Process. INTERNET-DRAFT "draft-christey-wysopal-vuln-disclosure-00.txt". The Internet Society (2002)
17. D'Ambros, M., Lanza, M., Robbes, R.: Evaluating defect prediction approaches: a benchmark and an extensive comparison. Empirical Software Eng. **17**, 531–577 (2012). https://doi.org/10.1007/s10664-011-9173-9
18. Zimmermann, T., Nagappan, N., Williams, L.: Searching for a needle in a haystack: Predicting security vulnerabilities for windows vista. In: Proceedings of Third International Conference on Software Testing, Verification and Validation (ICST), SVM, pp. 421–428 (2010)
19. Joshi, C., Singh, K.U., Tarey, K.: A review on taxonomies of attacks and vulnerability in computer and network system. Int. J. Adv. Res. Comput. Sci. Software Eng. **5**, 742–747 (2015)
20. Sabetta, A., Bezzi, M.: A practical approach to the automatic classification of security-relevant commits. In: 34th International Conference on Software Maintenance and Evolution. IEEE Computer Society, Sept. 2018, pp. 1–5 (2018)
21. Weber, S., Karger, P.A., Paradkar, A.: A software flaw taxonomy: Aiming tools at security. In: Proceedings of the 2005 Workshop on Software Engineering for Secure Systems—Building Trustworthy Applications, St. Louis, Missouri, pp. 1–7 (2005)
22. Li, X., Chang, X., Board, J.A., Kishor, S.: A novel approach for software vulnerability classification. In: IEEE Annual Reliability and Maintainability Symposium (RAMS), (2017). https://doi.org/10.1109/ram.2017.7889792
23. Weka 3–Data Mining With Open Source Machine Learning Software in Java. Available: http://www.cs.waikato.ac.nz/ml/weka/. Accessed Aug 2019
24. Zhang, J., Zulkernine, M., Haque, A.: Random-forests-based network intrusion detection systems. IEEE Trans. Syst., Man, Cybernetics Part-C, Appl. Rev. **38**(5), 649–659 (2008)
25. Neuhaus, S., Zimmermann, T.: Security trend analysis with CVE topic models. In: IEEE International Symposium on Software Reliability Engineering, pp. 111–120 (2010). https://doi.org/10.1109/issre.2010.53
26. Perl, H., Dechand, S., Smith, M., Arp, D., Yamaguchi, F., Rieck, R., Fahl, S., Acar, Y.: VCCFinder: Finding potential vulnerabilities in open source projects to assist code audits. In: 22nd CCS'15, Denver, Colorado, USA, ACM, pp. 426–437 (2015). https://doi.org/10.1145/2810103.2813604
27. Rangwala, M., Zhang, P., Zou, X., Li, F.: A taxonomy of privilege escalation attacks in Android applications. Int. J. Secure. Network **9**, 40–55 (2014). https://doi.org/10.1504/IJSN.2014.059327
28. Raheja, S., Munjal, G., Shagun: Analysis of linux kernel vulnerabilities. Ind. J. Sci. Technol. **9**, 12–29 (2016). https://doi.org/10.17485/ijst/2016/v9i48/138117
29. Haibo, H., Garcia, E.A.: Learning from imbalanced data. IEEE Trans. Knowl. Data Eng. **21**, 1263–1284 (2009). https://doi.org/10.1109/TKDE.2008.239
30. Piessens, F.: A taxonomy of causes of software vulnerabilities in Internet software. In: Supplementary Proceedings of the 13th International Symposium on Software Reliability Engineering, pp. 47–52 (2002)
31. Pothamsetty, V., Akyol, B.A.: A vulnerability taxonomy for network protocols: Corresponding engineering best practice countermeasures. In: Communications, Internet, and Information Technology, IASTED/ACTA Press, pp. 168–175 (2004)
32. Takahashi, T., Miyamoto, D., Nakao, K.: Toward automated vulnerability monitoring using open information and standardized tools. In: IEEE International Conference on Pervasive Computing and Communication Workshops (PerCom Workshops), pp. 1–4 (2016). https://doi.org/10.1109/percomw.2016.7457049
33. Massacci, F., Nguyen, V.H.: Which is the right source for vulnerability studies? An empirical analysis on mozilla firefox. I:n Proceedings of the 6th International Workshop on Security Measurements and Metrics (MetriSec 2010), pp. 15–15 (2010). https://doi.org/10.1145/1853919.1853925
34. Michael, G., Kishore, S.T.: Software faults, software aging and software rejuvenation. J. Reliab. Eng. Assoc. Jpn **27**, 425–438 (2005)
35. National Vulnerability Database. https://nvd.nist.gov/

36. Igure, V.M., Ronald, D.W.: Taxonomies of attacks and vulnerabilities in computer systems. IEEE Commun. Surv. Tutorials **10**, 6–19 (2008)
37. Khazai, B., Kunz-Plapp, T., Büscher, C.: Int. J. Disaster Risk Sci. **5**, 55 (2014). https://doi.org/10.1007/s13753-014-0010-9
38. Wijayasekara, D., Manic, M., McQueen, M.: Vulnerability identification and classification via text mining bug databases. In: IECON 2014-40th Annual Conference of the IEEE Industrial Electronics Society, pp. 3612–3618 (2014). https://doi.org/10.1109/iecon.2014.7049035
39. Witten, I.H., Frank, E.: Data Mining: Practical Machine Learning Tools and Techniques, 2nd edn (Morgan Kaufmann Series in Data Management Systems). Morgan Kaufmann Publishers Inc., San Francisco, CA (2005)
40. Torkura, K.A., Meinel, C.: Towards cloud-aware vulnerability assessments. In: 11th International Conference on Signal-Image Technology Internet-Based Systems (SITIS), pp. 746–751 (2015). https://doi.org/10.1109/sitis.2015.63
41. Zhou, Y., Sharma, A.: Automated identification of security issues from commit messages and bug reports. In: Proceedings of the 2017 11th Joint Meeting on Foundations of Software Engineering, pp. 914–919. ACM, New York (2017). https://doi.org/10.1145/3106237.3117771
42. Bowes, D., Gray, D: Recomputing the Confusion Matrix for Prediction Studies Reporting Categorical Output, Technical Report 509, Univ. of Hertfordshire (2011)
43. Wijayasekara, D., Manic, M., McQueen, M.: Vulnerability identification and classification via text mining bug databases. In: IECON—40th Annual Conference of the IEEE Industrial Electronics Society, pp. 3612–3618 (2014)
44. Ghaffarian, S.M., and Shahriari, H.R: Software vulnerability analysis and discovery using machine-learning and data-mining techniques: A survey. ACM Comput. Surv. **50**(4), 56:1–56:36 (2017). https://doi.org/10.1145/3092566
45. Morrison, P., Herzig, K., Murphy, B., Williams, L.: Challenges with applying vulnerability prediction models. In: Proceedings of the 2015 Symposium and Bootcamp on the Science of Security, pp. 1–9. ACM, New York (2015)

Computational Neuroscience Models and Tools: A Review

Parampreet Kaur and Gurjot Singh Gaba

Abstract A detailed and robust mathematical theory and modeling paradigms with discrete functions and relationships within diverse regions of the brain are essential to analyze the operations of neural system. Although the complete actions and events of nervous systems, with its causal entities, cannot be revealed by mathematical or statistical systems only, a requirement of dynamic and powerful virtual modeling paradigms is crucial for processing them. Artificial Intelligence techniques such as machine learning and deep learning have proven to be the unrivaled approaches for solving problems pertaining to numerous recognition systems. Computer vision systems integrated with enhanced neural network methods can provide qualitative and better results in the medical and healthcare sector to a large extent. The study of multifaceted models is therefore organized by discussing the current state-of-the-art in the field of computational neuroscience by focusing on the currently available tools and software resources for effectively carrying out the research work in this domain. The sources of relevant databases and online repositories consisting of brain-related imaging data for detailed research and investigation have also been comprehensively described.

Keywords Neuroscience · Healthcare · Computational tools · Cognitive science · Artificial intelligence

1 Introduction

The current neuroscience domain in academics originally initiated with the "Neurosciences Research Program (NRP)" in 1962 [1] at MIT. Among the diverse set

P. Kaur (✉)
School of Computer Science and Engineering, Lovely Professional University, Phagwara, Punjab 144411, India
e-mail: parampreet.18758@lpu.co.in

G. S. Gaba
School of Electronics and Electrical Engineering, Lovely Professional University, Phagwara, Punjab 144411, India
e-mail: gurjot.17023@lpu.co.in

A. K. Bhoi et al. (eds.), *Bio-inspired Neurocomputing*, Studies in Computational Intelligence 903, https://doi.org/10.1007/978-981-15-5495-7_22

of disorders and malfunctioning which occurs in the human body, neurological diseases specifically related to brain, amounts to approximately 13% of the globally found healthcare problems [2]. It is cautionary and alarming than other types of disorders such as heart diseases which amount to nearly 5% of the worldwide ailment set. Distinctive categories of cancers also, however, constitute 10% of the global illness. Existing neuroscience investigators belong to the groups of physiologists, experimentalists, physicists, statisticians, computer experts, engineers, molecular researchers, biology scientists, physicians, environmentalists, clinicians, clinicians, bio-informaticians, psychologists, and neurosurgeons, philosophers, etc. Such a varied fraternity of research groups provides an indication of the amusing vibrancy of this subject as depicted in Fig. 1. The nervous system computes and performs processing of information very quickly [3]. Computation and analysis of neurological signals is not simply processed digitally or using some analog techniques rather takes into account relatively sophisticated computational schemes based on varied modeling strategies [4].

There exist multiple levels of categorization in which the human nervous system can be organized. Every level is composed of different subsystems that are further classified into low-level smaller systems. In order to tackle such highly decomposed and detailed levels of systems, the field of neuroscience has been categorized into several related sub-domains, one of which is broadly known as "*computational neuroscience*". As the data to be processed by these tools and applications is extremely large

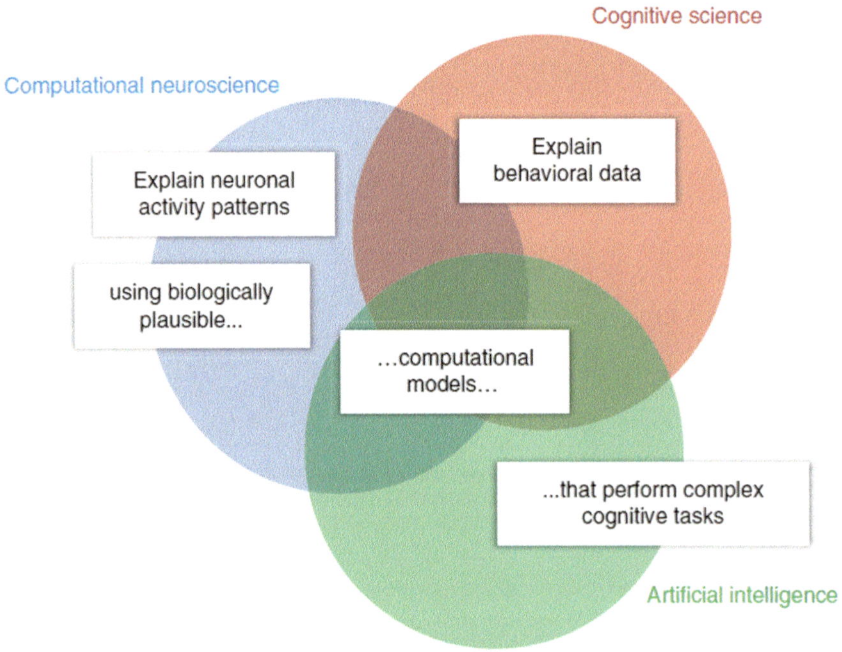

Fig. 1 Areas of neuroscience computational models [5]

and complicated, specialized storage techniques and high-speed capable processers pose a challenging task. Hence, automated and efficient modeling of neural informative data, its effective management, and then analysis of neurobiological information is ultimately a necessary phase. The severe complication of the brain's internal structure imparts significantly huge challenging task to neuroscience researchers for computational modeling, and for scientific data-analysis introspection.

Neurological researchers and scientists have investigated the brain's compositional and functional aspects from different dimensions, spanning from electron-microscopy examinations of synaptic interfaces [6] to axon-level details [7], and then later to fully connected neurons, by further extending towards brain-slices, and even live animal/human brain studies [8]. The scope of several experiments ascends from smallest to high-level magnitudes of brain data. Computational neuroscience is regarded as one of the dedicated disciplines of biological computational theories [9]. Figure 2 represents the illustration of the automated mechanism used by scientists today for capturing the cognitive abilities of the brain's functional areas. This is analogous to the fully connected neurons which pass information from sensory organs

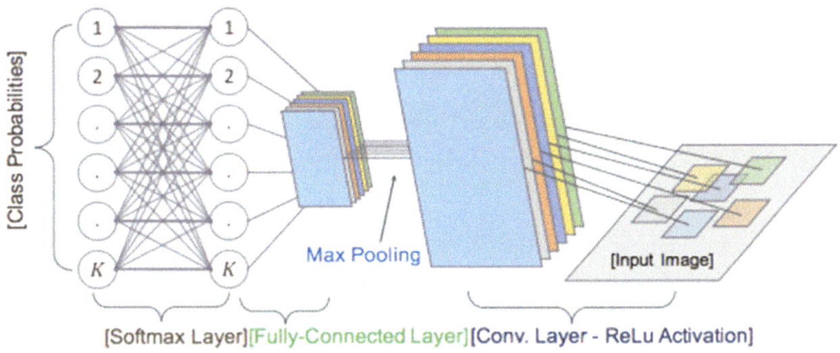

(a)

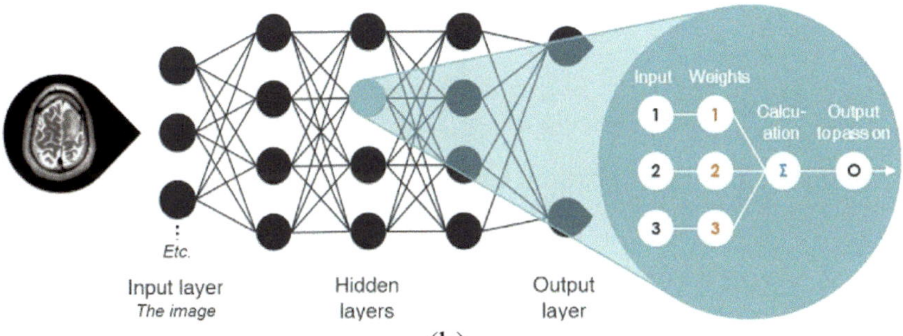

(b)

Fig. 2 **a, b** Fully connected layers mapping the neurological connections via the brain's neural network. Also, in Fig. 2, probabilities of actual real-world data are linked after the system has learned from its training datasets.

It delivers unambiguous multi-scale models that can collate experimental annotations and shape innovative hypothetical frameworks. Traditionally, a few years back, research carried by [11] directed towards the progress of "integrate-and-fire models". Approximately, around 50 years later, [7] provided a comprehensive multilevel biophysical prototype of the axon, overlaying the means for consequent advances in computational neuroscience. Initially, simulations and virtual prototype engines were very limited and specific to a particular domain only, with minimum scope for scalability or replicability. However, recently in the past, many important and critical analytical tools of neuroscience have been introduced. These can now assist in providing high-quality standards and scope of improved efficiency [12].

The flow chart is shown in Fig. 3 provides a representation of biological information to modeling of data for analysis. Three categories of realistic model analysis have

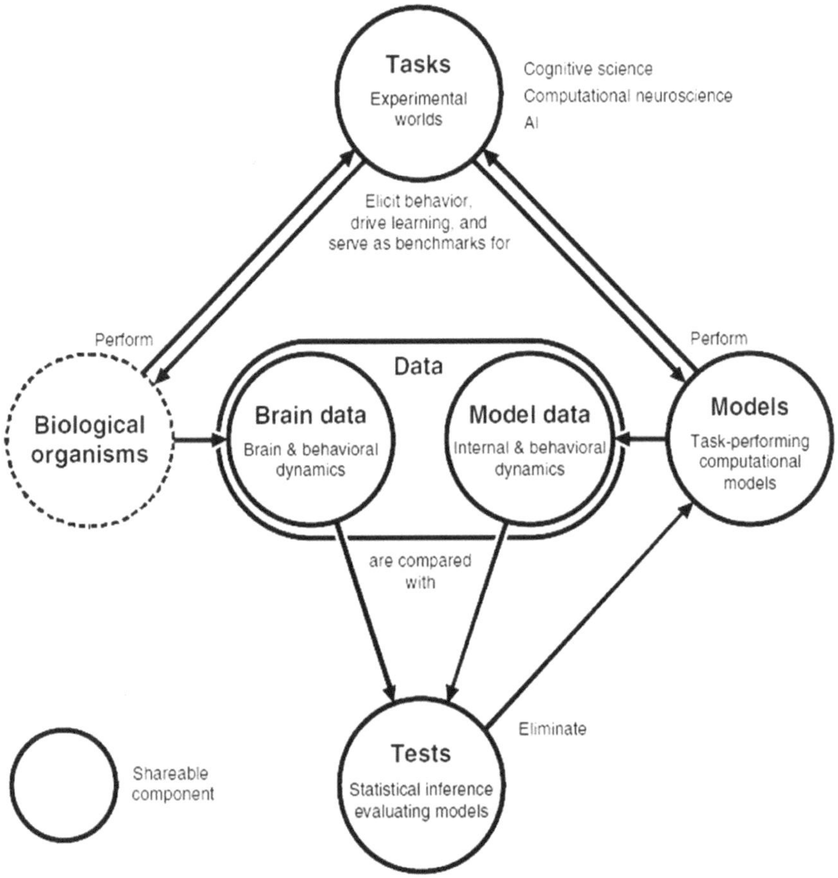

Fig. 3 Representation of biological information to modeling data for analysis flow diagram [5]

been presented in the research works. These are named as (a) encoding based models [13–15], pattern analysis models [16] and representative comparison study models as highlighted in [17–19]. These three approaches test theories about the sample space and are built on multivariate imageries of the trial situations, for instance, a semantic picture of a conventional structure of stimuli, or the motion patterns through different layers of neural networks [20].

With the beginning of human brain images and scanned dataset availability, researchers started relating cognitive systems with the person's intelligence and mind's thinking capability. This endeavor was given the name of neural cognition [21]. The scientific community belonging to this research focus group started by representing reasoning psychology's data handling components and their connections between these units against the brain interactions. This was one step forward in characterizing brain-related activities, but a drawback in the computational power of the automated tools [22].

Cognitive neuroscience has represented the comprehensive functional design of the human's brain and the ape's brain [23]. But it cannot attain a complete computational interpretation of brain's way of data organization and handling activities. The next level challenge is to construct computational models of brain's data handling activities which are conforming to the brain's structural as well as behavioral patterns for implementing and realizing multifaceted composite thinking capabilities. The subsequent current progresses in fields of cognition sciences, computational neuroscience put forward the success of AI (Artificial Intelligence) in their respective domains.

1.1 Cognitive Science

Cognitive science has progressed in a top-down manner, by disintegrating multifaceted cognitive functions into processing units. In order to derive sensible information from brain data, it has developed functional computational models at the logically rational level. One of the popular network models, i.e., Bayesian cognitive models, which intelligently associates previous knowledge about the world with physical objects for verification and testing [24–26]. Originally practiced on the basic perceptive and mechanical functions [26, 27], Bayesian models were applied to encompass intricate cognition, together with the way our observances exhibit the physical and outside world [28].

The advancements in the field of machine learning and artificial intelligence have led to commendable discoveries and provided effective solutions in a vast variety of domains. The state-of-the-art techniques in the area of deep learning methodologies to study the imagery as well as video-based data is immensely growing at a very fast pace where the computers automatically analyses the patterns of neurological images by comparing it with the stored databases records and finding the appropriate solutions to it [29]. Moreover, it makes existing processes and rules for fairly accurate interpretation of propagative models that can expand in convolution with the accessible data which is essentially needed for actual intelligence [30].

1.2 Computational Neuroscience

Computational neuroscience works on the basis of bottom to the top manner, representing how different connections between biological neurons can execute computational operations. The field of computational neuroscience has designed and generated multiple mathematical models since the last twenty years by utilizing the thrust of neurobiological science [31, 32]. These comprise of mechanisms for sensory coding [33, 34], normalization [35], working memory [36], signal growth and result/output techniques [37, 38]. Generally, these modular functionalities are processed easily by delivering structural theories for the problem of cognition. Computational neuroscience also instigated to examine multifaceted models that can explicate high-level perceptive and reasoning brain depictions [39].

Artificial intelligence has presented by what means module operations can be conjoined to generate intellectual behavior. Initial AI methods could not succeed due to lack of mapping of several features of real-world data into automation [40], as well as several difficult cognitive challenges [41, 42]. Many developments were maneuvered by reasoning based representational models. The majority of significant current developments have been steered by deep convolution networks, comprised of elements that calculate linear sequences of their inputted images, and then applying non-linear functions [29]. These models use a limited set of features to extract the neural abilities and provide approximately correct reasoning power to the machines.

The predominant task is to figure out solid connections between theoretical and experimental knowledge. The initial part of this analysis defines hierarchal advances that originate from the investigational data and make an effort to build links from the data in the track of philosophical concepts. Specified with brain activity statistics, linking models target divulging out the complicated dynamic forces of brain's stimulation; deciphering and encrypting representations intends to disclose the specifications and planned structures of neural system depictions [43]. The modeling strategies described in this work stipulate constrictions applicable in the domain of neuro-computational theories, however, in general, cannot accomplish the cognition-based abilities due to which there arises limitations in this theory.

1.3 Models in Neuroscience for Mathematical Computation

The term model can have different meanings in relation to the activities associated with brain and its structural studies. One of the models, i.e., **Data-analysis models** are basic mathematical models that assist in understanding the pre-determined variables and their relationships. These models are helpful in establishing connect for brain representations using linear correlation, single variable regression, multi-variable regression, etc.

Unlike a **box-and-arrow model**, this is a data management and analysis model in the shape of labeled squares which signifies perceptive and reasoning based elements along with directional edges which signify flow of data. In perceptive psychological sciences, these modeling constructs delivered suitable although imprecise, drawings for concepts of brain computation. An oracle model is a structure of brain's reactions that depends on data not available to the organism whose brain is being modeled. An oracle model could deliver a valuable description of the facts present in a brain section and its realistic format, with no specification of any philosophy about the brain's logical representation. A brain-computational model (B.C.M.) comparatively is a typical model that imitates the brain data and information management demonstrating the operation of a task at a particular level o f perception.

1.4 Neural Network Models

Neural networks inspired from the brain's logical functioning manner have been the trending areas of investigation from the past several years. Inputs are supplied into this processing network which analyses the data and generates relevant information in automated way. These network models fetch and represent biological details using neuron's methodology.

Neural networks provide a description of different cognition-based functionalities that can perform reasoning as well as to detect various objects. These models offer excellent capability of representing how the neurological system in humans operate. Some of the distinctive components calculate a sequential pattern of the inputted data and route the resulting outcome over a static non-linear path. The outcome of this process is inferred as equivalent to that of the message sending and passing a neuron. Moreover, the normal feedforward networks where only one layer is present as a hidden layer can fairly perform this action well with less probability rates [44], additional engineering advances [45], and better capturing of recurrent signaling in brains [46].

Neural networks method finds a solution to a problem in a distinctive way. The principle behind these is to take a huge dataset of n-images which are collectively called a training set, and then progress to design an algorithmic computational system that can relate those thousands of images and can acquire information from the training dataset. Putting it differently, the neural network utilizes the samples to automatically gather and deduce directions for recognizing patterns. Also, by enlarging the quantity of the training examples, the network can learn further about related patterns, and so as to enhance its accurateness [47, 48].

The feedforward nets are general model structures where the inputs are fed only in a serial forward manner without any cyclic function. However, the recurrent neural systems are dynamical representations of the activities taking place in biological neuroscience [49]. Such networks offer flexible although the computationally expensive approach of fulfilling the required deficiencies timely in a dynamic way. Due to this reason, recurrent networks can identify, distinguish, envisage, and produce effective patterns.

1.5 Deep Neural Network Models

Deep neural network models (DNNs) offer image processing models for visual analysis [10]. However, supervised deep neural networks depend on labeled images for training. Since labeled samples are not obtainable in large numbers for diverse sources in biological studies and systems, these prototypes are BCMs of photographic processing. Reinforcement learning models use environmental responses which lead to more real-time processing of imagery data. As can be seen in Fig. 4, the model wise complexity gradually rises in DCNN as well as recurrent neural networks in top-down as well as bottom-up strategies. The complexity is least for cognitive models.

A neurological training model is a B.C. Model in which the calculations which convert sensed inputs into a detailed low-level description. The behavioral interpretation modeled structure is a B.C.M. of the conversion from certain interior representation to a behavioral result. It is noteworthy that the B.C.M. simply designates that the model is proposed to encapsulate the brain's processed information at a particular level of abstraction. Figure 5 provides an overview of the methodology using the deep neural network tools such as Keras and Tensor-flow for implementing the deep convolution neural network-based training and testing functions.

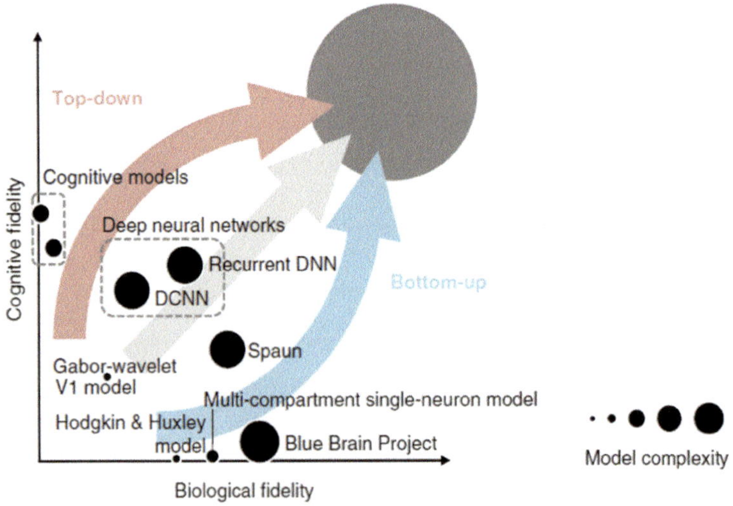

Fig. 4 Various models representing biological fidelity versus cognitive fidelity [5]

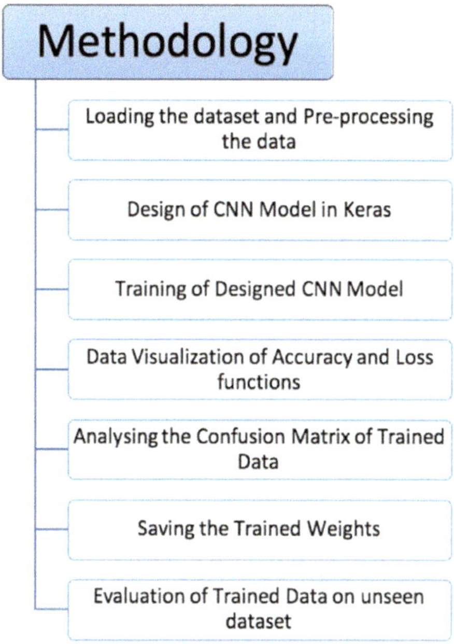

Fig. 5 Process workflow and methodology of training and validating data for deep neural networks in Keras

Both feedforward and recurrent networks are characterized by their structural design and the fixing of the connection weights. One technique of setting the required weights is done using iterative minor alterations that generate the output nearer to closely chosen output (supervised learning). Every weighing factor is adjustable relative to the errors that are reducing. This phenomenon is known as gradient descent. The backpropagation algorithm can be used to realize the gradient descent. But it is still unpredictable if the brain also utilizes a similar method for learning purposes and cognitive abilities.

Numerous biologically credible applications of backpropagation methods (gradient descent) and strongly associated practices of supervised learning are in existence [50–52]. Supervised signals may be created for linking outcomes with the trained datasets [53, 54]; It deals with the two most important fundamentals of classification and regression. Thousands of imagery or video-based datasets are used during training of samples, which provide probability values of whether or not any random unseen image belongs to the learned class or not. This way the model keeps learning the new information and adds to its repository [55].

The reinforced learning models based on hit and trial decision making [42] and unsupervised machine learning methodologies [55, 56] have become topics of the swift and recent development in academic as well as medical research. Neural net Technologies confirmed the ever-growing and endless motivation in the sciences and biological studies may lead to innovative results for a number of sectors. This seems potentially strong that the expedition for modeling paradigms that produce exactly similar and robust thinking and reasoning capabilities just like humans, derived from extensive biological structures can highly fascinate and entice the entire research community working in this direction [57].

Neurological scientific studies have defined a prolific assemblage of dynamic mechanisms, such as action abilities [58], recognizable patterns, nerve cell extensions [59], and network events such as alternations [31, 60], which can perform high scale computational operations. The field of biological sciences also has to offer some limitations on the universal structure while recommending other systems for inducing learning patterns by various automation machines [61]. Modeling of biological modules using neural network framework is considered to achieve significant success in the near future [50, 52].

1.6 Existing Software Applications and Frameworks

Numerous automated toolboxes and software applications exist to analyze computational neuroscience datasets and neuro-images. These applications are protected under the copyrights of the "Neuro-Informatics Tools and Resources (NITRC)" [62]. Some of the commonly used resources have been elaborated here with reference to their sources available online. Most of the software is accessible free of cost, i.e., they belong to open-source category of tools. Most of the toolboxes are easily accessible free of cost. The type of datasets utilized, their set of specific requirements to execute processing by these tools is described below.

- **FSL** [63]: FMRIB software Library (FSL) comprises of precompiled binaries and source codes for brain images useful for MRI primarily for non-commercial use. This tool is freeware.
- **SPM** [64]: This tool has been introduced by Math works Inc. It is connected to analysis of neuro-images of brain's multiple parts and its data classification samples for performing EEG, fMRI, MEG, PET, and SPECT. This is an open-source tool.
- **BS Mac** [65]: This is another MATLAB based numerical and pictorial imaging toolbox library for dealing with fMRI type of data accessible free of cost (open source).
- **Turbo-Brain-Voyager** [66]: This tool is accessible online using real-time image management system to operate fMRI information. The tool necessitates HASP authorization (open-source).
- **MRI-cron** [67]: MRI-CRON is a visual graphics inspecting system engine that performs actions on MRI data. It correspondingly offers provisions for numerical analysis (open-source).
- **Chronux** [68]: The recent software libraries of these tools can be realized as a MATLAB tool support apps. It pre-processes, searches, and examines neurological data.

- **itk-SNAP** [69]: It is a three-dimensional medicinal/healthcare image breakdown application software. It takes into account Magnetic Resonance Imaging and CT scan X-ray images and process it.
- **EEGLAB** [70]: It is an open-source software platform for EP signaling and analysis. This tool has collaborative MATLAB libraries for image creation, mimic cognitive thinking abilities, artifact elimination, and study of Electro Encephalon Gram (EEG), Magneto Encephalography (MEG) x-ray images.
- **Free-Surfer** [71]: It recreates the mind surface from MRI images, and superimposes fMRI information on the rebuilt portion.
- **Brain-Net Viewer** [72]: It is a visualization-based toolset that builds operational and functional network-diagrams after filtering and processing the original data (open source).
- **Brain-Voyager-QX** [73]: It can process "MRI, EEG, and MEG" datasets. The automated system integrates arithmetical, geometrical, and image processing libraries together.
- **CONN** [74]: It is a Math Works application for computation, demonstration, and fMRI statistics investigation (open source).

2 Conclusions

The Bayesian model can extract features from the available set of neurobiological features to any random degrees, but definitely predicts fewer operations of neurological activities. Psycho-physical modeling systems that forecast behavioral outcomes by the perceptive mechanisms and reasoning representations that might execute cognition responsibilities are B.C. Models outlined as higher degree modeling paradigms. Advancement in processing models is made by discarding candidate BCMs on an experimental basis. Figure 6 enlists a vast variety of domains, subdomains, and fields of research in the discipline of computational neuroscience. These are explicitly shown in the form of the human neural system constituting synapses, neurons, and dendrites (*represented in black*).

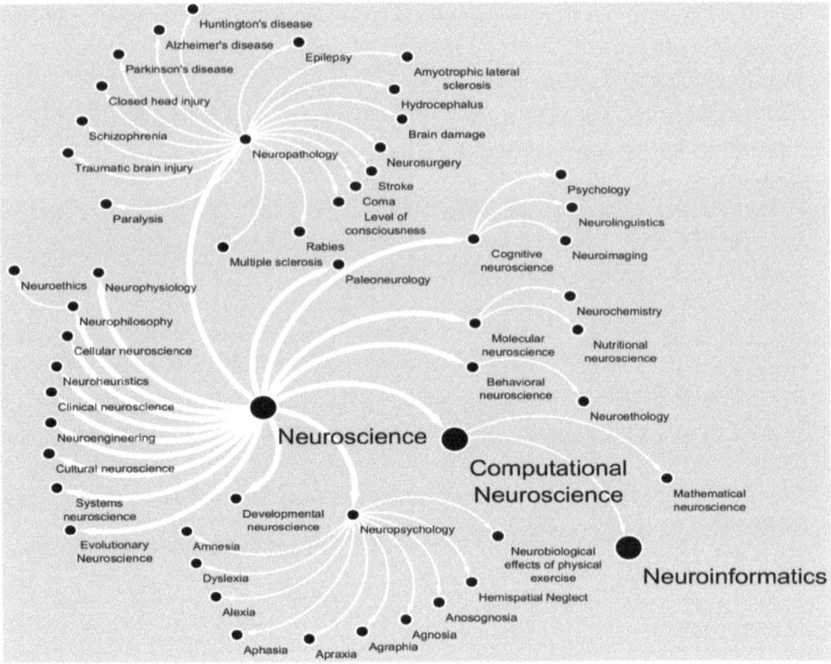

Fig. 6 Domains and fields under neuro-informatics discipline [5]

References

1. Adleman, G.: The neurosciences research program at MIT and the beginning of the modern field of neuroscience. J. Hist. Neurosci. **19**, 15–23 (2010)
2. Kiernan, M.C.: A fine neuroscience vintage. J. Neurol. Neurosurg. **86**(1), 1–2 (2015)
3. Piccinini, G., Shagrir, O.: Foundations of computational neuroscience. Curr. Opin. Neurobiol. **25**, 25–30 (2014)
4. Copeland, B.J., Shagrir, O.: Physical computation: how general are Gandy's principles for mechanisms? Minds Mach. **17**(2), 217–231 (2017)
5. Kriegeskorte, N., Douglas, P.K.: Cognitive computational neuroscience. Nat. Neurosci. **21**, 1148–1160 (2018)
6. Toni, N., Teng, E.M., Bushong, E.A.: Synapse formation on neurons born in the adult hippocampus. Nat. Neurosci. **10**(6), 727–737 (2007)
7. Hodgkin, A.L., Huxley, A.F.: A quantitative description of membrane current and its application to conduction and excitation in nerve. J. Physiol. **117**, 500–544 (1952)
8. Baddeley, A.: Working memory. Science **255**(5044), 556–559 (1992)
9. Spoerer, C.J., McClure, P., Kriegeskorte, N.: Recurrent convolutional neural networks: a better model of biological object recognition. Front. Psychol. **8** (2017)
10. Srinivas, S., Sarvadevabhatla, R.K., Mopuri, K.R., Prabhu, N., Kruthiventi, S.S.S., Babu, R.V.: A taxonomy of deep convolutional neural nets for computer vision. Front. Robot Ai, 1–13 (2016)
11. Drummond, C.: Replicability is not reproducibility: Nor is it good science. In: Proceedings of the 26th International Conference on Machine Learning (2009)
12. Piccinini, G., Bahar, S.: Neural computation and computational theory of cognition. Cogn. Sci. pp. 453–488 (2013)

13. Mitchell, T.M., et al.: Predicting human brain activity associated with the meanings of nouns. Science **80**(320), 1191–1195 (2008)
14. Kay, K.N., Naselaris, T., Prenger, R.J., Gallant, J.L.: Identifying natural images from human brain activity. Nature **452**, 352–355 (2008)
15. Dumoulin, S.O., Wandell, B.A.: Population receptive field estimates in human visual cortex. Neuroimage **39**, 647–660 (2008)
16. Diedrichsen, J., Ridgway, G.R., Friston, K.J., Wiestler, T.: Comparing the similarity and spatial structure of neural representations: a pattern component model. Neuroimage **55**, 1665–1678 (2011)
17. Kriegeskorte, N., Mur, M., Bandettini, P.A.: Representational similarity analysis—connecting the branches of systems neuroscience. Front. Syst. Neurosci. **2**(4) (2008)
18. Nili, H., et al.: A toolbox for representational similarity analysis. PLoS Comput. Biol. **10**, e1003553 (2014)
19. Kriegeskorte, R.A., Kievit, N.: Representational geometry: integrating cognition, computation, and the brain. Trends Cogn. Sci. **17**, 401–412 (2013)
20. Diedrichsen, J., Kriegeskorte, N.: Representational models: a common framework for understanding encoding, pattern-component, and representational-similarity analysis. PLoS Comput. Biol. **13** (2017)
21. Gazzaniga, M.S.: The Cognitive Neurosciences, pp. 33–43. MIT Press, Cambridge, MA (2004)
22. Fodor, J.A.: Précis of the modularity of mind. Behav. Brain Sci. **8**(1) (1985)
23. Van Essen, D.C., et al.: The brain analysis library of spatial maps and atlases (BALSA) database. Neuroimage **144**, 270–274 (2017)
24. Griffiths, T.L., Chater, N., Kemp, C., Perfors, A., Tenenbaum, J.B.: Probabilistic models of cognition: exploring representations and inductive biases. Trends Cogn. Sci. **14**, 357–364 (2010)
25. Ernst, M.O., Banks, M.S.: Humans integrate visual and haptic information in a statistically optimal fashion. Nature **415**, 429–433 (2002)
26. Weiss, Y., Simoncelli, E.P., Adelson, E.H.: Motion illusions as optimal percepts. Nat. Neurosci. **5**, 598–604 (2002)
27. Körding, K.P., Wolpert, D.M.: Bayesian integration in sensorimotor learning. Nature **427**, 244–247 (2004)
28. Lake, B.M., Ullman, T.D., Tenenbaum, J.B., Gershman, S.J.: Building machines that learn and think like people. Behav. Brain Sci. **40**, e253 (2017)
29. Lecun, Y., Bengio, Y., Hinton, G.: Deep learning. Nature **521**(7553), 436–444 (2015)
30. MacKay, D.J.C.: Information Theory, Inference, and Learning Algorithms, p. 2003. Cambridge University Press, Cambridge (2003)
31. Dayan, P., Abbott, L.F.: Theoretical Neuroscience: Computational and Mathematical Modeling of Neural Systems. MIT Press, Cambridge, MA (2001)
32. Abbott, L.F.: Theoretical neuroscience rising. Neuron **60**, 489–495 (2008)
33. Olshausen, B.A., Field, D.J.: Sparse coding of sensory inputs. Curr. Opin. Neurobiol. **14**, 481–487 (2004)
34. Simoncelli, E.P., Olshausen, B.A.: Natural image statistics and neural representation. Annu. Rev. Neurosci. **24**, 1193–1216 (2001)
35. Carandini, M., Heeger, D.J.: Normalization as a canonical neural computation. Nat. Rev. Neurosci. **13**, 51–62 (2011)
36. Chaudhuri, R., Fiete, I.: Computational principles of memory. Nat. Neurosci. **19**, 394–403 (2016)
37. Shadlen, M.N., Kiani, R.: Decision making as a window on cognition. Neuron **80**, 791–806 (2013)
38. Diedrichsen, J., Shadmehr, R., Ivry, R.B.: The coordination of movement: optimal feedback control and beyond. Trends Cogn. Sci. **14**, 31–39 (2010)
39. Yamins, D.L.K., DiCarlo, J.J.: Using goal-driven deep learning models to understand sensory cortex. Nat. Neurosci. **19**, 356–365 (2016)

40. Krizhevsky, A., Sutskever, I., Hinton, G.E.: ImageNet classification with deep convolutional neural networks. Adv. Neural Inf. Process. Syst. **25**, 1097–1105 (2012)
41. Silver, D., et al.: Mastering the game of Go with deep neural networks and tree search. Nature **529**, 484–489 (2016)
42. Mnih, V., et al.: Human-level control through deep reinforcement learning. Nature **518**, 529–533 (2015)
43. Cohen, J.D., et al.: Computational approaches to fMRI analysis. Nat. Neurosci. **20**, 304–313 (2017)
44. Uçar, A., Demir, Y., Guzelis, C.: Moving towards in object recognition with deep learning for autonomous driving applications. Innov. Intell. Syst. Appl. **93**(9) (2017)
45. Goodfellow, I., Bengio, Y., Courville, A.: Deep Learning. MIT Press, Cambridge, MA (2016)
46. Wang, X.-J.: Decision making in recurrent neuronal circuits. Neuron **60**(2), 215–234 (2008)
47. Wyatte, D., Curran, T., O'Reilly, R.: The limits of feedforward vision: recurrent processing promotes robust object recognition when objects are degraded. J. Cogn. Neurosci. **24**, 2248–2261 (2012)
48. Hunt, L.T., Hayden, B.Y.: A distributed, hierarchical and recurrent framework for reward-based choice. Nat. Rev. Neurosci. **18**, 172–182 (2017)
49. Schäfer, A.M., Zimmermann, H.G.: Recurrent neural networks are universal approximators. Int. J. Neural Syst. **17**, 253–263 (2007)
50. O'Reilly, R.C., Hazy, T.E., Mollick, J., Mackie, P., Herd, S.: Goal-driven cognition in the brain: a computational framework, p. 62 (2014)
51. Whittington, J.C.R., Bogacz, R.: An approximation of the error backpropagation algorithm in a predictive coding network with local Hebbian synaptic plasticity. Neural Comput. **29**, 1229–1262 (2017)
52. Schiess, M., Urbanczik, R., Senn, W.: Somato-dendritic synaptic plasticity and error-backpropagation in active dendrites. PLoS Comput. Biol. **12** (2016)
53. Marblestone, A.H., Wayne, G., Kording, K.P.: Towards an integration of deep learning and neuroscience. Front. Comput. Neurosci. **10**(94) (2016)
54. Shadlen, M.N., Shohamy, D.: Decision making and sequential sampling from memory. Neuron. Neuron **90**, 927–939 (2016)
55. Goodfellow, I., Pouget-Abadie, J., Mirza, M., Xu, B., Warde-Farley, D., Ozair, S., Courville, A, Bengio, Y.: Generative adversarial nets pp. 1–9 (2014)
56. Kingma, D.P., Welling, M.: Auto-encoding variational Bayes. In: International Conference on Learning Representation, Banff, Canada, pp. 1–14 (2014)
57. Kandel, E.R., Schwartz, J.H., Jessell, T.M., Siegelbaum, S.A., Hudspeth, A.: Principles of Neural Science. McGraw-Hill Professional, New York (2013)
58. Buesing, L., Bill, J., Nessler, B., Maass, W.: Neural dynamics as sampling: a model for stochastic computation in recurrent networks of spiking neurons. PLoS Comput. Biol. **7**(11), 1–22 (2011)
59. Larkum, M.: A cellular mechanism for cortical associations: an organizing principle for the cerebral cortex. Trends Neurosci. **36**, 141–151 (2013)
60. Bengio, Y., Scellier, B., Bilaniuk, O., Sacramento, J., Senn, W.: Feedforward initialization for fast inference of deep generative networks is biologically plausible (2016)
61. Kumaran, D., Hassabis, D., McClelland, J.L.: What learning systems do intelligent agents need? Complementary learning systems theory updated. Trends Cogn. Sci. **20**, 512–534 (2016)
62. NeuroInformatics Tools and Resources. Available: https://www.nitrc.org/
63. FSL, FMRIB Software Library. (2019). Available: http://fsl.fmrib.ox.ac.uk/fsl/fslwiki/. Accessed 05 Jan 2020
64. Mathworks. Inc. SPM12 (2014). Available: http://www.fil.ion.ucl.ac.uk/spm/
65. Mathworks. Inc, CBIS. Available: http://web1.sph.emory.edu/bios/CBIS/software.html
66. Goebel, R.: Turbo BrainVoyager 4.0 (2018). Available: http://www.brainvoyager.com/products/turbobrainvoyager.html. Accessed 06 Jan 2020
67. NITRC, MRICron (2018). Available: http://neuro.debian.net/pkgs/mricron.html
68. Chronux. Available: http://chronux.org/
69. ITK Snap. Available: http://www.itksnap.org/pmwiki/pmwiki.php

70. EEG Lab. Available: http://sccn.ucsd.edu/eeglab/
71. Free Surfer. Available: http://surfer.nmr.mgh.harvard.edu/fswiki/DownloadAndInstall
72. Brain Net Viewer. Available: https://www.nitrc.org/projects/bnv/
73. Goebel, R.: Brain Voyager (2019). Available: http://www.brainvoyager.com/products/brainvoyagerqx.html. Accessed 05 Jan 2020
74. NITRC, CONN. Available: https://www.nitrc.org/projects/conn/

Double Mediating Effects of Self-efficacy and Body Image in the Effect of Appearance Management Behavior on Life Satisfaction Among Old People

Jung-Soon Bae, Yun-Jeong Kim, and Sang-Jin Lee

Abstract *Background/Objectives* This study wants to test dual mediating effect of self-efficacy and body image in the relation between appearance management behavior and life satisfaction of the elderly in Korea. *Methods/Statistical analysis* The research objects of this study are 363 old men and women residing in South Korea. The survey was conducted in July and August in 2019. To examine relationship among variables like appearance management behavior, self-efficacy, body image, and life satisfaction, this study conducted correlation analysis. And to examine the relationship between appearance management behavior and life satisfaction, it uses Model 4 of SPSS PROCESS macro. *Findings* First, it was found that correlations among all the variables were significant. While the mean value (2.793) of appearance management behavior was a little lower than median, those of other variables were higher than median: self-efficacy (3.486), body image (3.070), and life satisfaction (3.312). Second, appearance management behavior has positive effect on life satisfaction. Third, in the relationship between appearance management behavior and life satisfaction, self-efficacy and body image were found to have full mediating effects. Based on such findings, this study suggested ways to improve life satisfaction by improving self-efficacy and body image. *Improvements/Applications* To improve life satisfaction of the elderly, it is more important to raise self-efficacy and have positive image of one's body than making efforts in appearance management. And it is necessary for old people to actively manage their aging bodies, which will positively change their body images, improving their life quality.

Keywords Double mediating effects · Self-efficacy · Body image · Appearance management behavior · Life satisfaction · Old people

---

J.-S. Bae
Department of Elderly Welfare, Hanseo University, Seosan-si, South Korea
e-mail: okskin4u@naver.com

Y.-J. Kim (✉)
Department of Health, Counseling and Welfare, Hanseo University, Seosan-si, South Korea
e-mail: twoyun21@hanmail.net

S.-J. Lee
Chungcheongnam-do Welfare Foundation, Yesan-gun, South Korea
e-mail: sjlee@cnwf.or.kr

A. K. Bhoi et al. (eds.), *Bio-inspired Neurocomputing*, Studies in Computational Intelligence 903, https://doi.org/10.1007/978-981-15-5495-7_23

1 Introduction

In 2018, Korea became an aged society with old people of 65 years old or over taking up 14.3% of the total population. It is expected that by 2026, the proportion of old people will be 20.8%, making Korean society a super-aged society [1]. Absolute and relative increase of old population means the increase of healthy and energetic old people enjoying leisure time. In particular, with rapid economic growth, the development of medical science and mass media has created various concepts of beauty, leading to high interest in one's body [2]. Interest in one's outward appearance does not escape old generations. Old people who are economically independent from their offspring are active [3] in their management of appearances to appear attractive, and taking care of appearance affects their life satisfaction. Appearance management is means to express oneself and the process of showing one's identity to others [4]. With change of social and cultural values, appearance management behavior is a way of realizing oneself. According to researches on the appearance management behavior of older women, the more interested they are, the more positive they are in appearance, the more positive they are in life, and the higher their self-esteem [2, 5–7]. And, with the change of society as one emphasizing appearance, not only women's but also men's appearance has important effects in their social activities and human relations [7, 8].

While interest in appearance can be universal to men and women in all ages, physical conditions become disadvantageous to old people, compared with other age groups. Accordingly, the aim of this research is to search for psycho-social means which can mediate effect of appearance management behavior on life satisfaction. Among various psycho-social variables, this study focuses on self-efficacy and body image.

First, self-efficacy is personal belief in one's ability to organize and perform behavior necessary in producing intended results in a specific scene [9]. It is an important element which affects decision of one's behavior and has effect on life in general [9, 10]. Accordingly, we can assume that appearance management behavior has positive effect on self-efficacy, and self-efficacy has effect on life satisfaction. That is, we can assume that in the relationship between appearance management behavior and life satisfaction, self-efficacy will have mediating effect.

Next, body image means consciousness or mental image of one's own body. It is feeling or attitude on one's appearance, health condition, function, sex [11], and social symbol to express oneself. In particular, to women, body image consists of physical self, femininity, and attractiveness and is considered as essential element to express one's natural being [12]. But, as women get old, they experience various experiences like physical aging, independence of offspring, and retirement of husband, and they are more likely to experience psychological atrophy and depression than in other age groups. Accordingly, in old age, the relationship between body image and life satisfaction gets greater than in other age groups [13, 14]. And, appearance management behavior like regular physical exercise [15] and appearance management behavior like clothing/make-up/accessories [16] make positive body image.

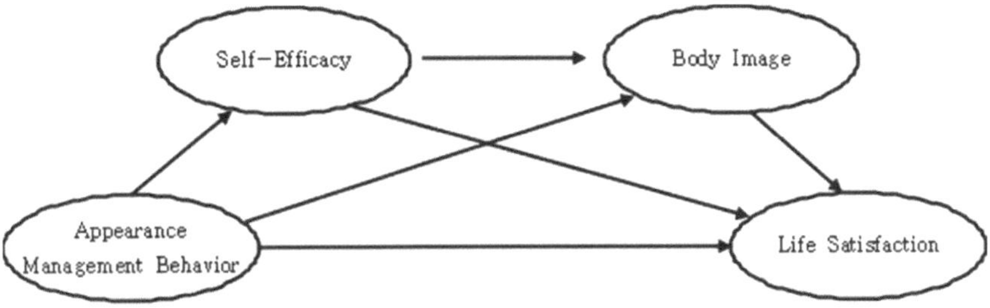

Fig. 1 Research model

That is, appearance management behavior has positive effect on body image, and positive body image will have positive effect on life satisfaction. That is, body image has mediating effect on the relationship between appearance management behavior and life satisfaction.

Based on such existing researches, this study wants to test dual mediating effect of self-efficacy and body image in the relation between appearance management behavior and life satisfaction, using old men and women as research objects.

2 Method

2.1 Research Model

Based on existing researches, this study sets the following research model (Fig. 1).

2.2 Research Objects and Data Collection

The research objects of this study are 363 old men and women residing in South Korea. The survey was conducted in July and August in 2019.

2.3 Measurement Tools

2.3.1 Appearance Management Behavior

The scale this study used to measure appearance management behavior is what Kim [17] revised. They are a set of ten questions, and they were measured with five-point Likert scale. The measure of reliability, Cronbach's alpha value is 0.856.

2.3.2 Self-efficacy

The scale of self-efficacy is what Park and Kim [2] developed. It consists of 20 questions, and they were measured with five-point Likert scale. Cronbach's alpha value is 0.930.

2.3.3 Body Image

The scale of body image in this study is the revised version of what Jang [8] developed. It consists of 14 questions, and they were measured with five-point Likert scale. Cronbach's alpha value is 0.847.

2.3.4 Life Satisfaction

The scale of life satisfaction in this study is what Diener et al. [18] developed. There are five questions, and they were measured with five-point Likert scale. Cronbach's alpha value is 0.811.

2.4 Socio-demographic Characteristics of Respondents

Respondents were composed of 36.6% males and 63.4% females. The average age of them was 67.33 years old. 78.2% of them have spouse. Among educational achievement groups, the largest proportion (35.0%) was those who had graduated from middle school or less. In religiosity, 65.6% did not have any religion. To the questions asking economic situation and health conditions, 48.5, 46.8% of them responded that they are 'similar' to other people in the same age groups. 68.6% of them were residing in urban areas and 31.4% in rural areas, which mean that the proportion of urban residents is two times larger than that of rural residents.

3 Results

3.1 Correlation Among Variables

To examine relationship among variables like appearance management behavior, self-efficacy, body image, and life satisfaction, this study conducted correlation analysis (Table 1) and found out that correlations among all the variables were significant. While the mean value (2.793) of appearance management behavior was a little lower

Table 1 Correlation among variables

	Appearance management behavior	Self-efficacy	Body image	Life Satisfaction
Appearance management behavior				
Self-efficacy	0.250***			
Body image	0.497***	0.385***		
Life satisfaction	0.143**	0.423***	0.336***	
Mean	2.793	3.486	3.070	3.312
SD	0.650	0.614	0.545	0.696

($N = 363$)
$^{**}p < 0.01$, $^{***}p < 0.001$

than median, those of other variables were higher than median: self-efficacy (3.486), body image (3.070), and life satisfaction (3.312).

3.2 *Double Mediating Effects of Self-efficacy and Body Image in the Relationship Between Appearance Management Behavior and Life Satisfaction*

3.2.1 Effect of Appearance Management Behavior on Life Satisfaction for Old People

To examine the relationship between appearance management behavior and life satisfaction, this study analyzed it using Model 4 of SPSS PROCESS macro as suggested by Hayes [19] (Table 2, Fig. 2). It was found that the more old people manage appearance, the higher their life satisfaction goes ($\beta = 0.154, p < 0.001$).

Table 2 Effect of appearance management behavior on life satisfaction among old people

Direct effect model (dependent variable: life satisfaction)						
	Coeffect	SE	t	p	LLCI	ULCI
Constant	2.883	0.160	18.025	0.000	2.568	3.197
Appearance management behavior	0.154	0.056	2.754	0.006	0.0440	0.263

($N = 363$)
$R^2 = 0.021, F = 7.584, p = 0.006$

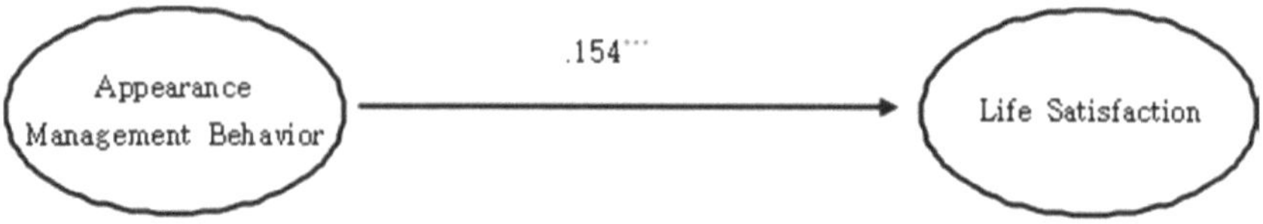

Fig. 2 Effect of appearance management behavior on life satisfaction among old people

3.2.2 Double Mediating Effects of Self-efficacy and Body Image in the Relationship Between Appearance Management Behavior and Life Satisfaction

To examine double mediating effect of self-efficacy and body image in the relationship between appearance management behavior and life satisfaction, this study analyzed using Model 6 of SPSS PROCESS macro suggested by Hayes [19] (Table 3, Fig. 3).

Mediating effect of each path was statistically significant. Appearance management behavior had positive effects on self-efficacy ($\beta = 0.192$, $p < 0.001$) and body image ($\beta = 0.359$, $p < 0.001$), which means that the more one is attentive to appearance management, the higher one's self-efficacy and body image get. Self-efficacy had positive effects on body image ($\beta = 0.247$, $p < 0.001$) and life satisfaction

Table 3 Double mediating effect of self-efficacy and body image in the relationship between appearance management behavior and life satisfaction

	Coeffect	SE	*t*	*p*	LLCI	ULCI
Mediating effect model (dependent variable: self-efficacy)						
Constant	2.826	0.138	20.474	0.000	2.554	3.097
Appearance management behavior	0.236	0.048	4.908	0.000	0.142	0.331
$R^2 = 0.063$, $F = 24.085$, $p = 0.000$						
Mediating effect model (dependent variable: body image)						
Constant	1.205	0.154	7.841	0.000	0.903	1.508
Appearance management behavior	0.359	0.038	9.531	0.000	0.285	0.433
Self-efficacy	0.247	0.040	6.201	0.000	0.169	0.326
$R^2 = 0.320$, $F = 84.728$, $p = 0.000$						
Dependent variable model (dependent variable: life satisfaction)						
Constant	1.206	0.228	5.283	0.000	0.757	1.654
Appearance management behavior	−0.063	0.058	−1.086	0.278	−0.177	0.051
Self-efficacy	0.395	0.058	6.859	0.000	0.282	0.508
Body image	0.295	0.072	4.082	0.000	0.153	0.437
$R^2 = 0.217$, $F = 33.062$, $p = 0.000$						

($N = 363$)

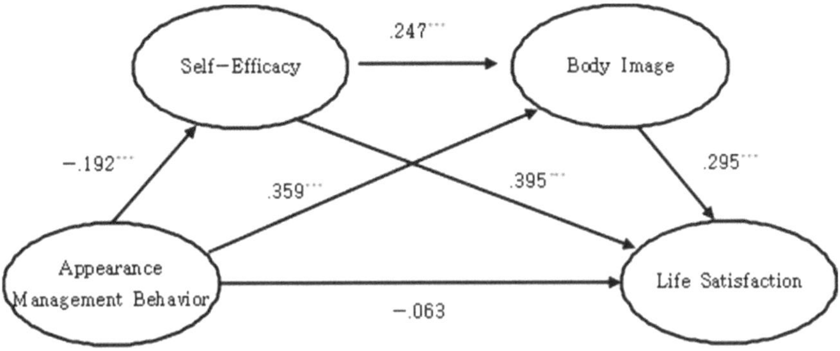

Fig. 3 Double mediating effect of self-efficacy and body image in the relationship between appearance management behavior and life satisfaction

Table 4 Verification of mediating effect of self-efficacy and body image

	Effect	Boot SE	BC 95.0% CI	
Appearance management → self-efficacy → body image → life satisfaction	0.017	0.007	0.005	0.033

($N = 363$)

($\beta = 0.295, p < 0.001$), which means that the higher one's self-efficacy is, the higher one's body image and life satisfaction become.

Though the effect of appearance management behavior on life satisfaction is statistically significant ($\beta = 0.154$, $p < 0.001$), when self-efficacy and body image are added to the equation, the effect of appearance management behavior on life satisfaction becomes insignificant ($\beta = -0.063$, $p > 0.05$), which means that self-efficacy and body image have full mediating effects on the relationship between appearance management behavior and life satisfaction.

Next, verification of mediating effect of self-efficacy and body image (Table 4) showed that the path of appearance management behavior → self-efficacy → body image → life satisfaction was significant because there is no 0 value between the lowest value and the highest value of bootstrap 0.017(0.005~0.033). Accordingly, self-efficacy and body image were proven to have mediating effects in the relationship between appearance management behavior and life satisfaction.

4 Discussion

Results showed the followings. First, appearance management behavior has positive effect on life satisfaction. That is, the more one make efforts to manage appearance such as fitness management, clothing procurement, accessories, hair-styling, cosmetic treatment, skin management, etc., the higher one's life satisfaction goes up. It is consistent with previous researches [2, 5–8] emphasizing importance of appearance and appearance management behavior in old life. In old age, managing appearance is an element to improve one's life satisfaction.

Second, this study proved full mediation effects of self-efficacy and body image in the effect of appearance management behavior on life satisfaction. That is, appearance management behavior does not have significant effect on life satisfaction without mediating effect of self-efficacy and body. To improve life satisfaction of the elderly, it is more important to raise self-efficacy and have positive image of one's body than making efforts in appearance management.

Self-efficacy is the ability to deal with the task one wants to achieve and is a kind of belief that one can overcome difficulties and create what one wants by one's behavior [9]. To improve self-efficacy among the elderly, it is important to help them to experience success and achievement through positive thoughts, behavior, and experiences in their daily lives. Next, as body image can be improved by appearance management behavior [16], it is necessary for old people to actively manage their aging bodies, which will positively change their body images, improving their life quality.

5 Conclusion

The aim of this study is to examine double mediating effects of self-efficacy and body image in the effect of appearance management behavior on life satisfaction among old people. This study conducted a survey to 363 old people from July to August, 2019.

First, it was found that appearance management behavior has positive effect on life satisfaction. Second, in the relationship between appearance management behavior and life satisfaction, self-efficacy and body image were found to have full mediating effects. Based on such findings, this study suggested ways to improve life satisfaction by improving self-efficacy and body image.

Acknowledgements This paper was supported by the Graduate Student Support Project of Hanseo University in 2019.

References

1. Kostat. Demographic Trend (2019). http://kostat.go.kr/portal/korea/index.action
2. Park, S.J., Kim, K.H.: The effects of appearance management behavior by women on psychological happiness and self-efficiency. J. Korean Soc. Knit Design **17**(1), 35–46 (2019)
3. Lee, Y.J., Seo, B.I.: Research on types of the old caring appearances according to their self-esteem and perception of physical attraction. J. Korean Acad. Health Welfare Elderly **29**(2), 59–75 (2010)

4. Cox, C.L., Glick, W.H.: Resume evaluations and cosmetics use: When more is not better. Sex Roles **14**(1–2), 51–58 (1986)
5. Park, Y.R., Son, Y.J.: Relationship of satisfaction with appearance, self-Esteem, depression, and stress to health related quality of life in women across the lifespan. Korean J. Fund. Nursing **16**(3), 353–361 (2009)
6. Ryoo, H.H., Jung, Y.: The relationship between body-consciousness and appearance management behavior of older women. J. Beauty Industry **3**(1), 1–19 (2011)
7. Lee, S.J., Park, K.S.: Appearance management behaviors of female university students by self-efficacy and self-esteem. Costume Culture Assoc. **19**(5), 1075–1087 (2011)
8. Jang, H.S.: Influences of body image and physical appearance management behavior on psychological stability. Doctorial Thesis. SeoKyung University (2014). Available from: https://scholar.google.co.kr/scholar?hl=ko&as_sdt=0%2C5&q=Influences+of+body+image+and+physical+appearance+management+behavior+on+psychological+stability&btnG=
9. Bandura, A.: Self-efficacy: The exercise of control. W. H. Freeman, New York (1997)
10. Jo, K.Y., Yoo, T.S.: The mutual relations of self–efficacy proximity of cosmetics the self and proximith of clothing to self. J. Korean Soc. Costume **32**, 183–200 (1997)
11. Fobair, P., Stewart, S.L., Chang, S., D'Onofrio, C., Banks, P.J., Bloom, J.R.: Body image and sexual problems in young women with breast cancer. Psycho-Oncol. **15**(7), 579–594 (2006)
12. Cohen, S.: Psychosocial models of the role of social support in the etiology of physical disease. Health Psychol. **7**(3), 269 (1988)
13. Kim, E.G., Lim, S.H., Nam, I.S.: Effect of sociocultural attitudes toward appearance of depression and psychological well-being of fitness participating adult woman. J. Korea Acad.-Ind. Cooperation Soc. **16**(1), 354–364 (2005)
14. Song, M.K., Ha, J.H., Park, D.H., Ryu, S.H., Oh, J.H., Yu, J.H., et al.: Effect of body image and eating attitude on depressive mood and suicide ideation in female adolescents. Korean J. Psychosom. Med. **18**(1), 40–47 (2010)
15. McAuley, E., Konopack, J.F., Motl, R.W., Morris, K.S., Doerksen, S.E., Rosengren, K.R., et al.: Physical activity and quality of life in older adults: influence of health status and self-efficacy. Ann. Behav. Med. **31**(1), 99 (2006)
16. Cash, T.F., Pruzinsky, T.: Future challenges for body image theory, research, and clinical practice. Body image: A handbook of theory, research, and clinical practice, pp. 509–516 (2002)
17. Kim, J.W.: The correlation of the body image satisfaction and the appearance caring behavior of mid-aged women regarding the propensity to narcissism. [Doctorial Thesis]. DaeGu University (2010). Available from: https://scholar.google.co.kr/scholar?hl=ko&as_sdt=0%2C5&q=The+correlation+of+the+body+image+satisfaction+and+the+appearance+caring+behavior+of+mid-aged+women+regarding+the+propensity+to+narcissism&btnG=
18. Diener, E.D., Emmons, R.A., Larsen, R.J., Griffin, S.: The satisfaction with life scale. J. Pers. Assess. **49**(1), 71–75 (1985)
19. Hayes, A.F.: Partial, conditional, and moderated moderated mediation: quantification, inference, and interpretation. Commun. Monographs **85**(1), 4–40 (2018)